Resources for the Future

Resources for the Future

An International Annotated Bibliography for the 21st Century

ALAN J. MAYNE *MA BSc FIMA FSS*

GREENWOOD PRESS
Westport, Connecticut • London

Library of Congress Cataloging-in-Publication Data

Mayne, Alan J. (Alan James), 1927-
 Resources for the future : an international annotated bibliography
/ Alan J. Mayne.
  p.   cm. – (Bibliographies and indexes in economics and
economic history, ISSN 0749-1786 ; no. 13)
 Includes bibliographical references and index.
 ISBN 0-313-28911-5 (alk. paper)
 1. Forecasting – Bibliography.   I. Title. II. Series.
Z5990.M39 1993
[CB161]
016.30349 – dc20   92-35820

British Library Cataloguing in Publication Data is available.

Copyright © 1993 by Adamantine Press Limited

Library of Congress Catalog Card Number: 92-35820
ISBN: 0-313-28911-5
ISSN: 0749-1786

First published in 1993

Greenwood Press, 88 Post Road West, Westport, CT 06881
An imprint of Greenwood Publishing Group, Inc.

Printed in the United States of America

English language edition, except the United States and Canada,
published by Adamantine Press, England

The paper used in this book complies with the
Permanent Paper Standard issued by the National
Information Standards Organization (Z39.48-1984).

10 9 8 7 6 5 4 3 2 1

CONTENTS

RESOURCES FOR THE FUTURE

Introduction: How to use this book

This book will be invaluable for those engaged in fundamental thinking about the future of the planet. It should prove essential for policymakers and decision makers who want to find out what is already being done in their field – and indispensable for practical men and women of action. It is designed to address the overriding priority of anticipating future developments – economic, environmental, political, social, scientific, technological, and conceptual – in order to keep pace with, and preferably one step ahead of, this headlong, accelerating era.

Within one volume, it offers a source of extremely up-to-date information on: books; periodicals; newsletters; non-print media including conventional and multi-media CD-ROMs, online databases, and computer software; networks; organisations; trusts, foundations, and charities; and future-oriented 21st century projects. It provides some information on almost any type of activity which is focussed on the empowerment of human and global prospects for the next millennium.

Chapter 1 describes 1,000 books in the English language, and includes many extensive entries on titles published this year and last. There are many entries for earlier key books and other important books. There are also occasional short but informative entries for early classics, by such authors as H. G. Wells and Lewis Mumford. It was not possible to achieve comprehensive coverage of all important recent books in English especially relevant to the 21st century, *and* keep the book to a manageable size. The books included are 'the tip of an iceberg', but I have endeavoured to identify and review as many as possible of the key books in depth. Numerous entries have been kept short, because longer reviews will appear in the issues of *New Paradigms* (entry No. 1110) and in other forthcoming Adamantine Press publications; longer reviews of some books also appeared in the Annotated Bibliography of *Into the 21st Century* (No. 144).

Chapter 2 indicates the subject matter and aims of nearly 200 relevant serial publications. In selecting them, I have given priority to interdisciplinary coverage and to breadth of interest. I have tended not to include highly specialist periodicals, unless the topics covered by them are really relevant to global problems. The chosen periodicals range from learned and scientific journals, through serials adapted to the needs of managers and decision makers and other people of action, to popular magazines with many readers.

Chapter 3 covers nearly 50 titles, representing the very large number of newsletters which address the human and planetary situation and look towards the future in various ways.

Chapter 4 begins by reviewing books and a few periodicals *about* non-print media, and then gives examples of important sources of information using several of these media: conventional CD-ROMs, providing text, diagrams and illustrations, sometimes in colour; multimedia CD-ROMs which also include moving pictures from films and TV programmes, animated diagrams, and music, speeches and other sound recordings; online databases; databases held on personal computer disks; and computer software. I have not, in this first edition, included examples of television programmes, radio broadcasts, films, video cassette recordings, or audio cassette tapes.

Chapter 5 gives a few examples of social networks more or less actively working for a better future; new networks of this kind are now continually emerging. Chapter 6 outlines the aims and activities of a varied but representative selection of relevant organisations, ranging from learned societies and professional associations to global, national, regional, local and grassroots movements, societies, and groups. Chapter 7 gives brief information about some of the more relevant trusts, foundations, and charities. Chapter 8 introduces twenty 21st century projects, oriented to preparing for and improving our long-term future; many more such projects are now appearing.

The end matter includes a directory of the publishers whose books and other products have been cited, together with a publisher's note. There is then a subject index, giving the reference numbers of all items corresponding to each of a very wide range of subjects and topics. Finally, there is an index by publisher, indicating which publishers have contributed most books relevant to the 21st century. I have chosen the subjects and topics so that as many readers as possible, whatever their walk of life, ranging from the academic to the highly practical, will be able to find easily many of the items that can help answer their questions and address their problems.

For ease of reference, all the items in the bibliography have been numbered serially. For each chapter, there is an alphabetical listing, and the serial numbers in the successive chapter listings follow in sequence.

This book is intended to be a starting point for exploration in greater depth of the complex web of human and planetary problems, and the intricate network of contemporary human knowledge and new and old paradigms. Both the publisher and I would be most interested and grateful to receive information from readers about relevant literature and non-print media items that I have *not* cited in this first edition, together with updatings or corrections to existing entries. We hope to remedy the more important of these omissions in the second edition, due in the latter part of 1994.

We would like to thank many friends who have provided additional information and encouragement for this book. I am especially grateful to Jeremy Geelan, Editorial Director of Adamantine Press, for his continued practical support and valuable advice throughout this project, to Robert Brown of Saxon Graphics for his invaluable help in the final make-up and design of the book, and to Brian Burrows and Peter de la Cour for especially useful help in several respects. In addition, we wish to thank Lloyds Bank for its generous contribution of sponsorship money.

Alan Mayne (26 December 1992)

Resources for the Future

1

BOOKS

List of Books

1 Abbott, John
Politics and Poverty: A Critique of the Food and Agriculture Organisation of the United Nations
This book provides a detailed history of the FAO, and analyses how far it has been successful in achieving its objectives.
1991, Routledge, London & New York, ISBN 0-415-06650-6 (hbk) 224 pp.

2 Aber, John D. & Jerry M. Melillo
Terrestrial Ecosystems
This book aims to integrate information from several disciplines, and to present a holistic view of ecosystem function. It considers various applications, and uses some simple computer models.
1991, Holt, Rinehart & Winston, Philadelphia & London, ISBN 0-03-047443-4 (pbk) xiv + 430 pp.

3 Abercrombie, Nicholas & Alan Warde (Eds.)
Social Change in Contemporary Britain
This book provides a lively and up-to-date account of major developments in British society over the last 25 years.
1992, Polity Press, Cambridge & Cambridge, MA, ISBNs 0-7456-0782-9 (hbk) & 0-7456-0783-7 (pbk) xi + 189 pp.

4 Abraham, John Colin Lacey & Roy Williams (Eds.)
Deception, Demonstration & Debate: Towards a Critical Environment & Development Education
This book brings together perspectives on education from many viewpoints, relating to environmental and development issues.
1990, Kogan Page, London ISBN 0-7494-0198-2 (pbk) 197 pp.

5 Abrams, Malcolm & Harriet Bernstein
Towards 2001: A Consumer's Guide to the 21st Century
This book briefly describes various recent inventions in communications and information technology.
2nd. Ed. 1990, Angus & Robertson, London, ISBN 0-207-16657-9 (pbk) 209 pp.; 1st. Ed. 1989, Penguin Books, New York, with the title "Future Stuff".

6 Abrams, Malcolm & Harriet Bernstein
More Future Stuff: Over 250 Inventions that Will Change Your Life by 2001
This book briefly describes products that may become available in the 1990s, estimating their probability and time of occurrence.
1991, Penguin Books, New York, ISBN 0-14-014523-0 (pbk) xvi + 336 pp.

7 Adams, W. M.
Wasting the Rain: The Ecology and Politics of Water Resources in Africa
The author offers constructive and informed criticism of the development of water resources in Africa, and advocates an alternative approach, based on working with local people using local knowledge.
1992, Earthscan, London, ISBN 1-85383-089-5 (pbk) 240 pp.

8 Adams, W. M.
Green Development: Environment and Sustainability in the Third World
This book analyses the evolution of the concept of 'sustainable development', and assesses how to apply it to the real world.
1992, Routledge, London & New York, ISBNs 0-415-0043-8 (hbk) & 0-415-08050-9 (hbk) 272 pp.

9 Agnew, Clive & Ewan Anderson
Water Resources in the Arid Realm
This book considers how well water management strategies can work in the hostile environments of arid lands, and refutes some fallacies. Because of the highly variable arid land environment, there are no simple universal solutions and the key to success is flexibility.
1992, Routledge, London & New York, ISBNs 0-415-04346-8 (hbk) & 0-415-07969-1 (pbk) 320 pp.

10 Aitchison, Jean
Teach Yourself Linguistics
This introduction to linguistics outlines and explains its scope, basic concepts, and terminology, with examples mainly from English.
4th. Ed. 1992, Hodder & Stoughton, London, ISBN 0-340-55938-1 (pbk) viii + 232 pp; 1st. Ed. 1972, with the title "General Linguistics".

11 Albery, Nicholas & Mark Kinzley (Eds.)
How to Save the World: A Fourth World Guide to the Politics of Scale
This book has over 60 articles on a broad range of subjects, by a wide range of alternative activists and thinkers, all supporting decentralisation and independence from the giantism of mass societies.
1984, Turnstone Press, Wellingborough, Northants., ISBN 0-85500-209-3 (pbk) 319 pp.

12 Albery, Nicholas (Ed.), assisted by Valerie Yule
The Book of Visions: An Encyclopedia of Social Inventions
This book brings together over 500 imaginative ideas and project proposals for handling social problems, that have been formulated by social inventors from most parts of the world. They range from global ideas to reverse our planet's crisis, through new proposals for reforms at the national level, to small ideas to improve everyday life simpler.
3rd Ed 1992, Institute for Social Inventions, London, ISBN 0-86369-601-5 (pbk) xiii + 338 pp.; 1st Ed 1989.

13 Albrow, M.& E. King (Eds.)
Globalization, Knowledge and Society
This collection of readings from *International Sociology* addresses key questions in the development of sociology.
1990, Sage Publications, London & Newbury Park, CA, USA, ISBN 0-8039-8324-7 (pbk) viii + 280 pp. .

14 Allaby, Michael
Green Facts: The Greenhouse Effect and Other Key Issues
This book outlines ecology and human ecology, examines some major problems and issues, and suggests how to reverse destructive trends.
2nd. Ed. 1989, Hamlyn, London, ISBN 0-600-51618-8 (pbk) 192 pp.; 1st. Ed. 1986, with the title "Ecology Facts".

15 Allaby, Michael
Guide to Gaia
This book explains the Gaia Hypothesis, assesses the implications of human activity on our planet, and questions our perceptions.
1989, Macdonald Optima, London, ISBN 0-356-17535-9 (pbk) ix + 181.

16 Allen, John, Peter Braham & Paul Lewis (Eds.)
Political and Economic Forms of Modernity
This introductory textbook examines the political and economic diemnsions of 20th century industrialised societies, mainly the UK.
1992, Polity Press, Cambridge, ISBNs 0-7456-0962-7 (hbk) & 0-7456-0962-7 (pbk) viii + 440 pp.

17 Allen, Peter
Off the Rocking Horse: How Local Councils Can Promote Your Health and Environment
This book explores the local health care agenda, viewing people and their communities as a source of enrichment for local government.
1992, Green Print, London, ISBN 1-85425-066-3 (pbk) v + 151 pp.

18 Allen, Robert
Waste Not, Want Not: The Production and Dumping of Toxic Waste
This book describes the campaigns surrounding several toxic waste disposal plants in the British Isles, many of which have been successful. It also examines the increasing trade in toxic waste, criticises its regulations and operating codes, and advocates legislative changes.
1992, Earthscan, London, ISBN 1-85383-095-X (pbk) 256 pp.

19 Allen, Tim & Alan Thomas (Eds.)
Poverty and Development in the 1990s
This book introduces some major issues in understanding poverty and development in the 1990s. It presents Third World problems, analyses their historical context, explains some current issues and concepts of development in the 1990s, and considers future options for development.

1992, Oxford University Press, Oxford & New York, ISBNs 0-19-877330-7 (hbk) & 0-19-877331-5 (pbk) vii + 421 pp.

20 Almond, Brenda & Donald Hill (Eds.)
Applied Philosophy: Morals and Metaphysics in Contemporary Debate
This selection from the *Journal of Applied Philosophy* (No. 1078) examines subjects of continued concern and debate, such as the environment, personal relationships, terrorism, and medicine.
1991, Routledge, London, ISBN 0-415-06015-X (pbk) x + 334 pp.

21 Alston, Philip (Ed.)
The United Nations and Human Rights: A Critical Appraisal
This book reviews the functions, procedures, and performance of each of the major United Nations organisations for human rights. It evaluates their effectiveness, and recommends improvements.
1992, Oxford University Press, Oxford & New York, ISBN 0-19-825450-4 (hbk) xiii + 765 pp.

22 Altbach, Philip G. & Gail P. Kelly (Eds.)
New Approaches to Comparative Education
This book illuminates the very varied perspectives on comparative education, provides some new visions, and indicates future trends.

1986, University of Chicago Press, Chicago, IL, USA & London, ISBNs 0-226-01525-4 (hbk) & ISBN 0-226-01526-2 (pbk) vi + 356 pp.

23 Anderson, Dennis
The Energy Industry and Global Warming: New Roles for Industrial Aid
This booklet discusses global warming in relation to world energy output and demand and different types of energy source.
1992 Overseas Development Institute, London, ISBN 0-85003-156-7 (pbk) 70 pp.

24 Anderson, Victor
Alternative Economic Indicators
This book seeks a new way of quantifying 'social well-being', to replace the normal national income measures. Part 1 reviews the growth debate, and discusses national income accounting. Part 2 describes 16 proposed alternative indicators, giving their values for 14 countries.
1991, Routledge, London & New York, ISBNs 0-415-04163-5 (hbk) & 0-415-04164-3 (pbk) 106 pp.

25 Andrews, David
The IRG Solution: Hierarchical Incompetence and How to Avoid It
This book criticises the inefficiency of hierarchical bureaucracies, and proposes a solution based on much more efficient laterally organised 'Information Routeing Groups' ('IRGs').
1984, Souvenir Press, London, ISBN 0-285-62662-0 (hbk) 272 pp.

26 Angell, David J. R., Justyn D. Comer & Matthew L. N. Wilkinson (Eds.)
Sustaining Earth: Response to the Environmental Threats
This book is a collection of chapters by distinguished scientists, industrialists, politicians, and environmentalists on various aspects of the environmental threat to the future of our planet. They present their responses to this threat in the context of the recommendations of *Our Common Future* (No. 988) and sustainable development. The book's six parts cover: the background, specific threats to the environment, individuals and society and sustainable development, national governments and sustainable development, the international community and sustainable development, and conclusions.
2nd Ed 1990, Macmillan, London & New York, ISBN 0-333-52492-6 (pbk) xvii + 226 pp.

27 Ansell, Jake & Frank Wharton
Risk: Analysis, Assessment and Management
This book highlights some of the more important and difficult aspects of risk management. It covers a wide range of applications of risk management, from methodology to ethics, with

examples from finance, the environment, industry, transport, and health.
1992, Wiley, Chichester & New York, ISBN 0-471-93464 (hbk) x + 270 pp.

28 Arestis, Philip
The Post-Keynesian Approach to Economics: An Alternative Analysis of Economic Theory and Policy
This book is concerned with the behaviour of advanced capitalist economies. It applies the approach of 'Post-Keynesian economics', and attempts to extend J. M. Keynes' ideas (see No. 537) to current problems.
1992, Edward Elgar, Aldershot & Brookfield, VT, USA, ISBN 1-85278-154-8 (hbk) xv + 316 pp.

29 Argyle, Michael
The Social Psychology of Everyday Life
This book attempts to apply social psychology to some of the most interesting and important aspects of human life. It presents briefly some up-to-date examples of the good research that has been done.
1992, Routledge, London & New York, ISBNs 0-415-01071-3 (hbk) & 0-415-01072-1 (pbk) xi + 319 pp.

30 Ash, Maurice
New Renaissance: Essays in Search of Wholeness
This books looks at things as a whole, and explores the relationships between different problems and concepts from a philosophy of wholeness. It discusses especially wholeness in plannng, the rural economy, and education.
1987, Green Books, Hartland, Bideford, Devon, ISBN 1-87009-8005 (pbk) x + 196 pp.

31 Ashton, John (Ed.)
Healthy Cities
This book presents the World Health Organisation's Healthy Cities Project, involving several hundred cities around the world which have adopted a programme intended to produce policies for a better environment and improvements in life styles and public health.
1992, Open University Press, Buckingham & Philadelphia, ISBNs 0-335-09477-5 (hbk) & 0-335-09476-7 (pbk).

32 Asland, Anders (Ed.)
The Post-Soviet Economy: Soviet and Western Perspectives
The contributors to this book, who include leading scholars from the West and key policy makers from the former USSR, analyse the current economic situation of the Commonwealth of Independent States and attempt to propose realistic and viable structures for the future.
1992, Pinter Publishers, London, ISBN 1-85567-039-9 (hbk) ix + 222 pp.

33 Assagioli, Roberto
Psychosynthesis: A Manual of Principles and Techniques
This book is a collection of the author's basic writings on 'psychosynthesis', which is concerned with the synthesis of the separate and separative elements of the human psyche.
1965, Hobbs, Dorman & Co., New York, (hbk) viii + 323 pp.

34 Atkinson, Adrian B.
Principles of Political Ecology
This book aims to unearth the basis of our cultural attitudes towards nature, and lay philosophical foundations for a sustainable relationship between society and the environment.
1991, Belhaven Press, London, ISBN 1-85293-183-3 (pbk) ix + 251 pp.

35 Attali, Jacques
Millennium: Winners and Losers in the Coming World Order
This book describes how the world is now undergoing continual change and upheaval, as a new geopolitical order is emerging. The author concludes that people have never had more power to shape their future than today, and never had to make so many urgent decisions.
2nd. Ed. 1991, Times Books, New York, ISBN 0-8129-1913-0 (hbk) x + 130 pp.; 1st. (French) Ed. 1990.

36 Attenborough, Sir David
The Living Planet: A Portrait of the Earth
This book explores each type of habitat on Earth, showing how it supports its own community of specially adapted plants and animals.
2nd. Ed. 1985, The Reader's Digest Association, London, ISBN 0-00-219169-5 (hbk) 368 pp.; 1st. Ed. 1984, Collins & BBC, London.

37 Avebury, Lord, Ron Coverson, Brian Humphries & Brian Meek (Eds.)
Computers and the Year 2000
This book presents a broad survey of the social implications and impact of computer technology, which will shape our future, for government, industry, organisations, the environment, and human values. Other topics include political and legal issues, the responsibilities of industry and computer professionals, and management aspects.
1972, NCC Publications, Manchester, ISBN 0-85012-074-8 (hbk) 319 pp.

38 Bahm, Archie J.
The Specialist: His Philosophy, His Disease, His Cure
This book, full of interesting insights, takes a close look at specialisation, its nature, its universality, and its 'cure'.
1977, World Books, Albuquerque, NM, USA, ISBN 0-911714-08-1 (hbk) ix + 119 pp.

39 Bahm, Archie J.
The Philosopher's World Model
The author of this book advocates the development of radically new, more complicated 'quantum-leap' concepts and paradigms to understand the present world situation. New multidisciplinary teamwork will be needed to provide expertise and understanding required for solutions.
1979, Greenwood Press, Westport, CT, USA, ISBN 0-313-21198-1 (hbk) 328 pp.

40 Baines, Chris
The New Pollution Handbook: An Action Guide to a Cleaner World
This book argues that our congested and wasteful *way* of life creates pollution problems. It describes many practical projects, that will help young people to gather the facts, assess the problem, find the source of pollution, and tackle it in their neighbourhood.
1991, Bedford Square Press, London, ISBN 0-7199-1263-6 (pbk) xiv + 130 pp.

41 Baker, Richard St. Barbe
My Life: My Trees
This book is an autobiographical account by the founder of 'Men of the Trees' of his worldwide work on tree conservation and planting.
3nd Ed. 1985, The Findhorn Press, Forres, Scotland, ISBN 0-905249-63-1 (pbk) xv + 167 pp.; 1st. Ed. 1970, Lutterworth Press.

42 Balasubramanyam, V. N. & Sanjaya Lall (Eds.)
Current Issues in Development Economics
This book discusses the key contemporary issues in development economics. Chapters by well-known international authors concisely survey recent theoretical developments and analyses of policy, and provide a valuable guide to current issues in development economics.
1991, Macmillan, Basingstoke & London, ISBNs 0-333-51323-1 (hbk) & 0-333-51324-X (pbk) xii + 287 pp.

43 Baldwin J. (Ed.)
Whole Earth Ecolog: The Book of Environmental Tools and Ideas
This compendium provides an authoritative review of essential books, hardware, tools, and ideas for Earth-conscious living. It also celebrates pioneering individuals and groups.
1990, Crown Publishers, New York, ISBN 0-317-57568-9 (pbk) 128 pp.

44 Banks, Ronald (Ed.)
Costing the Earth
This book proposes a survey of world resources as a first step to ensure the proper use of natural resources. It shows how this will have both environmental and economic benefits.

1989, *Shepheard-Walwyn, London, ISBN 0-85683-111-5 (pbk) 208 pp.*

45 Barbour, Ian (Ed.)
Technology, Environment, and Human Values
The author of this interdisciplinary book is concerned with environmental choices, especially how to combine conservation with distributive justice.
1980, Prager, New York, ISBN 0-03-05586-7 (hbk) xi + 331 pp.

46 Barde, Jean-Philippe & Kenneth Button (Eds.)
Transport Policy and the Environment: Six Case Studies
This book examines transport policies in the UK, USA and five continental European countries. Each case study shows the problems produced by transport policies' failure to take social and environmental costs into account, due to departmental division.
1990, Earthscan, London, ISBN 1-85383-075-5 (pbk) 224 pp.

47 Barenberg, Sumner et al. (Eds.)
Degradable Materials: Perspectives, Issues and Opportunities
This book aims to: (1) define the state of the art; (2) propose and agree standard terminology; (3) define characterisation techniques and protocols; (4) define mechanisms of degradation; (5) propose standard reference materials; (6) identify issues and needs.
1990, CRC Press, Boca Raton, FL, USA, ISBN 0-8493-4274-0 (pbk) xi + 761 pp.

48 Barker, Terry (Ed.)
Green Futures for Economic Growth: Britain in 2010
This book considers how far economic growth is compatible with environmental protection, and aims to examine the implications of concern for the environment for businesses and consumers. It attempts to assess possible futures for business and government, and their cost.
1991, Cambridge Econometrics, Cambridge, ISBN 0-9516638-0-1 (pbk) xii + 137 pp.

49 Barnaby, Frank (Ed.)
The Gaia Peace Atlas: Survival into the Third Millennium
This book is a call to action. It states the facts about violent conflicts and their extensive harmful effects, and presents positive approaches to conflict resolution and the development of a peaceful sustainable future. Peace goes beyond absence of war and violent conflict, and true security cannot be gained by arms. The development of a harmonious relationship between humankind and the planet is being recognised as the only path to a secure future. The decisions made at this stage will be vital to our survival and future prospects. The book has

three parts: Past – The Lessons of Experience, Present – Between Two Worlds, Future – Our Fate in Our Hands. There are many coloured maps and illustrations, with extensive accompanying text.
1988, Pan Books, London, ISBN 0-330-30151-9 (pbk) 271 pp.

50 Barney, Gerald O. (Ed.)
The Global 2000 Report to the President: Entering the Twenty-First Century
This book reprints unabridged the first two of the three volumes of the original 1980 *Global 2000 Report*, reporting the findings of the "Global 2000 Study", commissioned by US President Carter in 1977. Although the Study did not claim to be more than a reconnaissance of the future, its conclusions were similar to those of several other studies using global models, but saw some signs of hope.
5th. Ed. 1991, Seven Locks Press, Cabin John, MD, USA, ISBN 0-932020-96-8 (pbk) (Vol. 1 only, 51 pp; 3rd. Ed. 1982, Penguin Books, Harmondsworth, Midd., ISBN 0-14-022441-6 (pbk) 1 + 813 pp. (Vols. 1 & 2); 1st. Ed. 1980, US Government Printing Office, Washington.

51 Barr, John (Ed.)
The Environmental Handbook: Action Guide for the UK
This book, written for Friends of

the Earth, is adapted for British readers. Part 1 discusses the meaning of ecology, Part 2 is concerned with threats to the environment and alternative approaches to the environment, and Part 3 is an action guide for British people.
2nd Ed. 1971, Pan Books, London, SBN 345-02137-1 (pbk) xvi + 333 pp.; 1st. Ed. 1970, Ballantine, New York (Garrett de Bell (Ed.)).

52 Barrett, Brendan F. D. & Riki Therivel
Environmental Policy and Impact Assessment in Japan
The authors of this book show that the Japanese environmental policy making process has important implications, not only for the West but also for the newly industrialised countries of Asia. They examine the framework for environmental policy making, and outline the contemporary system for environmental management. They discuss the practical realities of environmental impact assessment in Japan.
1991, Routledge, London & New York, ISBN 0-415-03852-9 (hbk) 320 pp.

53 Barry, Brian J. L.
Long-Wave Patterns in Economic Development and Political Behavior
The author of this book argues strongly for the rehabilitation of long-wave theory in economics,

and aims to establish a sound empirical basis for it as a useful approach to economics and history. He thinks that the next step in controlling our economy is to control its longer swings, something that is beyond the range of Keynesian analysis.

1991, The Johns Hopkins University Press, Baltimore, MD, USA & London, ISBN 0-8018-4036-8, xiii + 241 pp.

54 Bartz, Bettina, Helmut Opitz & Elisabeth Richter
World Guide to Libraries

This book gives details for many countries of: national libraries, university and college libraries, government libraries, corporate and business libraries, public libraries, special libraries, etc.

10th. Ed. 1991, K. G. Saur, Munich, London & New York, ISBN 3-598-20541-4 & ISSNs 0939-1959 & 0936-0085 (hbk) xxvii + 1039 pp.

55 Baumol, William J. & Alan S. Blinder
Economics: Principles and Policy

This latest edtion of a leading economics textbook has an increased international emphasis, highlights ideas likely to be of lasting significance, and discusses many 'real world' examples.

5th. Ed. 1991, Harcourt, Brace, Jovanovich, New York & London, ISBN 0-15-518863-1 (pbk) xxxi + 912 pp.; 1st. Ed. 1979.

56 Bayliss, John & N. J. Rengger (Eds.)
Dilemmas of World Politics: International Issues in a Changing World

World politics has experienced more changes in the past two years than in the 45 years before that. In this book, leading academics identify key problems and dilemmas arising from these changes. They are concerned mainly with theories, institutions, and issues.

1992, Oxford University Press, ISBNs 0-19-827351-7 (hbk) & 0-19-827350-9 (pbk) xii + 438 pp.

57 Becker, Robert O.
Cross Currents: The Promise of Electromedicine, the Perils of Electropollution

This book describes the rapid rise of electromedicine, which could unlock the secrets of healing, and the parallel increase of man-made electromagnetic fields, which could become another source of widespread pollution. Its author is a top researcher in biological electricity.

1990, Tarcher, Los Angeles, CA, USA, ISBN 0-874-72609-0 (pbk) 352 pp.

58 Becker, Theodore, L. (Ed.)
Applying Quantum Theory to Political Phenomena

The contributors to this book argue that a Newtonian world view can no longer explain political phenomena, and is related to

classical liberal democracy and thereby to indirect, representative democracy. On the other hand, quantum thinking is linked to participatory democratic thinking, a more direct, appropriate, and pure form of democracy.
1991, Praeger, New York, ISBN 0-275-93310-5 (hbk) 248 pp.

59 Beer, Stafford
Platform for Change
This book provides a broad perspective of the approach of cybernetics, the science of communication and control, including applications to government and the achievement of social change. It is a good starting point for the scientific formulation of planet management principles and philosophy.
1975, Wiley, London & New York, ISBN 0-471-06189-1 (pbk) vii + 457 pp.

60 Beer, Stafford
The Heart of Enterprise
This book details how to apply the cybernetic approach to business management, and explores many models for improved management systems.
1979, Wiley, Chichester & New York, ISBN 0-471-27599-9 (hbk) xiv + 582 pp.

61 Beer, Stafford
Brain of the Firm
This book is about large and complicated systems, such as animals, computers, and economies, but mostly about the control of enterprises.
2nd. Ed. 1981, Wiley, Chichester & New York, ISBN 0-417 pp. (hbk) xiii + 417 pp.; 1st. Ed. 1972, Allen Lane, London.

62 Beer, Stafford
Diagnosing the System for Organisations
This book presents a new way of looking at organisational structure. It aims to show how to design such a structure, and how to diagnose a faulty organisational structure. It should be *worked through* while being read, using the author's exercises.
1985, Wiley, Chichester & New York, ISBN 0-471-90675-1 (hbk) 152 pp.

63 Beets, William C.
Raising and Sustaining Productivity of Smallholder Farming Systems
This book is based on 22 years of fieldwork, especially in Africa, Asia and the Caribbean. It brings together most relevant knowledge about the various tropical farming systems. It uses a multidisciplinary approach, stressing the interactions between agrotechnical, economic, environmental, sociological, institutional and political aspects. It aims to provide a framework for agricultural development in the tropics, emphasising sustainable increase of the farming

productivity.
1990, Ag Be Publishing, Alkmaar, Holland, ISBN 974-85676-4-8 (pbk) xvi + 740 pp.

64 Begg, David, Stanley Fisher & Rudiger Dornbusch
Economics
This is a leading introductory textbook on economics, introducing clearly both the theory and the reality of contemporary economics. This new edition includes material on European integration in the 1990s, environmental issues, unemployment, and local government finance.
3rd. Ed. 1991, McGraw-Hill, London & New York, ISBN 0-07-707245-6 (pbk) xx + 667 pp.; 1st. Ed. 1984.

65 Behrman, Jack N. & Robert E. Grosse
International Business and Governments: Issues and Institutions
This book shows how governments relate to transnational corporations (TNCs) at international, national and local levels; it provides a comprehensive survey of the relevant issues. TNCs and governments need to cooperate on several groups of issues.
1990, University of South Carolina Press, Columbia, SC, USA, ISBN 0-87249-696-1 (hbk) 434 pp.

66 Bell, Daniel
The Coming of Post-Industrial Society: A Venture in Social Forecasting
This book presents its author's view of a possible 'post-industrial society', making extensive use of high technology.
3rd Ed. 1976, Basic Books, New York, ISBN 0-465-09713-8 (pbk) 532pp.; 1st Ed. 1973.

67 Bellah, Robert N. et al.
The Good Society
The authors of this book hope that Americans can transform their society into a good society that actively involves all its people. There is a need for a new experiment in democracy, a newly extended and improved set of democratic institutions, within which we as citizens can know better what we want and what we ought to want to sustain life on Earth for ourselves and future generations. The author proposes a 'communitarian' approach, especially for American politics.
1991, Knopf, New York, ISBN 0-679-40098-2 (hbk) vii + 347 pp.

68 Bellamy, David J.
How Green Are You?
This book provides information and projects about ecology and environmental concerns, and shows children and families how to conserve energy, protect wildlife, and reduce pollution.
1991, Frances Lincoln, London & Potter Publishers, distributed by Crown Publishers, New York, ISBN 0-517-58429-8 (pbk) 31 pp.

69 Bello, Walden
Brave New Third World: Strategies for Survival in the Global Economy
This book discusses the reasons for the sharp increase in poverty and inequality throughout the Third World in the 1980s. It shows how only some developing countries have shown some success in breaking out of the vicious circle of their poverty, and outlines the key elements of a strategy to enable developing countries to become viable economically and begin their sustainable development.
1990, Earthscan, London, ISBN 1-85383-086-0 (pbk) 96 pp.

70 Bender, Arnold E.
Dictionary of Nutrition
This book provides information about different foods and food technologies, and includes tables of recommended daily intakes of different components of diet.
1990, Butterworth, London, ISBN 0-408-0373-9 (hbk) 352 pp.

71 Benedict, Ruth
Patterns of Culture
This book discusses culture in relation to human society, showing how the complex relationship between society and the individual is at the core of any viable society.
1935, Routledge & Kegan Paul, London, 104 pp.

72 Bennett, Graham
Dilemmas: Coping with Environmental Problems
This book considers six very different examples, from Europe and North America, of the conflicts of interest and policy arising from environmental issues. Its author shows the dilemmas that policy makers and environmentalists almost always face. Some of them arise from muddle and incompetence, but there are also the more serious problems presented by vested interests and the issues' complexity, which cannot always be resolved without compromise.
1992, Earthscan, London, ISBN 1-85383-021-6 (pbk) 288 pp.

73 Bennett, Robert & Robert Estall (Eds.)
Global Change and Challenge
This book examines some of the crucial challenges facing society in the 1990s, shows how geography can contribute to their understanding and management, and presents various views on the geography of change.
1991, Routledge, London & New York, ISBNs 0-415-00142-0 (hbk) & ISBN 0-415-00143-9 (pbk) 288 pp.

74 Bergesen, Helge Ole, Magnar Norderhaug, & Georg Parmann (Eds.)
Green Globe Yearbook 1992: An Independent Publication on Environment and Development

from the Fridtjof Nansen Institute, Norway

This book contains eleven articles on various aspects of environment and development, together with details of 31 international agreements on environment and development. It lists and describes 18 inter-governmental and 24 non-governmental organisations.

1992, Oxford University Press, Oxford & New York, ISBN 0-19-823322-1 (hbk) 305 pp.

75 Berkhaut, Frans
Radioactive Waste: Politics and Technology

This book analyses the radioactive waste management process, viewed as a technological system for which strategies have to be selected, ranging from dispersion and dilution to long-term containment by storage and disposal; each method has its own costs and benefits.

1991, Routledge, London & New York, ISBN 0-415-05492-3 (hbk) & 0-415-05493-1 (pbk) x + 256 pp.

76 Bernstein, Henry, Ben Crow & Hazel Johnson (Eds.)
Rural Livelihoods: Crises and Responses

This book discusses many rural issues arising in the Third World, such as: (1) securing a livelihood; (2) access to resources and employment; (3) organisation and changes in rural life; (4) how

much people respond to crisis and how they are helped or hindered by others.

1992, Oxford University Press, London & New York, ISBNs 0-19-877334-X (hbk) & 0-19-877335-8 (pbk) xi + 324 pp.

77 Berry, Peter S. & Catherine Lydford (Eds.)
Guide to Resources in Environmental Education

This guide contains about 40 sections, each on a given topic or aspect of the environment. Each section has subsections on: books and textbooks; slides, filmstrips and cassettes; films and videos for hire; packs, kits and games; charts and posters; simulations and computer software; organisations. There is a brief introduction about the arrangement of the entries and abbreviations used.

1990, The Conservation Trust, Reading, ISBN 0-907153-37-2 (semi-hardback) about 300 pp.

78 Berry, Thomas
The Dream of the Earth

This book considers spiritual approaches to resolving the environmental crisis. It analyses the present crisis in its historical context, and describes the new ecological consciousness evolving alongside and through current ideologies.

1990, Sierra Club Books, San Francisco, ISBN 0-87156-622-2 (pbk) 224 pp.

79 Biehl, Janet
Rethinking Ecofeminist Politics
This book raises several crucial problems that an adequate ecofeminist theory must address.
1991, South End Press, Boston, ISBN 0-89608-391-8 (pbk) pp.

80 Bignell, Victor & Joyce Fortune
Understanding Systems Failures
This book outlines a common approach to the understanding of many kinds of failures of machines, individuals, groups and businesses.
1984, Manchester University Press, Manchester, ISBN 0-7190-0973-1 (pbk) viii + 216 pp.

81 Bilderbeek, A. S. E. & A. van Buitenen (Eds.)
The Effectiveness of International Environmental Law
This book is the result of global consultation on the development and enforcement of international environmental law, with special reference to preserving biological diversity.
1992, IOS Press, Amsterdam, Oxford & Burke, VA, USA, ISBN 90-5199-014-6 (hbk) 200 pp.

82 Blinder, Alan S.
Growing Together: An Alternative Economic Strategy for the 1990s
This book argues that economic growth is a good thing, but that a radical transformation in its nature is needed.
1991, Whittle Direct Books, Knoxville, TN, USA, (hbk) 87 pp.

83 Bloom, William
The New Age
This book covers: the nature of the 'New Age', 'inner voice', human potential, healing, Gaia, and practical implications.
1991, Channel 4 Books, Rider, London, ISBN 0-7126-4804-6 (pbk) xix + 235 pp.

84 Bloomfield, Brian P.
Modelling the World: The Social Constructions of System Analysts
This book discusses the philosophy and application of system dynamics to world problems, with special reference to global modelling and to the motivations of those who have developed global models.
1986, Blackwell, Oxford & New York, ISBN 0-631-14163-4 (hbk) xv + 222 pp.

85 Blount, P. Clavell
Ideas into Action
The author of this book introduces a proposal to fuel higher UK productivity by 'Ideas Coordination', which would include a nationwide network of coordinated suggestion schemes, so that people's good ideas could be put in to action with much less delay. It is amazing that the author's proposals, or something

like them, have *still* not been implemented in the UK; he continues to promote them actively.
1962, The Clair Press, London, (hbk) 180 pp.

86 Blowers, Andrew (Ed.)
Planning for a Sustainable Environment
The contributors to this book consider how to plan for a sustainable environment, especially a sustainable urban environment, in the face of rising environmental pressures. They analyse present and future sustainability levels, while considering energy, natural resources, pollution and waste, building, livelihood, and transport.
1992, Earthscan, ISBN 1-85383-145-X (pbk) 240 pp.

87 Blowers, Andrew, David Lowry & Barry D. Solomon
The International Politics of Nuclear Waste
This account of the nuclear waste problem surveys its origins, describes the dramatic battle of the dumps, compares the search for 'nuclear oases' across five countries, and explains their differences. It suggests how to handle this ultimately insoluble problem.
1991, Macmillan, London & New York, 0-333-49363-X (hbk), 0-333-49364-8 (pbk) xxii + 362 pp.

88 Blunden, John & Nigel Curry
A Future for Our Countryside
This book examines various possible outcomes of the radical transformation now being undergone by the British countryside. With the help of revealing photographs, it presents the views of a wide range of people concerned with and interested in the future of the countryside.
1988, Blackwell, Oxford, ISBN 0-631-16272-0 (pbk) xx + 220 pp.

89 Blundon, John & Alan Reddish
Energy, Resources and Environment
This book analyses the impact of current systems of energy and mineral supply and use, from acid rain to radioactive pollution. It explores possible alternatives, including: mineral substitution and recycling, energy conservation, and sustainable energy supplies.
1991, Hodder & Stoughton, London, ISBN 0-340-53361-7 (pbk) v + 340 pp.

90 Bochuan, He
China on the Edge: The Crisis of Ecology and Development
This is the English translation of a book widely circulated in China until its printing was abruptly stopped by the Chinese Government. It discusses at least 12 urgent Chinese problems. The author sees little sign of improvement in the near future, because

of a prevailing short-term
outlook.
1991, China Books and Periodicals,
San Francisco, CA, USA, (hbk) &
(pbk) 208 pp.

91 Bocock, Robert & Kenneth
Thompson (Eds.)
Social and Cultural Forms of
Modernity
This introductory textbook examines the social and cultural dimensions of 20th century
industrialised societies. It is
designed as an introduction to
modern societies and modern
sociological analyses.
1992, Polity Press, Cambridge,
ISBNs 0-7456-09643-5 (hbk) &
0-7456-0964-3 (pbk) viii + 484 pp.

92 Boden, Margaret
The Creative Mind: Myths &
Mechanisms
The theme of this book is the
human mind, how it works, and
how it can surpass itself. It considers how human intuition and
creativity operate, and uses concepts drawn from 'artificial intelligence' theory.
1990, Weidenfeld and Nicholson,
London, ISBN 0-297-82069-9 (hbk)
xiii + 303 pp.

93 Boden, Margaret (Ed.)
The Philosophy of Artificial
Intelligence
The papers in this book provide a
wide range of commentary on the
philosophy of 'artificial intelligence' and represent various

viewpoints. They show that this
science of intelligence in general
embraces significantly different
methodologies, including computer science, logic, philosophy,
and neurobiology.
1990, Oxford University Press,
Oxford & New York, ISBNs
0-19-824855-5 (hbk) &
0-19-824854-7 (pbk) vii + 452 pp.

94 Bodmer, Frederick
The Loom of Language: A Guide
to Foreign Languages for the
Home Student
This book is one of the most
readable, useful, and wide-ranging introductions to human languages ever published.
1944, Allen & Unwin, London, (hbk)
669 pp.

95 Boehmer-Christiansen, Sonja
& Jim Skea
Acid Politics: Environmental and
Energy Policies in Britain and
Germany
This book argues that acid rain is a
key environmental issue in
Europe. It explores the reactions
to British and German policies in
relation to their international
context.
1991, Belhaven Press, London &
New York, ISBN 1-85293-116-7
(hbk) xiii + 296 pp.

96 Bohm, David
Wholeness and the Implicate
Order
This book develops a theory of

quantum physics, which treats all existence, including matter and consciousness, as an unbroken whole. It introduces a concept of 'implicate order', complementary to the 'explicate order' of the manifest physical universe.
1980, Routledge & Kegan Paul, London & Boston, ISBN 0-7100-0366-8 (hbk) xv + 224 pp.

97 Bohm, David
Unfolding Meaning: A Weekend of Dialogue with Bohm, David
In this book, the author's ideas and the dialogue developing from them present the basis for an exciting set of new possibilities for the release of creativity, and for more harmonious and fruitful relationships between individuals, groups, and even nations.
2nd Ed 1987, Ark Paperbacks, London & New York, ISBN 0-7448-0064-1 (pbk), xiii + 177 pp.; 1st Ed 1985, Routledge & Kegan Paul.

98 Bohm, David & F. David Peat
Science, Order and Creativity
The authors of this thought-provoking book propose a return to a greater creativity and better communication in the sciences. They offer insights into how scientific theories come into being, how to eliminate blocks to creativity, and how science can lead to better understanding of society, the human situation, and the human mind.

2nd. Ed., 1989, Routledge, London, ISBN 0-415-03079-X, v + 280 pp.; 1st. Ed., 1987, Bantam Books, New York.

99 Boland, Lawrence A.
The Principles of Economics: Some Lies My Teachers Told Me
This book is written for those who wish to understand neoclassical economics, especially for those who wish to develop a critical understanding of whether to improve it or only criticise it.
1992, Routledge, London & New York, ISBN 0-415-06433-3 (hbk) xv + 223 pp.

100 Bookchin, Murray
The Modern Crisis
This book presents the author's thinking on social ecology, moral economics, and solutions to the worsening social and ecological crisis.
1986, New Society Publishers, Philadelphia, ISBN 0-86571-084-8 (pbk) xi + 167 pp.

101 Bormann, F. Herbert & Stephen R. Kellert (Eds.)
Ecology, Economics, Ethics: The Broken Circle
This collection of reprinted articles can be used as an introductory text on aspects of ecology, economics, and ethics by general readers, students, and professionals.
1991, Yale University Press, Providence, RI, USA & London, ISBN 0-300-04976-5 (hbk) 236 pp.

102 Boserup, Este
Women's Role in Economic Development
This book investigates what happens to women during the process of economic and social growth throughout the Third World.
2nd. Ed. 1989, Earthscan, London, ISBN 1-85383-040-2 (pbk) iv + 283 pp; 1st. Ed. 1970.

103 Boulding, Elise (Ed.)
New Agendas for Peace Research: Conflict and Security Reexamined
This book presents papers about emerging best-case possibilities for peace building; they are derived from the 25th Anniversary Conference of the International Peace Research Association.
1992, Lynne Rienner Publishers, Boulder, CO, USA, ISBN 1-55587-290-5 (hbk) 250 pp.

104 Boulding, Kenneth
The World as a Total System
This book examines how far the Earth is a total system, of interacting parts, and how far a 'great mosaic' of isolated subsystems with little or no mutual impact. It describes some of the more important of these 'static' and 'dynamic' systems.
1985, Sage Publications, Beverly Hills, CA, USA & London, ISBN 0-8039-1648-8-5 (hbk) 183 pp.

105 Boulding, Kenneth E.
Towards a New Economics: Critical Essays on Ecology, Distribution, and Other Themes
This book is the seventh volume of the author's collected papers, representing his work during the 1980s; the previous six volumes, published by Colorado Associated Universties Press, are now out of print. The book covers about 125 themes, several of which, including viewing the world as a total system, are considered in detail.
1992, Edward Elgar, Aldershot, Hampshire & Brookfield, VT, USA, ISBN 1-85278-568-3 (hbk) xi + 344 pp.

106 Bowles, Samuel, David M. Gordon & Thomas E. Weisskopf
After the Waste Land: A Democratic Economics for the Year 2000
This book argues for an 'economics as if people mattered'; its authors see an socially and economically deteriorating USA, behind its mask of superficial economic progress and prosperity. They advocate a much less wasteful democratic and more egalitarian economy, that guarantees all citizens basic rights to an economic livelihood, offers them all opportunities to participate in economic decisions affecting their lives, and ends working peoples' dependence on their employers.
1991, M. E. Sharpe, Armonk, NY,

ISBNs 0-87332-644-X (hbk) & 0-87332-645-8 (pbk) 352 pp.

107 Boyd, Andrew
An Atlas of World Affairs

This extensively revised edition of an important atlas describes the events, people, and sectional interests, which have shaped history from 1945 into the 1990s. With its clear text and excellent maps, it presents facts about the past and contemporary world situation, and provides a compact guide to the world's troubled areas.
9th. Ed. 1992, Routledge, London & New York, ISBNs 0-415-06624-7 (hbk) & ISBN 0-415-06625-5 (pbk) 240 pp.

108 Boyd, Gavin
Structuring International Economic Cooperation

This policy-oriented book is also an advanced text on international economic organisations and the politics of international trade and direct investment. It examines the effects of political changes in industrialised democracies and the activities of international firms.
1991, Pinter Publishers, London, ISBN 0-86187-821-3 (hbk) viii + 163 pp.

109 Boyle, Stewart & John Ardill
The Greenhouse Effect

This book outlines clearly and objectively the reasons for current climate changes. It provides a practical guide to the greenhouse effect for non-experts, answers many important questions about it, refers to energy policy choices, and proposes a range of solutions.
1989, Hodder and Stoughton, London, ISBN 0-450-50638-X (pbk) xi + 298 pp.

110 Brackley, Peter (Compiled)
Energy and Environmental Terms: A Glossary

This book attempts to explain simply the technical terms met in the day-to-day management of a large energy enterprise, or when considering environmental and energy policy issues.
1988, Gower, Aldershot, Hampshire & Brookfield, VT, USA, ISBN 0-566-05759-X (pbk) ix + 189 pp.

111 Bramwell, Anna
Ecology in the 21st Century: A History

This book identifies the roots of ideas in the modern ecology movement from the biological and economic strands of the scientific community. A practical theory of ecology is contrasted with the biological approach, which has a more holistic outlook. There is a useful chapter on the relationship between energy and economics.
1989, Yale University Press, New Haven, CT, USA & London, ISBNs 0-300-04343-0 (hbk) & 0-300-04521-2 (pbk) xii + 292 pp.

112 Brandt Commission, The North-South: A Programme for Survival
This books reports the findings of the Independent Commission on International Development Issues on the urgent problems of world inequality and poverty and the world economic system's failure to solve them. It presents bold recommendations and a programme of priorities.
1980, Pan Books, London, ISBN 0-330-26140-1 (pbk) 304 pp.

113 Brandt Commission, The Common Crisis: North-South: Co-operation for World Recovery
This book is a sequel to *North-South: A Programme for Survival* (No. 112). It describes the different elements of the crisis in trade, energy, and food, and concentrates on the overriding problems of how to provide finance, reverse the decline in world trade, and revive the economy. It provides a message of hope that the world can eventually cooperate and expand to become safer and more prosperous.
1983, Pan Books, London, ISBN 0-330-28130-5 (pbk) ix + 174 pp.

114 Brennan, M. (Compiled), Dooge, J. C. I. et al. (Ed.)
An Agenda of Science for Environment and Development into the 21st Century
This book brings together the understanding and judgement of the world's scientific community on issues of the highest priority for the future of the environment and development. It looks beyond the current state of the art, formulates a research agenda for the environment and development, and identifies the scientific knowledge base that will be needed for rational policy discussions during the coming decades.
1992, Cambridge University Press, Cambridge & New York, ISBNs 0-521-43174-3 (hbk) & 0-521-43761-X (pbk) vii + 331 pp.

115 Breton, Denise & Christopher Largent
Spiritual Evolution Goes to the Marketplace
The authors of this book explore the frontier between religion and economics, and analyse the spiritual transformation that is needed to support a just and ethical economic system.
1992, Green Print, London, ISBNs 1-85425-075-2 (hbk) & 1-85425-075-2 (pbk) 378 pp.

116 Briggs, John P. & F. David Peat
Looking Glass Universe: The Emerging Science of Wholeness
This book presents the ideas of several well-known scientists, whose startling new theories could revolutionise our understanding of the universe. Although they work in very different fields, these scientists have

reached similar concepts and principles, which may fit together into one new paradigm giving scientific meaning to the ancient mystical idea that the universe is One.
2nd Ed 1985, Fontana Books, London, ISBN 0-00-636929-4 (pbk) 320 pp.; 1st Ed 1984, Simon & Schuster, New York.

117 British Medical Association Hazardous Waste & Economic Health
This report provides a comprehensive guide to all aspects of hazardous waste. It presents clearly its nature, existing methods of for its treatment and disposal, and evidence linking exposure to waste with ill health. It addresses the increasing public concern about waste, the problems of determining its effects, methods of controlling it by legislation, and how research can affect future developments.
1991, Oxford University Press, Oxford & New York, ISBNs 0-19-217782-6 (hbk) & 0-19-286142-5 (pbk) xiv + 242 pp.

118 Brotchie, John, Michael Batty, Peter Hall & Peter Newton (Eds.)
Cities of the 21st Century: New Technologies and Spatial Systems
This book explores various aspects of changing cities and new lifestyles. It examines the effects and impacts of new forms of transport, communications and information technologies. There is a discussion of the emergence of designer cities, deliberately created for the sharing and development of new technologies and the sustainability of current economic development.
1991, Longman Cheshire, Melbourne, Australia & Wiley/ Halsted, New York, ISBNs 0-582-87126-3 (pbk) & 0-470-21742-1 (hbk) xii + 446 pp.

119 Brown, Lester R. (Ed.)
The Worldwatch Reader on Global Environmental Issues
In response to the need for timely information about the urgent human and planetary situation, this book presents an anthology of papers written by the Worldwatch Institute's environmental research team, and previously published in *World Watch* magazine (No. 1176). It offers an indepth diagnosis of the Earth's problems, and a practical vision of how to create an environmentally responsible future, in a highly readable style. It presents both the bad news, for example about pollution and environmental damage, and the good news, for example about countering the greenhouse effect, developing renewable energy, and leading sustainable and satisfying lives. Altogeher, it provides a valuable interdiscipinary perspective.

1991, W. W. Norton & Co., New York & London, ISBN 0-393-03007-5 (pbk) 336 pp.

120 Brown, Lester R. et al. (Linda Starke (Ed.))
State of the World 1984: A World-watch Institute Report on Progress towards a Sustainable Society
This book is the first in a series of annual assessments of the world situation.
1984, W. W. Norton & Co., New York & London, ISBN 0-393-30176-1 (pbk) 272 pp.

121 Brown, Lester R. et al. (Linda Starke (Ed.))
State of the World 1985: A World-watch Institute Report on Progress towards a Sustainable Society
This book is the second in a series of annual assessments of the world situation.
1985, W. W. Norton & Co., New York & London, ISBN 0-393-30218-0 (pbk) 301 pp.

122 Brown, Lester R. et al. (Linda Starke (Ed.))
State of the World 1986: A World-watch Institute Report on Progress towards a Sustainable Society
This book is the third in a series of annual assessments of the world situation.
1986, W. W. Norton & Co., New York & London, ISBN 0-393-30255-5 (pbk) xvii + 263 pp.

123 Brown, Lester R. et al. (Linda Starke (Ed.))
State of the World 1987: A World-watch Institute Report on Progress towards a Sustainable Society
This book is the fourth in a series of annual assessments of the world situation.
1987, W. W. Norton & Co., New York & London, ISBN 0-393-30381-6 (pbk) xvii + 268 pp.

124 Brown, Lester R. et al. (Linda Starke (Ed.))
State of the World 1988: A World-watch Institute Report on Progress towards a Sustainable Society
This book is the fifth in a series of annual assessments of the world situation.
1988, W. W. Norton & Co., New York & London, ISBN 0-393-30440-X (pbk) xvii + 237 pp.

125 Brown, Lester R. et al. (Linda Starke (Ed.))
State of the World 1989: A World-watch Institute Report on Progress towards a Sustainable Society
This book is the sixth in a series of annual assessments of the world situation.
1989, W. W. Norton & Co., New York & London, ISBN 0-393-30567-8 (pbk) xvi + 256 pp.

126 Brown, Lester R. et al. (Linda Starke (Ed.))
State of the World 1990: A Worldwatch Institute Report on Progress towards a Sustainable Society
This book is the seventh in a series of annual assessments of the world situation.
1990, W. W. Norton & Co., New York & Unwin Paperbacks, London, ISBN 0-04-440711-4 (pbk) xvi + 253 pp.

127 Brown, Lester R. et al. (Linda Starke (Ed.))
State of the World 1991: A Worldwatch Institute Report on Progress towards a Sustainable Society
This book is the eighth in a series of annual assessments of the world situation.
1991, W. W. Norton & Co., New York & Earthscan, London, ISBN 1-85383-114-X (pbk) xvii + 254 pp.

128 Brown, Lester R. et al. (Linda Starke (Ed.))
State of the World 1992: A Worldwatch Institute Report on Progress towards a Sustainable Society
This book is the ninth in a series of annual assessments of the world situation.
1992, W. W. Norton & Co., New York & Earthscan, London, ISBN 1-85383-117-4 (pbk) xv + 256 pp.

129 Brown, Lester R. et al. (Linda Starke (Ed.))
Saving the Planet: How to Shape an Environmentally Sustainable Economy
This book was written especially for the June 1992 UNCED Conference, and for World Watch magazine's many thousands of regular readers. UNCED presented a tremendous challenge, to go beyond viewing environmental issues as single problems, and begin moving towards the basic economic and social reforms needed to save our planet and ourselves. The book is the first in the Worldwatch Institute's new Environmental Alert Series, an important addition to the Worldwatch Paper Series, (1975 on, No. 1177), the State of the World Series (1984 on, Nos. 120-128 & 130), and World Watch magazine (No. 1176).
The new series aims to provide relatively short but concise and lively books, assessing urgent contemporary problems and issues, and written by the Institute's experienced research staff. This first book takes an exceptionally broad view of the main issues of concern to the world, while the other books will focus on much more specific topics. They aim to provide comprehensive, up-to-date information and fresh insights into their subject matter. The authors hope that the books will help to raise

environmental literacy to the point where the process of reform becomes self-sustaining.

The book puts together a vision of a global economy that does not compromise the prospects of future generations, and confronts head-on the changes necessary to realise that vision. Part 1 considers energy efficiency, solar energy and other forms of renewable energy, recycling, protecting biodiversity, feeding a rising world population, and how to stabilise world population. Part 2 specifies the major restructuring of economic systems, tax systems, industrial and developing economies, international aid, etc., that would promote the shift to a sustainable and altogether more satisfying economy. Part 3 presents the challenge ahead.

2nd. Ed. 1992, Earthscan, London, ISBN 1-85383-133-6 (pbk) 224 pp.; 1st. Ed. 1991, W. W. Norton & Co., New York.

130 Brown, Lester R. et al. (Linda Starke (Ed.))
State of the World 1993: A Worldwatch Institute Report on Progress Toward a Sustainable Society
This book is the special tenth anniversary edition of this annual series, and examines the options for restoring our planet's health. It exposes the links between the environment and the global economy, and shows how we can use

water, energy, and other resources in a sustainable way, not consuming the resource base for future generations. It explores the pivotal role of private enterprise in developing a sustainable economy. Topics covered by its chapters include: the state of the world's indigenous peoples, the future of the military, the role of women in development, water scarcity, international trade and the environment, and energy for development.
1993, W. W. Norton & Co., New York & Earthscan, London, ISBN 1-85383-135-2 (pbk) 280 pp.

131 Brown, Lester R., Christopher Flavin & Hal Kane
Vital Signs: The Trends That Are Shaping Our Future
This book considers the real trends, many of them environmental, that are determining the kind of world in which we and our children will live in the future. Vital indicators like these have until recently often been hard to obtain, but this book, the first edition of an annual publication, will fill the gap. Part 1 describes 36 key trends in food, agricultural resources, energy, the atmosphere, the global economy, society, and the military. Each trend is presented on facing pages, of text and data and graphs. Part 2 contains essays on important trends for which no comparative, historical global

data are available. The book presents many key statistics and underlying trends.

1992, W. W. Norton & Co., New York & Earthscan, London, ISBN 1-85383-141-7 (pbk) 144 pp.

132 Bruyn, Severyn T.
A Future for the American Economy: The Social Market
This book is about the many contradictions in the American economy today, especially the central contradiction of government control versus market freedom. The resolution of this contradiction is important for the whole world, not only the USA. The author's main theme is that social factors, rather than purely economic factors, are at its root; he views the social organisation of the private sector of the economy as the key to the free market system.

1991, Stanford University Press, Stanford, CA, USA, ISBN 0-8047-1872-5 (hbk) viii + 424 pp.

133 Buarque, Cristovam
The End of Economics: Etihcs and the Disorder of Progress
The author of this book argues that, if economic theory is to command either public respect or intellectual confidence in relation to the modern world, it must incorporate the need for an ethical system. It must accept that there must be limits to growth and that the increasing inequality between classes and countries is neither morally tolerable nor politically sensible. The concept of 'economic progress' needs to be rethought; technological advances must respect nature, and economic theories must be applied with regard for human consequences.

1993, Zed Books, London, ISBNs 1-85649-079-1 (hbk) & 1-85649-098-6 (pbk) 192 pp.

134 Buatsi, Sosthenes
Technology Transfer: Nine Case Studies
This book presents nine case studies in technology transfer, which demonstrate experience of different technologies in different nations.

1988, Intermediate Technology Publications, London, ISBN 0-946668-29-X (pbk) 64 pp.

135 Buchholz, Roger A., Alfred A. Marcus & James E. Post
Managing Environmental Issues: A Casebook
This book presents about 20 recent case studies, from various parts of the world, which examine many environmental issues and corporate responses to them. It presents changing environmental perspectives, discusses public policy and economics in relation to the environment, and considers the role of business in the 'new environmentalism'.

1992, Prentice-Hall, Englewood Cliffs, NJ, USA & London, ISBN 0-13-563891-7 (pbk) xviii + 286 pp.

136 Budd, Stanley A. & Alun Jones
The European Community: A Guide to the Maze
This book mostly looks briefly at the European Community (EC) and its various policies. Details are given of many topics relating to the EC, ranging from its origin and achievements to current and likely future developments. There are references to more detailed information.
4th. Ed. 1991, Kogan Page, London, ISBN 0-7494-0310-1 (pbk) 218 pp.; 1st. Ed. 1985.

137 Buergenthal, Thomas
International Human Rights in a Nutshell
This book shows how international human rights law has evolved in the institutional context of international organisations, and provides a self-contained introduction to the international law of human rights.
1988, West Publishing Co., St. Paul, MN, USA, ISBN 0-314-43046-6 (pbk) xii + 283 pp.

138 Bullock, Alan, Oliver Stallybrass & Stephen Trombley (Eds.)
The Fontana Dictionary of Modern Thought
This book is a comprehensive compendium of information about contemporary (20th century) science, philosophy, ideas and thought, as well as current affairs, the humanities, literature, music and the visual arts. However, its coverage of ideas and new paradigms, potentially significant for the 21st century, is less complete. Its entries are arranged in alphabetical order, and written by many distinguished contributors. It is more discursive than an ordinary dictionary, more compact than an encyclopedia, and more selective than either. It is an invaluable reference book, also useful for browsing.
2nd Ed. 1988, Fontana Paperbacks, London, ISBN 0-00-686129-6 (pbk) xxvi + 917 pp.; 1st Ed. 1977.

139 Bulmer, Simon, Stephen George & Andrew Scott (Eds.)
The United Kingdom and EC Membership Evaluated
This book examines the gains and losses of European Community membership for the UK. Aspects covered include: EC membership, economic policy, government and legal system, foreign relations, social and educational policies, and weighing the gains and losses.
1992, Pinter Publishers, London, ISBN 0-86187-867-1 (hbk) xxi + 271 pp.

140 Bunyard, Peter & Fern Morgan-Grenville (Eds.)
The Green Alternative: Guide to Good Living

This book discusses what we can do as individuals to prevent environmental destruction and other major threats to our planet. It poses and answers about 500 questions on a wide range of subjects, including: conservation and nature, pollution, agriculture, nutrition, health, energy, transport, land use and urbanisation, arms races, the Third World, the Green philosophy, education, and politics. It not only shows what is wrong, but makes many practical and positive suggestions about what we can do to bring about a 'Green alternative'.

1987, Methuen, London, ISBNs 0-413-60280-X (hbk) & 0-413-42440-5 (pbk) xvi + 368 pp.

141 Burall, Paul
Green Design

This booklet provides a realistic and balanced introduction to Green design and the range of information available on environmentally safe technology. It includes a variety of industrial examples, which show how to use new opportunities and avoid pitfalls. It contains a list of information services on materials, processes and legislation.

1991, The Design Council, London, ISBN 0-85072-281-5 (pbk) 81 pp.

142 Burger, Julian
The Gaia Atlas of First Peoples: A Future for the Indigenous World

This book provides extensive information, with many maps and colour illustrations, about the world's more than 250 million indigenous peoples, about 5% of the global population. It describes the ways of life of many of their 5000 distinct groups, and outlines the often harmful effects of Western civilisation on them. It presents alternative options and visions for the indigenous peoples, including the indigenous movement (already with over 1,000 organisations worldwide), treaties for their rights, action by international organisations, and urging a UN Declaration of Indigenous Rights.

1990, Anchor/Doubleday, New York, ISBN 0-385-26653-7 (pbk) 191 pp.

143 Burling, Robbins
Patterns of Language: Structure, Variation, Change

This comprehensive introduction to linguistics integrates the variation and change in languages with more structural and static topics. It covers words, sounds, sentences, and the growth and evolution of languages, and surveys modern linguistics.

1992, Academic Press, San Diego, CA, USA, New York & London, ISBN 0-12-144920-3 (pbk) xiv + 461 pp.

144 Burrows, Brian C., Alan J. Mayne & Paul A. R. Newbury Into the 21st Century: A Handbook for a Sustainable Future

This book is intended to be a starting point for exploration of the complex web of human and planetary problems in greater depth. Its authors argue that what matters above all is an integrated solution, which can be achieved by approaching environmentalism in a rational manner and within a framework of coordinated action. This framework includes: a new economics of sustainable development, a responsible attitude to pollution and global environmental hazards, proposals for an outline of the politics of cooperation and conflict resolution, and guidelines to the appropriate use of science and technology.

Part 1 discusses and examines possible solutions for important specific problem areas. It does not consider human problems and world problems as isolated issues, but explores the linkages between them and shows that they all form parts of an interconnected whole. Tracing them back to their root causes, it shows that they seem to be caused mainly by fundamental flaws in human nature. Part 2 introduces new paradigms offering holistic approaches to world problems; it shows that unified approaches are much better than piecemeal solutions, and advocates the use of integrated planning methods at all levels of human activity. The emergence of a 'new social paradigm', based on a unified practical philosophy, could have a vast potential for alleviating, then solving world problems. Part 3 examines the consequences of the alternative approaches to world problems in terms of three scenarios resulting from continued neglect, piecemeal solutions, or a holistic approach. It shows that the holistic approach, together with an appropriate planet management system, is the only way in which we can hope to solve our growing man-made problems.

The final chapter summarises the author's main arguments, and presents their conclusions. It is followed by an annotated bibliography of over two hundred important, sometimes key, books and periodicals. In addition, each chapter ends with an 'Exploring Further' section, which provides annotated references to relevant literature, and suggests specific subproblems or aspects of problems that readers can investigate further. The authors wish to encourage feedback from the book's readers.

1991, Adamantine Press, London, ISBNs 0-7449-0041-7 (hbk) & 0-7449-0031-X (pbk) (ISSN 0954-6103) vi + 442 pp.

145 Burstein, Daniel
Euroquake: Europe's Explosive Economic Challenge Will Change the World

This book presents the results of interviews with many leaders worldwide, resulting in 100 predictions for the year 2000 about the 'battle of the capitalisms' of the 1990s.

1991, Simon & Schuster, NY, ISBN 0-671-09033-7 (hbk) 384 pp.

146 Busch, Lawrence, William B. Lacy, Jeffrey Burkhardt & Laura R. Lacy
Plants, Power and Profit: Social, Economic and Ethical Consequences of the New Biotechnologies

This book discusses the potential impact of new biotechnologies on plant breeding, agricultural science, and the structure of agriculture.

1991, Blackwell, Cambridge, MA, USA & Oxford, ISBN 1-55786-088-3 (hbk) 275 pp.

147 Butler, Kate, Sally Carr & Frances Sullivan
Citizen Advocacy: A Powerful Partnership

This book presents the principles of the 'citizen advocacy' approach to supporting disadvantaged people, which is rapidly attracting interest in the UK, USA, Canada, Australia, and Scandinavia.

1988, National Citizen Advocacy, London, ISBN 0-9514025-0-1 (pbk) ii + 62 pp.

148 Button, John (Ed.)
The Green Fuse: The Schumacher Lectures 1983-8

This book covers a wide variety of topics, including: Green philosophy, deep ecology, Gaia, Green spirituality, ecology, forests as a source of life, and a nopoverty society.

1990, Quartet Books, ISBN 0-7043-0121-0 (pbk) xiii + 198 pp.

149 Button, John
How to be Green

This book provides an important guide to how each of us can help to conserve and maintain planet Earth, and prevent environmental destruction. It contains hundreds of practical suggestions on many aspects of life. It shows how we can press for practical environmental action, and lists Green organisations and suppliers in the UK.

2nd Ed 1990, Century-Hutchinson, London, ISBN 0-7126-395-2, 237 pp.; 1st Ed 1989.

150 Button, John
New Green Pages: A Directory of Natural Products, Services, Resources and Ideas

This book aims to inspire its readers to find out more about what is being done, who is doing

it, and how to become more involved. It is an important reference book for the home, including many names and addresses, book reviews and quotations. It has a subject index, and an index of suppliers and organisations.
2nd. Ed. 1990, Macdonald Optima, London, ISBN 0-356-19190-5 (pbk) 352 pp.; 1st. Ed. 1988, with the title Green Pages.

151 Button, John (Ed.)
The Best of Resurgence Magazine
This book contains a selection of the best articles from *Resurgence* (No. 1135), with contibutions by distinguished writers.
1991, Green Books, Hartland, Bideford, Devon,
ISBN 1-870098-27-7 (pbk) 288 pp.

152 Bystydzienski, Jill M. (Ed.)
Women Transforming Politics: Worldwide Strategies for Empowerment
This book presents 13 case studies which examine women's culture and movements in various parts of the world, and identifies strategies and social conditions producing successful empowerment for women.
1992, Indiana University Press, Bloomington, IN, USA, (hbk) & (pbk) 229 pp.

153 Cadman, Martin & Geoffrey Payne (Eds.)
The Living City: Towards a Sustainable Future

This book analyses the critical issues facing cities in both the First and Third Worlds. Its editors conclude that city planning should respect social needs, ecological balance, and economic sustainability.
1990, Routledge, London & New York, ISBN 0-415-01250-3 (hbk) 288 pp.

154 Cairncross, Frances
Costing the Earth: What Governments Must Do; What Consumers Need to Know: How Businesses Can Profit
The author aims to explain how governments and the private sector can make people better off and at the same time improve their quality of life. She argues that the right government policies, combined with the innovative powers of industries, can unite ecological aims with industrial needs and political targets. She clearly explains widely debated issues and complex problems concerning us all, using worldwide illustrations. She proposes a range of radical but practical solutions. One major problem is how to put a market value on environmental benefits; the high cost of any environmental protection must be perceived as part of the costs for any product or service. Another is international agreement on legislation to make polluters pay. The book is full of useful facts, charts, and statistics, and

ends with a valuable checklist of suggestions for companies. It is an important contribution to the debate about how to develop sustainable economics, concluding that a truly Green economy is unlikely to be badly managed, and that a well-managed company can be Green relatively easily.

1991, The Economist Books, Business Books, London & (1992) Harvard Business School Press, Boston, MA, USA, ISBN 0-09-174918-2 (hbk) 256 pp.

155 Calder, Nigel
Spaceship Earth

This book presents many coloured illustrations of images from artificial satellites, which give us an unprecedented overview of our ever-changing planet. It shows how a global geography is emerging, which relates human life to natural systems worldwide and may lead to more exact answers to urgent planetary problems.

1991, Viking, London, & Viking Penguin, New York, ISBN 0-670-83628-1 (hbk) 208 pp.

156 Caldera, Anna Maria
Endangered Environments: Saving the World's Vanished Ecosystems

This book takes a close look at five of the Earth's most vital and threatened environments, through stunning colour plates

and an informative text. It explores how and why they are now seriously threatened, and describes current attempts to save them.

1991, Friedman Group, New York, ISBN 0-7924-5304-3 (hbk) 128 pp.

157 Calvocoressi, Peter
World Politics since 1945

This book is one of the most comprehensive and up-to-date analyses of world history and politics since the end of World War 2. *6th. Ed. 1991, Longman, London & New York, ISBN 0-582-07379-0 (pbk) xiv + 754 pp. ; 1st. Ed. 1968.*

158 Campbell, David
Writing Security: United States Foreign Policy and the Politics of Identity

This book offers a fundamental and unorthodox reappraisal of US foreign policy, now changing after the collapse of the Soviet threat.

1992, Manchester University Press, Manchester, ISBNs 0-7190-3417-8 (hbk) & 0-7190-3418-3 (pbk) ix + 269 pp.

159 Campbell, Eileen & J. H. Brennan
The Aquarian Guide to the New Age

This book contains a wide range of information about the 'New Age', a term widely used to cover a broad range of 'alternative' interests.

1990, The Aquarian Press, Well-ingborough, Northants., ISBN 0-85030-970-0 (pbk) 352 pp.

160 Capra, Fritjof
The Tao of Physics

This book links the views of modern physics, especially quantum theory, with Eastern mysticism.
3rd. Ed. 1991, Shambhala Publications, Boston, ISBN 0-87173-59-8 (pbk) 366 pp.; 2nd. Ed. 1975, Wildwood House, London; 1st. Ed. 1975.

161 Capra, Fritjof
The Turning Point: Science, Society and the Rising Culture

This book extends the holistic, systems-based approach of the new physics to other important areas of contemporary life, especially: medicine and health, psychology, economics, politics, and ecology. In all these areas, we are at a 'turning point' between old and new paradigms, whose combined effect will be to bring a major cultural transformation.
2nd. Ed. 1983, Flamingo Books, Fontana Paperbacks, London, ISBN 0-00-654017-1 (pbk) xxiii + 516 pp.; 1st. Ed. 1982, Wildwood House, London & Simon & Schuster, New York.

162 Capra, Fritjof
Uncommon Wisdom: Conversations with Remarkable People

This book records conversations of the author with leading exponents of new views of physics, spiritual freedom, psychology, ecology, economics, politics, futures, and feminism.
2nd. Ed. 1989, Flamingo Books, Fontana Paperbacks, London, ISBN 0-00-654341-3 (pbk) 352 pp.; 1st. Ed. 1988, Century Hutchinson, London.

163 Capra, Fritjof, David Steindl-Rast & Thomas Matus
Belonging to the Universe: Explorations on the Frontiers of Science and Spirituality

This book records a conversation between Fritjof Capra and two Benedictine monks. Capra presents a paradigm shift in science and technology towards a holistic, ecological viewpoint, which not only views something as a whole, but considers how this whole is embedded into larger wholes. He presents five criteria of new paradigms thinking in all of the sciences, expressed as shifts of viewpoint. Steindl-Rast and Matus present a parallel set of theological new paradigm shifts.
1991, HarperSanFrancisco, San Francisco, CA, USA, ISBN 0-06-250187-9 (hbk) 217 pp.

164 Carim, Enver et al. (Eds.)
Towards Sustainable Development

This book presents 14 reports on Third World development projects, presenting the lessons to be learned from both good and bad experiences of development, and showing how to make development more sustainable.

1987, Panos Publications, London & Alexandria, VA, USA, 1-870670-01-9 (pbk) xvi + 233 pp.

165 Carley, Michael & Ian Christie
Managing Sustainable Development

This book offers realistic guidelines on how to achieve genuinely sustainable development and sound environmental management at every level. The author argues for a new form of environmental management, linking government, business, and community groups in participatory 'action-centred networks', which have been shown to work in different contexts. Drawing on worldwide case studies, he provides a practical guide to better environmental management for professional people.

1992, Earthscan, London, ISBNs 1-85383-128-X (hbk) & 1-85383-129-8 (pbk) 256 pp.

166 Carlson, Richard & Bruce Goldman
2020 Visions: Long View of a Changing World

This book projects many aspects of the future of the world and especially North America up to 2020, and concludes that political and economic organisation are the keys to success of a modern society.

1990, Stanford Alumni Association/ The Portable Stanford, Stanford, CA, USA, (pbk) 252 pp.

167 Carr, Marilyn (Introduced by)
Sustainable Industrial Development

This book contains seven case studies of successful appropriate technology projects in Asia and Africa that have run for 5 to 20 years.

1988, Intermediate Technology Publications, London, ISBN 0-946688-89-3 (pk) 208 pp.

168 Carson, Rachel
Silent Spring

This book alerted the public to concern about the growing use of pesticides in agriculture. It led to rapid growth of public awareness of these problems and of the need for new methods of controlling pests.

3rd. Ed. 1965, reprinted 1991, Penguin Books, London, ISBN 0-14-013891-9 (pbk); 1st. Ed. 1962, Houghton Mifflin, Boston, USA.

169 Cartledge, Sir Brian (Ed.)
Monitoring the Environment: The Linacre Lectures 1990-91

In this book, eight leading British people in science and givernment examine environmental realities and future issues. They explore the complex questions of what science, technology, and human imagination can do to reverse damaging trends.

1991, Oxford University Press, Oxford & New York, ISBNs 0-19-858408-3 (hbk) & 0-19-858412-1 (pbk) viii + 216 pp.

170 Carwardine, Mark
The   WWF   Environment
Handbook
This book is a useful, scientifically
accurate overview of today's
major environmental issues,
drawing on the practical experience of the WWF (No. 1547) with
environmental and conservation
problems worldwide. Topics covered include: endangered species, Antarctica, atmospheric
pollution, tropical rain forests,
drought, famine, population, the
impact of war on the environment, and the work of the WWF.
*1990, Macdonald Optima, London,
ISBN 0-356-18839-6 (pbk) xii + 178
pp.*

171 Casti, John L.
**Searching for Certainty: What
Science Can Know about the
Future**
This book discusses how far contemporary science can predict
and explain everyday events in
natural and human affairs, for
example in weather and climate,
stock markets, and warfare.
*2nd. Ed. 1992, Scribners, London,
ISBN 0-356-20368-9 (hbk) 496 pp.;
1st. Ed. 1991, Morrow, New York.*

**172 Celente, Gerald with Tom
Milton**
**Trend Tracking: The System to
Profit from Today's Trends**
This book gives its readers easy-
to-use guidelines for setting up
one's own trend tracking system,
using readily available sources
such as newspapers and periodicals. It provides insights and surveys of the major trends changing
modern society, and helps its
readers to devise coprporate and
personal strategies to profit from
emerging trends.
*1990, Wiley, New York & Chiches-
ter, ISBN 0-471-50265-0 (hbk) xv +
303 pp.*

173 Cernea, Michael C. (Ed.)
**Putting People First: Sociological
Variables in Rural Development**
This book argues that success in
development projects is more
likely if sociological considerations have been included in their
planning, appraisal, implementation, and monitoring.
*2nd. Ed. 1991, Oxford University
Press, New York & Oxford, ISBN
0-19-520827-7 (pbk) xxi + 515 pp.;
1st. Ed. 1985.*

**174 Cetron, Marvin & Owen
Davies**
**Crystal Globe: The Haves and
Have-Nots of the New World
Order**
The authors of this book forecast
that the world will be a more
peaceful and prosperous place in
the 1990s than it has been during
the earlier decades since World
War 2. Appendices provide profiles of 37 countries and describe
90 major trends changing the
world.
*1991, St. Martin's Press, New York,
ISBN 0-312-06325-3 (hbk) 430 pp.*

175 Chaika, Elaine
Language: The Social Mirror
This book offers a comprehensive introduction to sociolinguistics, including the latest available research findings.
2nd. Ed. 1989, Newbury House Publishers, New York & London, ISBN 0-06-632613-3 (pbk) xxi + 374 pp.

176 Chandra, Rajesh
Industrialization and Development in the Third World
This book, provides a comprehensive introduction to Third World industrialisation, and includes a variety of case studies. It examines why the Newly Industrialised Countries have achieved spectacular growth in sharp contrast to most other developing countries.
1992, Routledge, London & New York, ISBN 0-415-01380-1 (pbk) 128 pp.

177 Chant, Sylvia (Ed.)
Gender and Migration in Developing Countries
This book considers various patterns of women's migration from the countryside to the cities and abroad in the Third World.
1992, Belhaven Press, London & New York, ISBN 1-85293-186-8 (hbk) xiii + 249 pp.

178 Chapman, J. D.
Geography and Energy: Commercial Energy Systems and National Policies
This book introduces the geography and functional structure of the commercial energy industry. It also considers possible energy futures, the analysis of energy policy, and public energy policy in Canada.
1989, Wiley & Longman, ISBN 0-470-21188-1 (?bk) xv + 260 pp.

179 Chapman, J. L. & M. J. Reiss
Ecology: Principles   and Applications
This comprehensive introductory textbook of ecology integrates studies of human ecology with biological ecology, and emphasises the links between ecology and related disciplines.
1992, Cambridge University Press, Cambridge & New York, ISBN 0-521-38951-8 (pbk) ix + 294 pp.

180 Charter, Martin (Ed.)
Greener Marketing: A Responsible Approach to Business
'Greener marketing' is essentially a strategic management process, that incorporates an integrated approach to corporate environmentalism. This book presents 20 original case histories to illustrate how proactive companies are examining environmental issues.
1992, Interleaf Productions, Sheffield, ISBN 1-874719-00-4 (?bk) ? pp.

181 Chase, Steve (Ed.)
Defending the Earth: A Dialogue between Murray Bookchin and Dave Foreman
This book presents a 'Great Debate' between social ecology theorist Murray Bookchin and deep ecology activist Dave Foreman. The editor considers that a more unified, holistic radical ecology movement may be emerging, to expand wilderness and create a humane, ecological society.
1991, South End Press, Boston, MA, USA, ISBNs 0-89608-383-7 (hbk) & 0-89608-382-9 (pbk) 147 pp.

182 Chorafas, Dimitris N. & Eva Maria Binder
Technoculture and Change: Strategic Solutions for Tomorrow's Society
This book surveys scientific breakthroughs like 'artificial intelligence' and biotechnology, and then goes beyond science into sociology, philosophy, epistemology, finance, and the fabric of the post-industrial era. Part 1 provides a simple, comprehensive view of technology's impact on our lives. Part 2 focusses on human capital and considers future prospects for business and career development.
1992, Adamantine Press, London, ISBNs 0-7449-0039-5 (hbk) & 0-7449-0038-7 (pbk) (ISSN 0954-6103) 216 pp.

183 Chossudovsky, Michel
Towards Capitalist Restoration? Chinese Socialism after Mao
This book describes how the emergence of a new Communist Party leadership in China has led to a major upheaval in the foundations of Chinese socialism, in both agriculture and industry.
1986, Macmillan, London & New York, ISBNs 0-333-38440-7 (pbk) & 0-333-38441-5 (pbk) xiv + 252 pp.

184 Christensen, Karen
Home Ecology
This book is a domestic and discursive but hard-hitting guide to the things that we can do in our personal lives, to ensure the well-being of our families and our environment.
1989, Arlington Books, London, ISBN 0-85140-725-0 (pbk) xii + 366 pp.

185 Christodoulou, Demetrios
The Unpromised Land: Agrarian Reform and Conflict Worldwide
This book is a compassionate discussion of the rural crisis in the Third World, describing the danger of famine and the marginalisation of the disadvantaged people there by inappropriate development policies.
1990, Zed Books, London, ISBNs 0-86232-778-4 (hbk) & 0-86232-779-2 (pbk) 256 pp.

186 Churchman, C. West & Richard O. Mason (Eds.)
World Modelling: A Dialogue
This collection of 19 papers discusses many aspects of global models and of the global problematique that they address. The final paper is a general assessment of global modelling in the mid-1970s.
1976, North Holland, Amsterdam & Oxford, & American Elsevier, New York, ISBNs 0-7204-0338-X (NH) & 0-444-11029-1 (AE) (pbk) xv + 163 pp.

187 Clark, John (Ed.)
Renewing the Earth: The Promise of Social Ecology: A Celebration of the Work of Murray Bookchin
This book explores the many dimensions of social ecology, one of the most fully developed expressions of contemporary Green thinking. Its contributors apply the ecological world view to a wide range of questions, and present social ecology as a comprehensive philosophy, that could give meaning and direction to the Green movement worldwide.
1990, Green Print, London, ISBN 1-85425-001-9 (pbk) xii + 219 pp.

188 Clark, John
Democratizing Development: The Role of Voluntary Organizations
This book defines a broad concept of democracy, based on more participation, freedom and equity, that is likely to lead to more efficient, effective and sustainable government. Applied to development, its principles lead to structural transformation to a new democratic order and new values based on the needs of the people. Voluntary and non-governmental organisations (NGOs) have a unique ability to advoocate and facilitate this transformation.
1991, Kumarian Press, West Hartford, CT, USA, ISBN 0-931816-91-2 (pbk) xii + 226 pp.

189 Clark, John, Sam Cole, Ray Curnow & Mike Hopkins
Global Simulation Models: A Comparative Study
This book outlines and assesses the earlier global models, considers the range of global modelling techniques, compares them with other forecasting techniques, and presents some conclusions on the future usefulness of global models, especially for policy making.
1975, London & New York, ISBN 0-471-15899-2 (hbk) x + 135 pp.

190 Clark, Mary E.
Ariadne's Thread: The Search for New Modes of Thinking
This book aims to present many aspects of the modern world in a holistic way, allowing students to begin to understand, criticise and evaluate it meaningfully. It attempts to provide enough background information, to evaluate

critically the ideas presented by each discipline and to interpret them in ways reflecting the contemporary scene. The author offers an alternative vision, that she sees as the surest road to survival. This way explores cooperation rather than competition, diversity rather than uniformity, social bonding rather than self-centredness, sacred meaning rather than personal consumption.

1989, Macmillan, London & New York, ISBNs 0-333-46599-7 (hbk) & 0-333-46600-4 (pbk) xxviii + 584 pp.

191 Clark, Mary E. & Sandra A. Wawrytko (Eds.)
Rethinking the Curriculum: Toward an Integrated, Interdisciplinary College Education
This book's key chapter proposes a core curriculum based on human communities, rather than on disciplines and departments, and urges a value system where students learn that knowledge should be directed for human purposes. Its other contributions are grouped around five themes.

1990, Greenwood Press, Westport, CT, USA, ISBN 0-313-27306-5 (hbk) 288 pp.

192 Clark, Michael, Dennis Smith & Andrew Blowers
Waste Location: Spatial Aspects of Waste Management, Hazards and Disposal
The contributors to this book provide a comprehensive survey of the waste location debate in the early 1990s. They discuss the basically spatial nature of waste disposal, raise several key issues, and present a varied selection of case studies. Although the chapters focus on the UK, they also concern the international trade in pollution exports.

1991, Routledge, London & New York, ISBN 0-415-04824-9 (hbk) 288 pp.

193 Clarke, Michael
British External Policy-Making in the 1990s
This book brings together much evidence and original research, to evaluate systematically the factors shaping British foreign and defence policy and policy making in the 1990s.

1992, Macmillan, London & New York, ISBNs 0-333-57055-3 (hbk) & 0-333-57056-1 (pbk) xi + 353 pp.

194 Cleveland, Harlan
The Global Commons: Policy for the Planet
This book presents the emerging concept of the 'Global Commons', including important parts of: the human environment, physical environment, biological and cultural diversity, and the information commons managing the flow of information. The book proposes 60 propositions about the Global Commons as a new policy frontier.

1990, University Press of America,

Lanham, MD, USA, ISBNs 0-8191-7835-7 (hbk) & 0-8191-7836-5 (pbk) 118 pp.

195 Clinch, Peter
Business Statistics: How to Find Them, How to Present Them
This booklet provides information about UK, European Community, international, and computer database sources of business statistics.
1990, Headland Press, Headland, Cleveland, ISBN 0-906889-23-5 (pbk) vii + 50 pp.

196 Coates, Joseph N. & Jennifer Jarratt
What Futurists Believe
This book presents the views of 17 futurists on important human, technological, and planetary issues, and analyses them in ways that will stimulate important insights into these issues. It summarises, compares, and contrasts their beliefs about the future, discusses the views of each futurist in turn, and presents the authors' overview.
1989, Bethesda, MD, USA, ISBN 0-912338-66-0 (hbk) xiii + 340 pp.

197 Coates, Joseph F. & Jennifer Jarratt
The Future: Trends into the Twenty-First Century
This periodical issue presents examples of subjects, that have received special attention from futurists and show the variety of future studies approaches. The editors' initial essay surveys 200 years of future studies, and provides eight lists about different aspects. In an article on methods, Theodore J. Gordon describes several forecasting techniques. In an article which provides useful information additional to that in *Resources for the Future*, Michael Marien recommends over 100 good recent books on: environmental issues and sustainability, global issues, US domestic issues, technology, and methods of shaping the future. The other articles cover a variety of topics.
July 1992, The Annals of the American Academy of Political and Social Science, Vol. 522, pp 1-151 (hbk) & (pbk).

198 Coates, Joseph F., Jennifer Jarratt & John B. Mahaffie
Future Work: Seven Critical Forces Reshaping Work and the Work Force in North America
This book identifies recent trends from an ongoing scan of the corporate environment, and analyses their implications for human resources planners and managers. It outlines 37 trends, and discusses counteracting forces, implications for human resource planning, possible futures, opportunities for action, and interactions with related trends. The trends are grouped under seven major themes.
1990, Jossey-Bass Publishers, San

Francisco, ISBN 1-55542-240-3 (hbk) 445 pp., together with workbook, ISBN 1-55542-246-2 (hbk) 139 pp.

199 Codlin, Ellen M. & Keith W. Reynard
Directory of Information Sources in the United Kingdom
This directory provides a comprehensive list of sources of information for research, reference, business development, and marketing, with an easily understood scheme of organisation. Areas covered include: business, finance, commerce, economics, the social sciences, the humanities, medicine, science, and technology. This edition has information on the European Community Single Market.
7th. Ed. 1992, Aslib, London, ISBN 0-85142-292-6 (hbk) 2 vols., viii + 1072 pp.

200 Coghill, Roger
Electro Pollution: How to Protect Yourself Against It
This book presents a pioneering study of new, pervasive and, until recently, unrecognised threats to human health, resulting from an unprecedented explansion in the use of electromagnetic energy for heating, lighting, telecommunications, and other applications.
1990, Thorsons, Wellingborough, Northants., ISBN 0-7225-2307-6 (pbk) 192 pp.

201 Cohen, Bernard L.
The Nuclear Energy Option: An Alternative for the 90s
The author of this book argues that concern about the harmful effects of nuclear power seems to have been reduced, at least in the USA, and presents the case for a revival of nuclear power in the USA.
1990, Plenum Press, New York & London, ISBN 0-306-43567-5 (hbk) x + 338 pp.

202 Cohen, Bernice
The Cultural Science of Man
This book surveys all aspects of human culture, and may be of considerable help in developing a wider historical approach to the process of understanding cultural activities. It develops a holistic view of our world and its social, economic and cultural development from prehistory to the present day. In the final chapter, the author formulates some guidelines for future developments and possibilities.
1988, Codek Publications, London, ISBNs 0-870386-00-0 (hbk) & 1-870386-04-3 (pbk) 3 vols., xliv + 1100 pp.

203 Cohen, L. Jonathan
The Probable and the Provable
The author of this book questions how far 'othodox' probability theory can provide a basis for legal proof. He aims to present an alternative, paradox-free approach

with legal and other applications.
1977, Oxford University Press, Oxford & New York, ISBN 0-19-824412-6 (pbk) xvi + 363 pp.

204 Cohen, L. Jonathan
An Introduction to the Philosophy of Induction and Probability
This book aims to introduce its readers to philosophical arguments and issues about probability and induction.
1989, Oxford University Press, London & New York, ISBN 0-19-875079-X (hbk) x + 217 pp.

205 Coker, Annabel and Cathy Richards
Valuing the Environment: Economic Approaches to Environmental Valuation
In this wide-ranging collection of essays, economists, geograohers, environmental managers, and philosophers examine and debate the necessarily complex question of what the environment is worth. They emphasise practical problems and the need to inform decision makers. The book is a useful, practical manual of conceptual, methodological, and procedural ideas of economic evaluation of environmental goals, especially though not only for wildlife resources at coastal sites.
1992, Belhaven Press, London & Boca Raton, FL, USA, ISBN 1-85293-212-0 (hbk) xiii + 183 pp.

206 Cole, H. A.
Understanding Nuclear Power: A Technical Guide to the Industry and Its Processes
This book aims to allay some of the fears about nuclear power by exploring its science and technology in everyday language. It should enable its readers to assess the pros and cons of nuclear power.
1988, Gower Technical Press, Aldershot, Hants. & Gower Publishing Co., Brookfield, VT, USA, ISBN 0-291-39704-2 (hbk) xv + 410 pp.

207 Cole, Ken, John Cameron & Chris Edwards
Why Economists Disagree: The Political Economy of Economics
This book should provide all students of economics with a useful comparative approach, which will help them to understand better the broad range of theoretical issues (economic, political, and social), underlying how we perceive and manage in the real world. The authors argue that there are several competing theoretical perspectives, each of which is logically consistent and offers a plausible explanation of economic phenomena. They explains why each theory offers different answers to essentially the same groups of economic problems.
2nd. Ed. 1991, Longman, London & New York, ISBN 0-582-06400-7 (pbk) xi + 295 pp.; 1st. Ed 1983.

208 Cole, Sam (H. S. D.), Christopher Freeman, Marie Jahoda & K. L. R. Pavitt
Thinking about the Future: A Critique of the Limits to Growth
This book presents the most important and comprehensive of the many critiques made of *The Limits to Growth* (No. 642) and its global models. It examines the structure and assumptions of the models, and discusses the ideological background to the debate on them.
1973, Chatto & Windus, London for Sussex University Press, ISBNs 0-85621-018-8 (hbk) & 0-85621-020-X (pbk) vi + 218 pp.

209 Coleman, Alice
Utopia on Trial: Vision and Reality in Planned Housing
This book presents evidence to establish links between social malaise and the design and layout of modern housing estates. It proposes practical measures to correct and avoid past mistakes.
1985, Hilary Shipman, London, ISBNs 0-948096-00-4 (hbk) & 0-948096-01-2 (pbk) viii + 219 pp.

210 Coleman, David & John Salt
The British Population: Patterns, Trends and Processes
This book describes the contemporary population of Britain, analyses possible trends, and discusses various demographic issues.
1992, Oxford University Press, Oxford & New York, ISBNs 0-19-874097-2 (hbk) & 0-19-874098-0 (pbk) xxvi + 680 pp.

211 Coleman, Vernon
The Health Scandal: Your Health in Crisis
This controversial book, written by a medical author with experience as a general practitioner, brings almost all aspects of the UK's health come under clinical and critical scrutiny, disclosing many abuses. The author is very concerned that, by 2020, the health of the UK may have reached a state of crisis leading to social breakdown.
1988, Sidgwick & Jackson, London, ISBN 0-283-99509-2 (hbk) 245 pp.

212 Collins, Mark
The Last Rain Forests (Ed.)
This book brings a deeper level of understanding of the world's rain forests and their associated problems and issues, on which to base opinions and take action. It provides a unique atlas of the conservation status of all the rain forests, and a blueprint for positive action to secure their future. It shows what rain forests are and why we need them, and includes many beautiful colour illustrations.
1990, Mitchell Beazley, London, ISBN 0-85533-789-3 (hbk) 200 pp.

213 Combs, Allan (Ed.)
Cooperation: Beyond the Age of Competition
The 20 chapters in this book

explore a remarkably growing awareness of the central role and importance of cooperation. They are written from various disciplinary perspectives, including new ideas emerging from systems studies, biology, psychology, and the social sciences. The essays question the long-held assumption that our existence depends on competition, and offer insights into far-reaching world challenges.
1992, Gordon and Breach Science Publishers, New York & Reading ISBN 2-88124-573-4 (hbk) 272 pp.

214 Conroy, Czech & Miles Litvinoff (Eds.)
The Greening of Aid: Sustainable Livelihoods in Practice
This book gives 33 case studies of agricultural, fisheries, and industrial projects, attempting to achieve sustainable development in the Third World. They show that there are forms of development that allow people to control their own resources, while improving their condition and enhancing their environment.
1989, Earthscan, London, ISBN 1-85383-016-X (pbk) xv + 302 pp.

215 Conway, Gordon & Edward Barbier
After the Green Revolution: Sustainable Agricuture for Development
This book examines the methods of improving agricultural sustainability in developing countries, and the necessary trade-offs between development goals, at international, national and local levels.
1990, Earthscan, London, ISBN 1-85383-035-6 (pbk) 208 pp.

216 Cooke, Belinda
The Trade Trap: Poverty and the Global Commodity Markets
This book shows how countries, depending on the export of primary commodities, are caught in a trap: the more they produce, the lower the prices of their goods fall on the world market. It recommends actions that could be taken by governments, shows how consumers could make a *real* difference by buying products promoted by the new 'fair trade' movement, and presents many case studies.
1992, Oxfam, Oxford, ISBNs 0-85598-134-2 (hbk) & 0-85598-135-0 (pbk) x + 214 pp.

217 Cooper, David E. & Joy A. Palmer (Eds.)
The Environment in Question: Ethics and Global Issues
The essays in this book discuss the ethical questions arising from various disciplines, including environmental science, geography, psychology, and philosophy. They introduce several key environmental issues and develop several practical

approaches. By addressing specific global problems within an ethical context and emphasising both scientific principles and ethical perspectives, they provide both theoretical and practical understanding of environmental issues.
1992, Routledge, London & New York, ISBNs 0-415-04967-9 (hbk) & ISBN 0-415-04968-7 (pbk) xii + 256 pp.

218 Cooper, David, Renee Velvee & Henk Hobbelink (Eds.)
Growing Diversity: Genetic Resources and Food Security
In this book, activist contributors present their experience of managing plant genetic resources, and show how the approaches used by indigenous farmers help to promote conservation and development.
1992, IT Publications, London, ISBNs 1-85339-123-9 (hbk) & 1-85339-119-0 (pbk) viii + 166 pp.

219 Cooper, Richard N.
Economic Stabilization and Debt in Developing Countries
This book discusses how developing countries have coped with persistent problems of debt, formulates some lessons based on their experience of these nations, and presents some proposed solutions.
1992, MIT Press, Cambridge, MA, USA & London, ISBN 0-262-03187-6 (hbk) xvii + 195 pp.

220 Coopers & Lybrand Deloitte
Your Business and the Environment: A D-I-Y Review for Companies
This report, with a foreword by HRH the Prince of Wales, provides a framework for completing an environmental review of a business.
1991, Business in the Community, London, ISBN 1-85271-193-0 (pbk) 122 pp.

221 Cornish, Edward (Ed.)
The 1990s and Beyond
This book reprints important papers from *The Futurist* (No. 1050). Its themes include: issues for the 1990s, values and societal change in terms of nine basic lifestyles, 'artificial intelligence', important social trends, the future of education, the future of cities, the 21st century economy, the future of health care, and the next 100 years.
1990, Transaction Books, New York, ISBN 0-930-292-37-8 (pbk) 160 pp.

222 Cornwall, Malcolm G. (Ed.)
Special Issue on Energy: *Physics Education*, Vol. 27, No. 4
This special issue contains articles on: energy priorities for the UK, energy and the environment, the prospects for renewable energy in the UK, the pros and cons of nuclear energy, the Centre for Alternative Technology, and approaches to the teaching of energy.

July 1992, I of P Publishing, Bristol, ISSN 0031-9120, 66 pp.

223 Costanza, Robert (Ed.)
Ecological Economics: The Science and Management of Sustainability
This book has three parts: (1) developing an ecological world view; (2) accounting, modelling and analysis; (3) institutional changes and case studies. It introduces the conceptual and modelling facilities of the newly emerging discipline of ecological economics.
1991, Columbia University Press, New York, ISBN 0-231-07562-6 (hbk) xiii + 525 pp.

224 Cottrell, Richard
The Sacred Cow: The Folly of Europe's Food Mountains
The author of this book, a British Euro-MP, argues that Europe's Common Agricultural Policy is leading towards a world food crisis. He presents a case for ending all forms of state support for agriculture.
1987, Grafton Books, London, ISBN 0-246-13224-8 (pbk) 192 pp.

225 Coulson-Thomas, Colin J. & Susan D. Coulson-Thomas
Managing the Relationship with the Environment: A Survey Sponsored by Rank Xerox (UK) Limited
This independent survey aims to answer some importamt questions about company approaches to the environment. Its results strongly suggest that commercial suppliers will lose customers if they do not satisfy the emerging new environmental criteria.
1990, Adaptation Ltd., London, ISBN 0-9516900-2-7 (pbk) iii + 42 pp.

226 Craig, Frances & Phil Craig
Britain's Poisoned Water
This book shows that millions of British families drink water, containing in excessive concentrations of toxic chemicals.
1989, Penguin Books, London & Viking Penguin, New York, ISBN 0-14-011050-X (pbk) xiv + 135 pp.

227 Crane, G. T. & A. Amawi (Eds.)
The Theoretical Evolution of International Political Economy
This book charts the historical and political evolution of international political economy, and includes classical works and leading contemporary papers. It outlines the development of Liberalism, Marxism, and Realism, and includes recent syntheses of them.
1991, Oxford University Press, New York & Oxford, ISBN 0-19-506012-1 (pbk) ix + 302 pp.

228 Cropley, Jacqueline
Directory of Financial Information Sources
This directory is designed to help busy financial professionals, business executives, and their

researchers to identify quickly and easily material appropriate for their areas of work.
1991, Woodhead-Faulkner, London & New York, ISBN 0-85491-564-3 (pbk) xii + 308 pp.

229 Crow, Ben, Mary Thorpe et al.
Survival and Change in the Third World
This book provides a framework of ideas useful to those interested in problems of development and underdevelopment in the Third World.
1988, Polity Press, Cambridge, ISBNs 0-7456-0332-7 (hbk) & 0-7456-0333-5 (pbk) xi + 365 pp.

230 Dahle, Kjell
On Alternative Ways of Studying the Future: International Institutions, an Annotated Bibliography and a Norwegian Case
This very useful overview of the field of future studies contains: (1) an annotated bibliography, giving a good idea of the field's present status; (2) a list of institutions, providing information about names, people, and institutional settings; (3) a description of the Alternative Future Projects in Norway (No. 1632), showing a special way of doing future studies, as an ongoing experiment.
1991, The Alternative Future Project, Oslo, Norway, ISBN 82-7480-008-7 (pbk) 189 pp.

231 Dakin, A. J.
Feedback from Tomorrow
This book is one of the most wide ranging books on future developments; it attempts to provide a synthesis of knowledge and action for addressing global problems, and aims to develop a basic background for planners. Its approach can be used as a basis for developing a planet management philosophy.
1979, Pion, London, ISBN 0-85086-071-7 (hbk) 489 pp.

232 Dalton, Russel J. & Manfred Kuechler (Eds.)
Challenging the Political Order: New Social and Political Movements in Western Democracies
This book examines the first wave of systematic, comparative research on new movements, concerned with environmentalism, women's rights, peace and other urgent issues facing advanced industrial countries, to determine the extent of their challenge. Its findings provide a key to understanding how such movements might contribute to their evolving political order.
1990, Polity Press, Cambridge, ISBNs 0-7456-0744-7 (hbk) & 0-7456-0750-0 (pbk) xii + 329 pp.

233 Daly, Herman E.
Steady-State Economics
This book is a classic of sustainable economics, originally considered too radical, but now

recognised as providing a key to concepts of sustainable development and economics. Various aspects of the steady-state economy are discussed, including its concepts, institutions, efficiency, application to developing economies, and alternative strategies for integrating economics with ecology. The current debate on growth versus environment is also discussed.

3rd. Ed. 1992, Earthscan, London, ISBN 1-85383-140-9 (pbk) 304 pp.; 2nd. Ed. 1991, Island Press, Washington, DC, USA; 1st. Ed., 1977.

234 Daly, Herman E. & John B. Cobb, Jr. with Clifford W. Cobb For the Common Good: Redirecting the Economy Towards Community, the Environment and a Sustainable Future

This book is a penetrating critique of orthodox economics, that shows how the present growth-oriented industrial economy leads to environmental disaster. It offers an exciting new paradigm for economics, public policy, and social ethics. It disagrees with the assumption of neoclassical economic theory that the 'market' is rational and organises the production and distribution of goods and services in the best interests of society. Its alternative economic model gives a moral priority to national and regional community, and attends to the

needs of future generations and non-human life.

Part 1 discusses economics as an academic discipline. Part 2 proposes an alternative approach to the economy, with new foundations for economics. Part 3 discusses policies for community in the USA, that follow from the alternative perspective. Part 4 considers how to achieve the proposed new economy. A long appendix develops and presents a proposed Index of Sustainable Economic Welfare. The authors propose an 'optimal size' of community, that can provide a good, sustainable life for its members, and applies the model of 'person-in-community' to a wide range of issues. They show how present policies are directly responsible for today's environmental and social problems, locate our economic difficulties in the spiritual failings of Western society, and find hope for the future in the human sense of community.

2nd. Ed. 1990, Green Print, London, ISBN 1-85425-039-6 (pbk) xii + 482 pp; 1st. Ed., Beacon Press, Boston, MA, USA.

235 Dammann, Erik The Future in Our Hands

This book is the English translation of the book which inspired the launching of the popular Future in Our Hands movement in Scandinavia. It dramatically reveals the injustices that have been and are suffered by most Third World people, proposes some radical adjustments to Western society, shows our responsibilities as individuals, provides guidelines for action, and presents the movement's manifesto.

2nd. Ed., 1979, reprinted 1983, Pergamon Press, Oxford & New York, ISBNs 0-08-024284-7 (hbk) & ISBN 0-08-024283-9 (pbk) xviii + 181 pp; 1st. (Norwegian) Ed. 1979.

236 Dammann, Erik
Revolution in the Affluent Society

This book describes the enormous gap between the rich and poor nations of the world, and urges the citizens of the affluent societies embark on a radical change. It calls for the recognition of a basic human solidarity as the foundation for a radical approach, which has inspired many people in the Future In Our Hands movement in Scandinavia, and has lessons for people elsewhere.

2nd. Ed. 1984, Heretic Books, London, ISBN 0-946097-06 (pbk) 173 pp.; 1st. (Norwegian) Ed. 1979.

237 Dammers, Horace
Lifestyle: A Parable of Sharing

This book presents the aims, philosophy, and early history of the LifeStyle movement, which invites us all to "live more simply so that all of us may simply live" and recall our sense of oneness with the whole human family. Adopting a simpler lifestyle defends us against overconsumption environmental damage, and is also an act of solidarity with the majority of humankind at present forced to live in poverty.

1982, Turnstone Press, Wellingborough, Northants, ISBNs 0-85500-159-3 (hbk) & 0-85500-160-7 (pbk) 224 pp.

238 Dammers, Horace
A Christian Lifestyle

This book affirms that a simple lifestyle should be the Christian response to a world of rapidly diminishing resources, where billions now live in poverty. Its realistic suggestions for simple living provide practical encouragement for all who want a more equal and fair distribution of the Earth's resources.

1986, Hodder and Stoughton, London & Toronto, ISBN 0-340-38176-1 (pbk) 224 pp.

239 Daniel, Joseph & Ilana Kotin (Eds.)
1992 Earth Journal: Environmental Almanac and Resource Directory

This volume is reviewed and updated annually, and sets the stage for active participation. Its contents range from updates on the major environmental challenges to directories of the resources needed to live an eco-lifestyle. It documents extensively what has happened, is happening, and is likely to happen, in the environmental arena.

1992, Buzzworm Books, Boulder, CO, USA, ISBN 0-9603722-9-6 (pbk) 448 pp.

240 Dankelman, Irene & Joan Davidson
Women for the Environment in the Third World: Alliance for the Future
This book describes the important activities of women in the Third World, and the problems that arise. Its case studies presents examples of how women are organising themselves to change their lives for the better, and show that an improved attitude of the male population is vital if the Third World is to escape from its poverty trap.
1988, Earthscan, London, 1-85383-003-8 (pbk) 224 pp.

241 Da Silva, E. J., C. Rutledge & A. Sasson (Eds.)
Biotechnology: Economic and Social Aspects: Issues for Developing Countries
This collection of papers reviews current trends in harnessing biotechnology, as an integral component of the labour and industrial sectors of society in developed and developing countries.
1992, Cambridge University Press, Cambridge & New York, ISBN 0-521-38473-7 (hbk) xiii + 388 pp.

242 Dauncey, Guy
After the Crash: The Emergence of the Rainbow Economy
This book traces the emergence of a new economy at the grassroots and community level in many parts of the world. It introduces a rainbow spectrum of values, expressing both personal and planetary wholeness: PURPLE for spiritual values, DARK BLUE for planetary values, LIGHT BLUE for economic values, GREEN for environmental values, YELLOW for values of personal creativity and wholeness, ORANGE for local community values, and RED for social values.
1988, Green Print, Basingstoke, Hants., ISBN 1-85425-004-3 (pbk) viii + 312 pp.

243 Davidson, Joan
How Green Is your City? Pioneering Approaches to Environmental Action
This book shows how local groups, pioneering environmental action, can provide new hope for the communities living in rundown and neglected areas. It describes a wide range of schemes involving city greening, energy conservation, and waste recycling.
1988, Bedford Square Press, London, ISBN 0-7199-1215-6 (pbk) x + 110 pp.

244 Davidson, Joan & Dorothy Myers with Manab Chakrabarty
No Time to Waste: Poverty and the Global Environment
20% of the world's people live in poverty, while 25% enjoy life styles of extravagant consumption; a deteriorating environment links both groups of people. This book examines issues of sustainable development from a 'Southern' viewpoint. It gives examples of how poor people are responding to safeguard and improve the environment, on which their livelihood and our common future depend.
1992, Oxfam, Oxford, ISBNs 0-85598-182-2 (hbk) & 0-85598-183-0 (pbk) v + 217 pp.

245 Davidson, Paul
International Money & the Real World
This book's Post-Keynesian approach provides foundations for developing proposals for civilised economic policies and approaches to major international problems. Post-Keynesian analysis is here presented as distinct from orthodox approaches, recommended by neoclassical economists and adopted by many nations in the 1980s. The author explains the defects of those approaches, and why they often failed.
2nd. Ed. 1992, Macmillan, London & New York, ISBNs 0-333-52153-6
(hbk) & 0-333-52154-4 (pbk) xiv + 288 pp.; 1st. Ed. 1982.

246 Davies, Paul (Ed.)
The New Physics
This is an up-to-date survey of all the major subject areas of modern physics at the frontiers of modern research.
2nd. Ed. 1992, Cambridge University Press, ISBN 0-521-43831-4 (pbk) 528 pp.; 1st. Ed. 1989, ISBN 0-512-30420-2 (hbk) ix + 516 pp.

247 Davies, Paul
The Cosmic Blueprint
This book presents some remarkable discoveries at the frontiers of modern mathematical and physical research.
1987, Heinemann, London, ISBN 0-434-17702-4 (hbk) viii + 224 pp.

248 Davies, Paul
The Mind of God: Science and the Search for Ultimate Meaning
This book comprehensively reviews the universe and the nature of scientific enquiry and rationality, and asks whether science can provide insights the existence of God and a meaning to our life.
1992, Simon & Schuster, London & New York, ISBN 0-671-71069-9 (hbk) 253 pp.

249 Davies, Paul & John Gribbin
The Matter Myth: Towards 21st Century Science
This book examines the current

revolutionary transformation of scientific thinking, and presents evidence of a massive paradigm shift.
1991, Viking, London & New York, ISBN 0-670-83585-4 (hbk) 314 pp.

250 Davis, John
The Greening of Business
The author of this book urges business to pursue a new path of discriminating sustainable development. The richer countries will have to consume very much less material and energy resources than before, to enable people in the developing countries to achieve decent standards of living, and at the same time try to increase the quality of life all round. Local communities will have to become more self-reliant. In Chapter 1, the author outlines the concepts of sustainable business development, which he develops and applies in the rest of the book. He discusses nine reorientations that are required in business life; meeting each of these requirements is a separate task for a company, from which a transformation programme should be produced. The new constraints will be more than offset by new opportunities for imaginative firms during the coming period of intense innovation. He addresses the specific problems and opportunities in each area of sustainable development, and provides guidelines for constructively handling these basic changes, whose outcome will determine the future of humankind and the planet. The final chapter states the need for global sustainable development and a new business philosophy.
1991, Blackwell, Oxford & Cambridge, MA, USA, ISBN 0-631-17202-5 (hbk) 215 pp.

251 Davis, Kingsley & Mikhail S. bernstein (Eds.)
Resources, Environment, and Population: Present Knowledge, Future Options
This book explores the complex relationships between population trends, resource use, and environmental impacts. It documents local, regional, and global processes, and examines policy options and obstacles to action, at various stages of development and under different economic systems. It considers: (1) estimating the future; (2) water, energy, and development; (3) the rise of global pollution; (4) world deforestation and its consequences; (5) limits to growth.
1991, Oxford University Press, New York & Oxford, ISBNs 0-19-507049-6 (hbk) & 0-19-507050-X (pbk) xii + 423 pp.

252 Dawson, Alastair
Global Climate Change
This booklet examines the evidence for global climate change,

the relevant factors, and the likely impacts of future climate changes.
Oxford University Press, Oxford & New York, ISBN 0-19-913366-2 (pbk) 48 pp.

253 de Garine, I. & G. A. Harrison (Eds.)
Coping with Uncertainty in Food Supply
This book addresses various questions concerned with availability of food resources to people, and fluctuations in their supply.
1988, Oxford University Press, Oxford & New York, ISBN 0-19-857643-9 (hbk) xiv + 483 pp.

254 de la Court, Thijs
Beyond Brundtland: Green Development in the 1990s
This book explores the main ideas in *Our Common Future* (No. 988), and considers their further development and practical application into the 1990s. It also outlines some criticisms of *Our Common Future* from leading environmental campaigners, especially in the Third World. The author suggests some detailed principles that 'sustainable development' should incorporate, if it is to constitute a rigorous benchmark against which to judge the economic policies and development proposals of governments and international agencies.

1990, Zed Books, London, ISBNs 0-86232-904-3 (hbk) & 0-86232-905-1 (pbk) 128 pp.

255 de la Court, Thijs
Different Worlds: Environment and Development Beyond the Nineties
The author praises the Dutch Government for taking sustainable development far beyond the limits of the Brundtland Report (No. 988). But he argues that even tougher policy instruments have to to be produced if a government is to succeed in creating sustainable development. He indicates the policy changes needed to achieve real sustainable development. The book is aimed at people who make decisions on environment or development, and at people who influence them.
1992, International Books, Utrecht, The Netherlands, ISBN 90-6224-996-5 (pbk) 144 pp.

256 Delgado, Jose M. (Ruth Anshen (Ed.))
Physical Control of the Mind: Toward a Psychocivilized Society
This book describes the basic principles of brainwashing and other forms of psychological control of the mind.
2nd. Ed. 1971, Irvington, USA, ISBN 0-8290-1765-8 (pbk) 280 pp.; 1st. Ed. 1969, Harper & Row, New York.

257 Dennis, Everitt & John C. Merrill
Media Debates: Issues in Mass Communication
This book examines some of the most significant arguments about the process of mass communication, whether in the media generally, media institutions, individual communicators, or communication contexts.
1991, Longman, New York & London, ISBN 0-8013-0436-9 (pbk) xi + 228 pp.

258 Department of Energy
Renewable Energy in the UK: The Way Forward (*Energy Paper*** No. 55)**
This report presents the UK Department of Energy's research, development and demonstration strategy for renewable energy, then summarises its research programme, considers institutional and environmental factors, and indicates future actions. It has nine appendices on various forms of renewable energy.
1988, HMSO, London, ISBN 0-11-412905-3 (pbk) vii + 31 pp.

259 Department of the Environment
Global Climate Change
This booklet explanains for the general public the principles of man-made climate change via the greenhouse effect, the scope of the threat, and the possible options for responding to it.
1989, HMSO, London, (pbk) 16 pp.

260 Department of the Environment
This Common Inheritance: Britain's Environmental Strategy
This book is the British Government's first White Paper on the environment, providing an extensive survey of many environmental issues. It describes what the Government has done and proposes to do, and sets out several steps that can be taken by local government, business, schools, voluntary bodies, and individuals, working together.
1990, HMSO, London, ISBN 0-10-112002-8 (pbk) 291 pp.; summary version, ISBN 0-11-752334-8 (pbk) 30 pp.

261 Department of the Environment
Recycling: A Memorandum Providing Guidance to Local Authorities on Recycling
This report aims to provide guidance to British local authorities on devising and implementing recycling strategies, and advice to waste collection authorities on how to draw up recycling plans.
1991, HMSO, London, ISBN 0-11-752445-X (pbk) iv + 86 pp.

262 Department of the Environment
A Review of Options: A Memorandum Providing Guidance on the Options Available for Waste Treatment and Disposal (*Waste Management Paper*** No. 1)**

This report provides a general guide to the wide range of methods of treatment and disposal of controlled wastes, showing how to match appropriate techniques to the specific types of waste arising.

2nd. Ed. 1992, HMSO, London, ISBN 0-11-752644-4 (pbk) viii + 75 pp.; 1st. Ed. 1976.

263 *Development*
The Wealth of Humanity: A People's Development Agenda
This issue explores the 'people's agenda, and examines issues of sustainablity, democracy, and alternative economic systems. It presents what may be the most important institutional challenge of the 1990s: how to strengthen the institutions of the civil society, to enable a true democracy where citizen's movements, are the social innovators.

1991, Development, Vol. 1991, No. 3/4, Special Issue, ISSN 1011-6370, 100 pp.

264 Dicken, Peter
Global Shift – The Internationalization of Economic Activity
This book shows how the world's economic activities are becoming increasinglylobalised, ħat is, functionally integrated across national boundaries in ways which are dramatically altering events and prospects in different countries and localities. Its main theme is that the patterns and processes of globalisation are created by the

interaction between transnational corporations and nations in the context of a volatile and evolving technological environment. Almost every part of the world is affected, as transnational corporations restructure their operations worldwide, national governments attempt to increase or preserve their economic strength, the pace of technological change accelerates and its nature changes.

2nd. Ed. 1992, Paul Chapman Publishing, London, ISBN 1-85396-142-6 (pbk) xv + 492 pp., 1st. Ed. 1986.

265 Dickens, Peter
Urban Sociology: Society, Locality and Human Nature
This urban sociology textbook develops the themes of 'space' and 'locality' through a detailed re-assessment of older forms of urban sociology, and combines them with Marxist, feminist and political economic perspectives to develop a new approach.

1990, Harvester Wheatsheaf, London & New York, ISBNs 0-7450-0642-6 (hbk) & 0-7450-0643-4 (pbk) xiii + 192 pp.

266 Dickens, Peter
Society and Nature: Towards a Green Social Theory
The author of this book criticises the established theoretical approaches of contemporary social science, and argues for a

new 'Green' social theory, which advocates the integration of social, biological and physical sciences. This approach allows for the development of a new understanding of the people-environment relationship.
1992, Harvester Wheatsheaf, London & New York, ISBNs 0-7450-0966-2 (hbk) & 0-7450-0967-0 (pbk) xx + 203 pp.

267 Dierkes, Meinolf, Hans N. Weiler & Ariane Berthoin Antal (Eds.)
Comparative Policy Research: Learning from Experience
This book assesses the state of comparative research in economic policy, social welfare policy, environmental policy and educational policy. It provides a full basis for future discussions and actions.
1987, Gower, Aldershot, Hants. & Brookfield, VT, USA, ISBN 0-566-05196-6 (hbk) vi + 531 pp.

268 Dilloway, James
Human Rights and World Order
This booklet provides an excellent summary of human rights in relation to the world situation. It gives in full three of the most important declarations of human rights, together with useful commentaries on them and on the way in which they evolved.
1983, H.G. Wells Society, London.

269 Dimitriou, Harry T.
Urban Transport Planning: A Developmental Approach
This book analyses the problems of contemporary urban transport systems and transport planning, including congestion and environmental pollution. Its author suggests some more appropriate solutions for the future, and questions the appropriateness of applying the Western approach to urban transport planning to the Third World.
1992, Routledge, London & New York, ISBN 0-415-03857-X (hbk) 320 pp.

270 Dobson, Andrew (Ed.)
The Green Reader: Essays Towards a Sustainable Future
This anthology of 'classics' of Green thinking has an introductory paragraph from the editor for each item, to set its context. It is a useful source on the origins of contemporary Green thought, and explains the aims and strategies of the Green movement. Green political thought is distinguished from socialism, liberalism, and conservatism.
1991, Deutsch, London & Mercury House, San Francisco, USA, ISBN 0-233-98653-7 (pbk) xii + 224 pp.

271 Dobson, Wendy
Economic Policy Coordination: Requiem or Prologue?
This book discusses how to strengthen the Group of Seven

(G-7) process for coordinating economic policy, which is the latest attempt by the industrialised nations to manage an increasingly integrated world economy, whose success has hitherto been rather limited.

1991, Institute for International Economics, Washington, DC, USA, ISBN 0-88132-102-8 (pbk) xii + 162 pp.

272 Dodds, Felix (Ed.)
Into the Twenty-First Century: An Agenda for Political Realignment
This book brings together articles by well-known socialists, Liberals, Greens, and others. What their ideas have in common is more important than their political differences. They explore the opportunities for a new and Green parliamentary politics, which will take us beyond contemporary politics into the 21st century.

1988, Green Print, Basingstoke, Hants., ISBN 1-85425-014-0 (pbk) xii + 203 pp.

273 Dogan, Mattei & John D. Kasarda (Eds.)
The Metropolis Era: Volume 1: A World of Giant Cities, Volume 2: Mega-Cities
Vol. 1 focusses on general aspects of giant cities worldwide, and defines a new type of city, the 'region-dominating city'. Vol. 2 presents in-depth case studies of ten very large cities, and examines their urgent problems and policy successes and failures.

1988, Sage Publications, Newbury Park, CA, USA & London, ISBNs 0-8039-2602-2 (Vol. 1) & 0-8039-2603-0 (Vol. 2) (pbk) 706 pp.

274 Dogan, Mattei & Dominique Pelassy
How to Compare Nations: Strategies in Comparative Politics
This book discusses strategies of comparison between nations, and explores some new perspectives in comparative research. It focusses on the general strategy of comparative research, and aims to present a critical appraisal of the state of the art of comparative politics.

2nd. Ed. 1990, Chatham House, Chatham, NJ, USA, ISBN 0-934540-79-9 (pbk) vii + 215 pp.; 1st. Ed. 1984)

275 Dornbusch, Rudiger & James M. Poterba (Eds.)
Global Warming: Economic Policy Responses
This book evaluates the key issues in the policy debate on global warming. Topics covered include: economic approaches, economic responses, carbon taxes, and specific initiatives.

1991, MIT Press, Cambridge, MA, USA & London, ISBN 0-262-04126-X (hbk) xi + 389 pp.

276 Dossett, Patti (Ed.)
Handbook of Special Librarianship and Information Work
The latest editon of this handbook contains sections on: (1) information management and an organisation's information needs; building an information store; (3) organising information; (4) providing information services; (5) technology trends.
6th. Ed. 1992, Aslib, London, ISBN 0-85142-269-1 (hbk) vii + 581 pp.

277 Dotto, Lydia
Planet Earth in Jeopardy: Environmental Consequences of Nuclear War
This book summarises in a convenient and readable form the SCOPE *Report on the Environmental Consequence of Nuclear War.*
1986, Wiley, New York & Chichester, ISBNs 0-471-90908-4 (hbk) & 0-471-99836-2 (pbk) vii + 134 pp.

278 Douglas, Mary
Risk: Acceptability According to the Social Sciences
This book examines the cultural bases of risk acceptance, and the social factors which influence our beliefs and moral judgements about what sort of society we should live in.
1986, Routledge & Kegan Paul, London & New York, ISBN 0-7102-1108-2 (pbk) 115 pp.

279 Douthwaite, Richard
The Growth Illusion: How Economic Growth Has Enriched the Few, Impoverished the Many, and Endangered the Planet
The author of this book argues, from experience in Ireland, the UK, and the USA, that negative effects of economic growth have outweighed any positive effects. Rather than strive for continual expansion to prevent collapse, nations must learn to build stable economies.
1992, A Resurgence Book, Green Books, Hartland, Bideford, Devon, in association with Lilliput Press, Dublin, ISBNs 0-870098-41-2 (hbk) & 0-870098-47-1 (pbk) xiii + 367 pp.

280 Doxiadis, Constantinos
Ekistics: An Introduction to the Science of Human Settlements
This is the author's introductory book on 'ekistics', the science of human settlements that he pioneered, and it presents an outline of his personal experiences.
1968, Hutchinson, London & New York, SBN 09-080300-0 (hbk) xxix + 527 pp.

281 Doxiadis, Constantin A. (Ed. Gerald Dix)
Ecology and Ekistics
This posthumous book presents the author's arguments for his belief that ecology and ekistics could work together to achieve global harmony and global balance. It provides a possible conceptual framework.
1977, Elek, London, ISBN 0-236-40045-2 (pbk) xxvii + 91 pp.

282 Doyal, Len & Ian Gough
A Theory of Human Need
Rejecting fashionable subjectivist and relativist approaches, the authors of this book argue that all people have common basic needs that cannot be reduced to individual cultural preferences. They develop cross-cultural indicators of need satisfaction, and use them to measure and compare levels of human welfare worldwide. They argue that the results support an innovative 'dual strategy' for optimising need satisfaction, which recognises the economic importance of both the market and central planning, and the political importance of both state intervention and a flourishing democracy.
1991, Macmillan, London & New York, ISBNs 0-333-38234-9 (hbk) & 0-333-38325-7 (pbk) xvi + 365 pp.

283 Drexler, K. Eric
Engines of Creation: The Coming Era of Nanotechnology
This is a remarkably original book about the consequences of new technologies. One revolutionary technology in the 21st century could be 'nanotechnology', the construction of machines with linear dimensions a thousand times smaller than those of the microcomputer 'chip'. This new technology could lead to changes far more profound than those caused by most other advances hitherto. For example, *everybody* could be wealthy and have a high quality of life, but, unless precautions are taken in good time, there could be weapons highly devastating to life. The book is an excellent attempt to prepare us to think of what might happen to us, if we continue to develop new technologies.
3rd. Ed., 1992, Oxford University Press, Oxford & New York, ISBN 0-19-286149-2 (pbk) xii + 298 pp.; 1st. Ed. 1986, Morrow, New York.

284 Drexler, K. Eric & Chris Peterson with Gayle Pergamit
Understanding the Future: The Nanotechnology Revolution
This comprehensive but very readable book explains the remarkable new technology of nanotechnology and how it could revolutionise life between ten and 50 years from now. If currently plausible possibilities are fulfilled, there could be dramatic advances in medicine, agriculture, solar energy generation, clearing up and preventing pollution, and alleviating the greenhouse effect and ozone layer depletion. The present 'limits to growth' could be swept right back, much of the presently devastated environment could be restored, and there could be abundance for all people on Earth. However, nanotechnology is *not* a panacea, and some uncertainties remain.

1991, Morrow, New York, ISBN 0-688-09124-5 (hbk) 304 pp.

285 Drucker, Peter F.
The Frontiers of Management: Where Tomorrow's Decisions Are Being Shaped Today

This book presents variations on one unfying theme, the challenges of tomorrow that face executives today. The nature of the future depends very much on the knowledge, insight, foresight and competence of today's decision makers. The book is addressed to *these* people especially, to enable them to see and understand the long-term implications and impacts of their immediate, everyday, urgent actions and decisions. It aims to create vision, and is full of apt lessons.

1987, Heinemann, London, ISBNs 0-434-90394-9 (hbk) & 0-434-90392-2 (pbk) xiii + 368 pp.

286 Duke of Edinburgh, HRH The
Down to Earth

This book is a collection of writings and speeches on man and the natural world from 1961 to 1987. It expresses clearly the thoughts and viewpoint of one of the world's leading conservationists, and describes the situation faced by the threatened areas of the natural world, both wildlife and environment. It introduces the problems, discusses exploitation of the natural system, assesses the population factor, describes what is happening to the victims, and states conclusions.

The Duke gives details of the World Conservation Strategy, and of the work of the International Union for the Conservation of Nature and Natural Resources (No. 1408) and the World Wide Fund for Nature (No. 1547). He discusses many aspects of the conservation of nature and wildlife and of the human impact on them, including: tropical rain forests, trees and plants, animals, natural habitats and genetic resources, culling and control of wild populations, captive breeding in zoos, and trade in wild species. The book is a powerful challenge to decision makers, industrialists, agriculturalists, and ordinary people, to face the moral responsibility of solving the present human and planetary crisis.

1988, Collins, London & (1989) Viking Penguin, New York, ISBNs 0-00-210947 (C) & 0-8789-0711-0 (VP) (hbk) 240 pp.

287 Dunleavy, Patrick & Brendan O'Leary
Theories of the State: The Politics of Liberal Democracy

This textbook is a carefully structured introduction to the five broad schools of political, social and economic thought, demonstrating current dicussions about

the modern state: Pluralist, New Right, Elitist, Marxist, and Neo-Pluralist. For each of them, there is a summary of its intellectual origins, methodology, and main strengths and weaknesses, presented to make comparison between them easy. The final chapter shows how the theories interlock and can be evaluated.
1987, Macmillan, London & New York, ISBN 0-333-38698-1 (pbk) xiii + 382 pp.

288 Durning, Alan B.
How Much Is Enough? The Consumer Society and the Future of the Earth
This book explores the roots of the consumer society, and its toll on Earth and its natural systems; it assesses the real benefits of consumption to consumers. If we are to achieve healthier and more rewarding lives, we need to understand what the Earth can achieve and when enough *is* enough. Otherwise, other efforts to avert environmental crisis will fail. The book assesses consumerism, then shows how we can achieve sufficient consumption and tame consumerism.
1992, Earthscan, London & W. W. Norton, New York, ISBN 1-85383-134-4 (pbk) 200 pp.

289 Durrell, Lee
State of the Ark: An Atlas of Conservation in Action
This book provides a wide ranging well-illustrated approach to

the state of our planet, concludes that we must think globally, and provides information on which to base a planet management strategy.
1986, Bodley Head, London.

290 Dvorak, Eldon J. (Ed.)
Social and Private Costs of Alternative Energy Technologies
This periodical issue is a collection of papers on the greenhouse effect, acid rain, the social costs of nuclear power, energy conservation, and private and external energy costs.
July 1990, Contemporary Policy Issues, Vol. 8, No. 3, Special Issue.

291 Dwyer, Denis J. (Ed.)
South East Asian Development: Geographical Perspectives
This book considers the major aspects of development problems and policies in South East Asia today, including the impact of rapid economic growth and capitalism. It assesses South East Asia's future prospects and the challenges that it will face in the 21st century. It shows how significant progress made by geographers can help. It is a companion book to *Latin American Development* (No. 761).
1990, Longman, London, & Wiley, New York, ISBNs 0-582-30149-1 (L) & 0-471-21665-4 (W) (pbk) xi + 315 pp.

292 Earth Works Group, The
50 Simple Things You Can Do to Save the Earth
This book is a practical guide to simple things that we can all do, especially in the home. It includes guidance on how to save energy, conserve the environment, be much less wasteful, recycle various forms of waste, stay involved, and spread the Green message.
2nd. Ed. 1990, Hodder and Stoughton, London, ISBN 0-450-54020-0 (pbk) 96 pp; 1st. Ed. 1989, Earth Works Press, Berkeley, CA, USA.

293 Easterby-Smith, Mark, Richard Thorpe & Andy Lowe
Management Research: An Introduction
This book presents a clear and accessible introduction to the practice and principles of management research. It considers the particular characteristics of research in the management area, and the factors likely to lead to successful research outcomes.
1991, Sage Publications, Newbury Park, CA, USA & London, ISBNs 0-8039-8392-1 (hbk) & 0-8039-8393-X (pbk) xi + 172 pp.

294 Eccles, Sir John C.
The Human Psyche: The Gifford Lectures University of Edinburgh
In this book, the author considers whether there is mind and what we know about it. He explores in depth the dualist-interactionism concept that he developed with Sir Karl Popper in the context of a wide variety of brain activities relating to self-consciousness.
2nd. Ed. 1992, Routledge, London & New York, ISBN 0-415-07222-0 (pbk) xv + 279 pp.; 1st. Ed. 1989, Springer, Berlin, etc.

295 Echikson, William
Lighting the Night: Revolution in Eastern Europe
This book tells the first inside story of the people, who exposed Communism's fraud in Eastern Europe and brought about its collapse. It shows how freedom came after more than 40 years of struggle.
1990, Pan Books, London, ISBN 0-330-31835-X (pbk) viii + 310 pp.

296 *Ecologist, The*
Whose Common Future? Reclaiming the Commons
This book diagnoses the social roots of the world's accelerating environmental crises, tracing them to the enclosure of the local commons. A community can live in harmony with its surroundings, when it has authority over its commons, but, everywhere, local commons are being eroded by enclosures. As a result, the ways of life and lands of whole communities are being forced into a restrictive framework, determined by global forces and state

and company bureaucracies. The only way to understand and handle this process is to face the issue of power, and find out what it consists of, who is exercising it, how, and in whose interests. The authors argue that the Earth Summit revealed an official detarmination to extend the process of enclosure by managing it at a global level. They make it clear that the communities need control over their own land, forests, water, and air.

1993, Earthscan, London, ISBN 1-85383-149-2 (pbk) 176 pp.

297
Economic Survey of Europe in 1991-1992
This report gives details, including statistical data, about the economies of Western Europe, North America, Eastern Europe, and the former USSR. It surveys and discusses various aspects of the transition of the Eastern European and Soviet economies from Communism to capitalism, and also has a chapter analysing migration from East to West. It provides statistical data on international trade and payments.

1992, United Nations, New York, distributed by HMSO, London, ISBN 92-1-116540-7 (ISSN 0070-8712) (pbk) xi + 323 pp.

298 Eden, Michael J.
Ecology and Land Management in Amazonia

This book discusses the Amazonian forest region and the exploitation of its renewable resources, and also considers the dry savannahs and wetlands lying within the Amazon forest.

1990, Belhaven Press, London & New York, ISBN 1-85293-118-3 (hbk) xi + 268 pp.

299 Edge, Hoyt L., Robert L. Morris, Joseph H. Rush & John Palmer
Foundations of Parapsychology
This authoritative textbook on parapsychology provides a clear, comprehensive survey with a good balance of topics, and includes a detailed discussion of the criticisms made of this important field.

1986, Routledge & Kegan Paul, Boston & London, ISBNs 0-7102-0726-1 (hbk) & 0-7102-0805-7 (pbk) xvi + 432 pp.

300 Edwards, Michael & David Hulme (Eds.)
Making a Difference: NGOs and Development in a Changing World
Non-governmental organisations (NGOs) are now a major feature of the international environmental and development scene, and their number and influence continue to grow. However, few of them have thought through the implications of recent rapid changes, and few of them have a clear idea how to increase their

impact in the future. This book is designed to help them make their work more effective, by examining how they can enhance their impact. Its contributors have considerable development experience, review the different options and opportunities for NGOs, and relate them to a range of specific case studies.

1992, Earthscan, London, ISBN 1085383-144-1 (pbk) 256 pp.

301 Eggleston, Sir Richard
Evidence, Proof and Probability
This book covers various topics on the legal implications of logic and probability theory.
2nd. Ed. 1983, Weidenfeld and Nicolson, London, ISBNs 0-297-78262-2 (hbk) & 0-297-78263-0 (pbk) xiv + 274 pp.; 1st. Ed. 1978.

302 Ehrlich, Paul R. & Anne H. Ehrlich
The Population Explosion
This book examines the world population problem in the wider context of impending environmental collapse. Even if every nation adopted zero growth population policies immediately, the world population would still reach about 10 billion people before it stabilised. In the authors' view, escalating world population has serious side-effects, including: world food shortages as agriculture is unable to keep up with rising population and soil starts to degrade, increased

threats to the global ecosystem, and pressures on public health services. However, the authors believe that catastrophe can be avoided by adopting appropriate birth control and population control policies. They suggest what individuals as well as governments can do.
1990, Hutchinson, London & Simon & Schuster, New York, ISBN 0-09-174551-9 (hbk) 320 pp.

303 Ehrlich, Paul R. & Anne H. Ehrlich
Healing the Planet: Strategies for Resolving the Environmental Crisis
This book is a companion volume to the *The Population Explosion* (No. 302), and warns that the human enterprise is rapidly outstripping Earth's carrying capacity to support civilisation, which is being threatened almost everywhere. The authors believe that the 1990s must be a decade of environmental politics to ensure human survival. They discuss human life-support systems, energy and the environment, meeting future energy needs, global warming, the ozone layer, air and water pollution, pressures on land resources, sustainable agriculture, and cost-benefit decisions. They urge the development of a new ecological-economic paradigm, make many specific proposals to move towards a sustainable society,

and describe tasks for professionals and ordinary people to help improve the human prospect.
1991, Addison Wesley, Reading, MA, ISBN 0-201-55046-6 (hbk) xv + 366 pp.

304 Ekins, Paul (Ed.)
The Living Economy: A New Economics in the Making
In this book, leading representatives of new economic thinking analyse the defects of conventional economic theory in the light of contemporary economic problems, and provide a preliminary formulation of a coherent, consistent economic framework. They propose a positive new economic theory, policy, and practice for wealth and well-being.
1986, Routledge & Kegan Paul, London & New York, ISBN 0-7102-0946-0 (pbk) xviii + 398 pp.

305 Ekins, Paul
A New World Order: Grassroots Movements for Global Change
This book assesses global problems, then extensively criticises conventional policy approaches to these problems, as exemplified by the reports of the Brandt, Palme and Brundtland Commissions (Nos. 112, 113 & 988). Much useful information is provided about organisations and individuals involved in policy and action at the grassroots level. Peace and security, human rights, economic

development, the environment, and human development are all discussed. The author considers that the root causes of crisis lie in Western scientism, developmentalism, and the concept of the nation state. He argues that, if this is so, it is only through the success of democratic mobilisation of peoples that a new world order, based on peace, human dignity, and ecological sustainability, can be created.
1992, Routledge, London & New York, ISBNs 0-415-07114-3 (hbk) & 0-415-07115-1 (pbk) xi + 248 pp.

306 Ekins, Paul, Mayer Hillman & Robert Hutchinson
Wealth Beyond Measure: An Atlas of the New Economics
The economics of consumption is full of hidden costs, and is drawing us deeper into economic, ecological, and social crisis. This book shows us a way out of this destructive preoccupation with economic growth. It describes a new economic synthesis between the market, state, families, and communities. It states what people and governments can do to build a sustainable society, to create prosperity and a free world in a healthy environment.
Part 1 exposes the nature of the economic crisis. Part 2 explains the concept of wealth, its nature, how it is created, and how it can be measured. Part 3 explores policies for wealth in many different

sectors. For most sectors, the problems and dominant trends are analysed, and proposed policy responses are outlined, with the objectives of economic justice, democracy, and sustainability.
1992, Gaia Books, London & Anchor Books, Doubleday, New York, ISBN 1-85675-050-7 (pbk) 191 pp.; the US Ed. has the title "The Gaia Atlas of Green Economics".

307 Ekins, Paul & Manfred Max-Neef (Eds.)
Real Life Economics
This book considers how economics can address the most urgent contemporary problems, and what are the most exciting new strands of thought in modern economics. Conventional economics has too often ignored environmental and social costs, or failed to analyse them appropriately. The book constructs a framework, within which the wider impact of economic activity can be both understood and improved. This framework emphasises an in-depth understanding of real-life processes, rather than mathematical formalism, to express the interdependence of the economy with the social, ecological and ethical dimensions of human life. The book has an extensive bibliography, and includes a list of names and addresses of 'new economics' and other organisations.
1992, Routledge, London & New

York, ISBN 0-415-07977-2 (pbk) xxi + 460 pp.

308 Elfstrom, Gerard
Moral Issues and Multinational Corporations
This book analyses the moral issues related to the existence and activity of multinational corporations, in the context of their increasing role in world economic integration.
1991, Macmillan, London & New York, ISBN 0-333-52690-2 (hbk) vi + 144 pp.

309 Elkington, John & Tom Burke
The Green Capitalists: Industry's Search for Environmental Excellence
This book discusses the impact of the Green movement on large corporations, and shows how they are changing as a result. It also describes the rise of Green entrepreneurs.
1987, Gollancz, London, ISBN 0-575-04163-3 (hbk) 258 pp.

310 Elkington, John, Tom Burke & Julia Hailes
Green Pages: The Business of Saving the World
This comprehensive reference book on Green business and enterprise describes Green opportunities for business, investment and employment, and shows how environmental pressures threaten major

markets.
*1988, Routledge, London, ISBN
0-415-00232-X (pbk) 256 pp.*

**311 Elkington, John & Julia
Hailes**
**The Green Consumer Guide:
From Shampoo to Champagne:
High-Street**
**Shopping   for   a   Better
Environment**
This guide brings together in a
convenient form important infor-
mation for consumers on a wide
range of products.
*1988, Gollancz, London, ISBN
0-575-04177-3 (pbk) x + 342 pp.*

**312 Elkington, John & Julia
Hailes**
**The Green Consumer's Super-
market Shopping Guide**
This sequel to *The Green Consumer
Guide (No. 311) makes it easier for
consumers to make appropriate
choices in supermarkets.*
*1989, Gollancz, London, ISBN
0-575-04582-5 (pbk) ix + 371 pp.*

**313 Elkington, John, & Peter
Knight with Julia Hailes**
**The Green Business Guide: How
to Take Up – and Profit From – the
Environmental Challenge**
This book is part of a series of
projects by SustainAbility Ltd.,
designed to explore the relationships between business and the
environment. It builds on work
carried out during the production
of *The Green Capitalists (No. 309).*

We are still learning about the
appropriate relationship between
business and the environment,
and the environmental agenda for
business is still emerging. All
businesses, large and small, need
to develop an environmental policy for every aspect of their business, if they are to retain
reasonable markets into the 21st
century. The authors provide
practical guidance on how to
achieve competitive environmental quality standards. They advise
on a wide variety of subjects,
including: investor relations,
environmental auditing, environmental law, financial management, marketing, research and
development, project planning,
manufacturing, the workplace,
personnel, corporate communications, and an agenda for sustainability. They have over 20
years' experience of advising
leading companies and other
organisations, and provide many
useful tips about sensible management. The book includes 15
short corporate case studies.
*2nd.Ed. 1992, Gollancz, London,
ISBN 0-575-05291-0 (pbk) 256 pp.;
1st. Ed. 1991, ISBN 0-575-04675-9
(hbk) 256 pp.*

**314 Elliott, John E., in collabora-
tion with Robert W. Campbell**
Comparative Economic Systems
This textbook compares theories
of economic systems, both the
classical capitalism and socialism,

and their contemporary variants.
1973, Prentice-Hall, Englewood Cliffs, NJ, USA & London, ISBN 0-13-153379-7 (hbk) xix + 540 pp.

315 Ellis, Derek
Environments at Risk: Case Histories of Impact Assessment
This book presents seven case studies of notorious pollution incidents, and three case studies of reducing risk by environmental audits, permitting and regulation, fact finding and social impact.
1989, Springer, London, New York, ISBNs 0-387-51180-6 & 3-540-51180-6 (pbk) xiv + 329 pp.

316 Ellison, Arthur
The Reality of the Paranormal
This book provides a general introduction to parapsychology and psychical research, taking account of recent scientific investigations.
1988, Harrap, London & Dodd Mead, New York, ISBN 0-245-54474-7 (hbk) 159 pp.

317 Ellman, Michael & Vladimir Kontorovich (Eds.)
The Disintegration of the Soviet Economic System
In this book, it is argued that, despite its deep-seated flaws, the Soviet economy was a viable system, whose performance was actually improving from 1983 to 1985; it could have been improved without endangering its principles and structure, but for the

misguided economic paradigm, policies and reforms of the Gorbachev period.
1992, Routledge, London & New York, ISBNs 0-415-06349-3 (hbk) & 0-415-07314-6 (pbk) xv + 281 pp.

318 Ellul, Jacques
The Technological Bluff
This book provides a wide-ranging critique of technology and where it is taking us, and makes extensive use of French and American sources of information. Its author argues that he is not against technology but is working out a theory of the technical society and system.
1990, Eerdmans Publishing Co., Grand Rapids, MI, USA, ISBN 0-8028-3678-X (hbk) 418 pp.

319 Elster, Jon
The Cement of Society: A Study of Social Order
This book addresses the question of what binds societies together and prevents them from disintegrating into chaos and war.
1989, Cambridge University Press, Cambridge & New York, ISBNs 0-521-37456-1 (hbk) & 0-521-37607-6 (pbk) viii + 311 pp.

320 Emms, Peter
Social Housing: A European Dilemma?
This book offers a comprehensive account of the changes which have affected social housing in five European countries.

1990, School for Advanced Urban Studies, Bristol, ISBN 0-86292-358-1 (pbk) xiii + 318 pp.

321 *Energy Policy*, Vol. 20, No. 6, Special Issue
Energy, Environment and Development
This issue opens with a lead article on the theme by Maurice Strong, and addresses various specific aspects and topics of the theme.
June 1992, Energy Policy, ISSN 0301-4215.

322 Engel, J. Ronald & Joan Gibb Engel (Eds.)
Ethics of Environment and Development: Global Challenge and International Response
This book presents contributions to informed debate on the values of sustainable development, and how this goal might be achieved within the foreseeable future. Its papers address these questions from the different perspectives of environmental ethics, developmental ethics, ecofeminism, religion, and humanistic philosophy.
1990, Belhaven Press, London & New York, ISBNs 0-85293-084-5 (hbk) & 0-85293-251-1 (pbk) xvi + 264 pp.

323 ENTEC Press, The
The ENTEC Directory of Environmental Technology: European Edition

This directory is the first European edition of the only comprehensive guide to environmental technology products and services, and over 20,000 companies that supply them, in Western Europe.
1993, Kogan Page, London & Earthscan, London, ISBN 0-7494-0853-7 (pbk) 900 pp.

324 Ereira, Alan
The Heart of the World
This book shows how the author visited the 'lost' Kogi civilisation in the Colombian mountains. Its people have preserved the way of life and philosophy of a culture lost everywhere else. After centuries of deliberate isolation, the Kogi people decided that the time had come to speak to us, convinced that our ignorance and greed would destroy the balance of life on Earth within a few years. They believe that the only hope is for us to change our ways, so that they have set out to teach us what they know about the balance between humankind, nature, and the spiritual world. This book is their warning and their message, attempting to make us understand the consequences of our way of living.
1990, Cape, London, ISBN 0-224-02908-8 (hbk) x + 243 pp.

325 Etzioni, Amitai
The Moral Dimension: Toward a New Economics

This book considers the paradigms that we use in trying to make sense of the social world surrounding us, of which we are an integral part. Out of the dialogue between the entrenched utilitarian, rationalist-individualist, neoclassical paradigm and the social-conservative paradigm, which considers that individuals require a strong controlling authority, arises a third paradigm, presented and discussed by the author, that sees individuals as able to act rationally and on their own, but rooted within a sound community and sustained by a firm moral and emotional personal foundation.

1988, Free Press, New York & Collier-Macmillan, London, ISBN 0-02-909900-5 (hbk) xvi + 314 pp.

326 Etzioni, Amitai (Ed.)
Socio-Economics: Toward a New Synthesis
This book is the published proceedings of the First International Conference on Socio-Economics, held at the Harvard Business School in 1989. Its editor points out that socio-economics has a more complex picture of economic reality than neoclassical economics.

1990, M. E. Sharpe, Armonk, NY, USA, ISBN 0-87332-685-7 (hbk) 289 pp.

327 Evans, Leonard
Traffic Safety and the Driver
This book examines figures for deaths, injuries and property damage from road accidents, and applies scientific method to throw light on their characteristics, nature and severity. It provides important information about the results of published research, answers many questions about traffic safety and driver behaviour, and discusses the current state and future direction of traffic safety.

1991, Van Nostrand Reinhold, New York, and Chapman & Hall, London, ISBN 0-442-00163-0 (hbk) xv + 405 pp.

328 Evans, Roger & Peter Russell
The Creative Manager
This showing managers how to draw on the creativity within them and use it to empower themselves and their organisation. It presents a practical approach to stimulating creative thinking and problem solving, and explores what is really involved in being creative in the face of new challenges under high pressure. It is based on the authors' extensive consultancy and training work on companies worldwide, but it is relevant to everyone, because it applies general principles.

1989, Unwin Hyman, London, ISBN 0-04-440144-2 (hbk) xxiv + 215 pp.

329 Faulks, Rex W.
Principles of Transport
This textbook provides an overall

understanding of transport, and relates theory to practicalities throughout, by including many case studies. It has a modern, international perspective.
4th. Ed. 1990, McGraw-Hill, Maidenhead & New York, ISBN 0-07-707229-4 (pbk) xii + 190 pp.; 1st. Ed. 1964, with the title "Elements of Transport".

330 Felder, David W.
How to Work for Peace
This introduction to practical peace studies discusses issues of security and peace, describes peace education and peace movements, and shows how to think and work positively for peace.
1991, A & M University Press, Gainesville, FL, USA, ISBN 0-8130-1071-3 (hbk) xv + 195 pp.

331 Ferencz, Benjamin B. (Ed.)
World Security for the 21st Century: Challenges and Solutions
The 20th century has seen two parallel but contradictory processes in international law: (1) the progressive development of the theory of international law; (2) the increasing violation of international law in practice. In this book, 12 experts from various parts of the world discuss the legal aspects of an improved world order, and consider how best to replace the law of force by the force of law.
1991, Adamantine Press, London, ISBN 0-7449-0050-6 (ISSN 0954-6073) (hbk) 228 pp.

332 Ferguson, Marilyn
The Aquarian Conspiracy: Personal and Social Transformation in the 1980s
This book discusses the transformation of human consciousness in the 1970s and 1980s, that will eventually transform human awareness and lead to a new social paradigm. It describes a renaissance occurring in *all* the disciplines, both scientific and human, breaking the boundaries between them, and converging at their farthest reaches.
3rd. Ed. 1987, Tarcher, Los Angeles, ISBN 0-87477-191-9 (hbk) 464 pp.; 2nd. Ed. 1982, Paladin, London (pbk).

333 Foley, Gerald & Charlotte Nassim
The Energy Question
This book surveys energy needs and resources, from technical, environmental, and political perspectives.
3rd. Ed., 1992, Penguin Books, London & Viking Books, New York, ISBN 0-14-015683-6 (pbk) 336 pp.

334 Foley, Gerald
Global Warming: Who Is Taking the Heat
This book summarises clearly and concisely the evidence for global warming so far. It examines the potential impacts of climate change worldwide, and outlines what can be done to reduce the risks.

1991, Panos Publications, London, ISBN 1-870670-22-1 (pbk) viii + 104 pp.

335 Foley, Patrick (Ed.)
Why Inflation? (Lloyds Bank Annual Review, Vol. 5)
Most of the articles in this volume discuss tha causes of inflation. More attention is devoted to analysing why inflation persists than why governments give its control such high priority; both the effects of inflation and policies to combat it are considered.
1992, Pinter Publishers, London & New York, ISBN 1-85567-011-9 (ISSN 9053-5004) (hbk) v + 273 pp.

336 Fontaine, Jean-Marc (Ed.)
Foreign Trade Reforms and Development Strategy
This book aims to clarify the objectives, constraints, and dangers of foreign trade reforms, and to contribute to formulating a positive case for flexible, targeted, reversible import controls. It aims to reintroduce long-term considerations into a debate dominated by the management of short-term constraints.
1992, Routledge, London & New York, ISBN 0-415-07294-8 (hbk) xiv + 304 pp.

337 Food and Agriculture Organisation
FAO in 1990
This report includes an overview by the FAO Director-General, an assessment of the world food situation, advice to governments, and a discussion of development assistance.
1991, FAO Information Division, Rome, (pbk) i + 25 pp.

338 Ford, Brian J.
The Cult of the Expert
This highly entertaining and ingenious, yet tremendously relevant, book attacks the 'Experts', who abuse their expert knowledge relevant to the scientific and administrative professions to which they belong. The author's message is that our blind subservience to Experts is leading to the decay and destruction of modern society.
1982, Hamish Hamilton, London, ISBN 0-241-10476-9 (hbk) 194 pp.

339 Foresta, Ronald A.
Amazon Conservation in the Age of Development: The Limits of Providence
This book attempts to expand our understanding of conservation's relationship to several other broad sociopolitical currents.
1991, University of Florida Press, Gainesville, FL, USA, ISBN 0-8130-1092-6 (hbk) x + 366 pp.

340 Forrester, Susan
Business and Environmental Groups: A Natural Partnership?
This book explores the developing partnerships between Greening businesses and

environmental groups and the practical and ethical issues involved. It shows how such relationships can develop, with advice on approach and good practice, and case studies from both sides.
1990, Directory of Social Change, London, ISBN 0-907164-58-7 (pbk) xi + 212 pp.

341 Foster, Philips
The World Food Problem: Tackling the Causes of Undernutrition in the Third World
This book provides an extensive array of facts about the world food problem, all important to understanding the issues. But it presents the facts around a conceptual framework that selects the data to be presented and weaves them into a coherent picture leading to useful, applicable conclusions and a set of policy reform recommendations.
1992, Lynne Riener, Boulder, CO, USA & Adamantine Press, London, ISBNs 1-55587-296-4 (LR) & 0-7449-0072-7 (AP) (hbk) & 1-55587-274-3 (LR) & 0-7449-0073-5 (AP) (pbk) xx + 367 pp.

342 Fox, Avril & Robin Murrell
Green Design: A Guide to the Environmental Impact of Building Materials
This book has been written to provide designers and their clients with clear information about the environmental impact of the materials that they specify and the construction techniques that they use.
1989, Architectural Design and Technology Press, London, ISBN 1-85154-200-1 (pbk) 138 pp.

343 Fox, Matthew
Original Blessing: A Primer in Creation Spirituality
This influential book introduces creation-centred spirituality, and takes us on a spiritual journey that has changed many people's lives. It puts humanity and the whole of nature at the centre of creation, and links Christianity and spirituality to ecology and modern science.
2nd. Ed. 1990, Mountain Books, London, ISBN 0-939680-79-3 (pbk) 355 pp.; 1st. Ed. 1983 Bear & Co., Santa Fe, NM, USA.

344 Fox, Matthew
Creation Spirituality: Liberating Gifts for the Peoples of the Earth
This book presents a framework for a far-reaching spirituality, and challenges people of every religious and political persuasion to unite in a new vision, through which we learn to honour the Earth and the people who inhabit it as the gift of a good and just creator.
1991, HarperSanFrancisco, Harper-Collins, New York, ISBNs 0-06-062916-9 (hbk) & 0-06-062197-7 (pbk) xv + 155 pp.

345 Fox, Warwick
Toward a Transpersonal Ecology: Developing New Foundations for Environmentalism
This book clarifies and reformulates deep ecology's metaphysics in psychological terms, using biography, comparison and review.
1990, Shambhala Publications, Boston & London, ISBN 0-87773-533-6 (pbk) 328 pp.

346 Freeman, Christopher (Ed.)
Design, Innovation and Long Cycles in Economic Development
This book surveys the state of knowledge in design and innovation, and discusses them in relation to 'economic long waves'.
2nd. Ed. 1986, Pinter Publishers, London, ISBN 0-86187-616-4 (hbk) vii + 248 pp.; 1st. Ed. 1984, Royal College of Art, London.

347 Freeman, Christopher, Margaret Sharp & William Walker
Technology and the Future of Europe: Competition and the Global Environment in the 1990s
This book surveys Europe's current technological capabilities across a range of economic sectors, and addresses the international context and policy agenda of technology. It concludes that technologies, which improve quality of life, protect the environment and save resources, could lead world technological competition.
1991, Pinter Publishers, London & Columbia University Press, New York, ISBN 0-86187-075-4 (hbk) 430 pp.

348 Freeman, R. Edward (Ed.)
Business Ethics: The State of the Art
This collection of essays on business ethics discusses four groups of research questions about its current state of the art.
1991, Oxford University Press, Oxford & New York, ISBN 0-19-506478-X (hbk) xi + 225 pp.

349 Friday, Laurie & Ronald Laskey (Eds.)
The Fragile Environment
This book reprints a series of environmental lectures. Topics covered include: the changing human impact on the environment, tropical and temperate forests, attitudes to animals, species diversity and the number of species, famine and its prevention, exhaustible resources, changing climates, and observing Earth's environment from space.
1989, Cambridge University Press, Cambridge & New York, ISBN 0-521-36337-3 (pbk) 188 pp.

350 Friedman, John
Empowerment: The Politics of Alternative Development
This book argues that development policies, guided by mainstream economics, offer little

promise of a better life for most of the world's poor people. It provides a theoretical foundation for an alternative approach to development, rooted in a politics of empowerment, and an informed framework for appropriate new policies that could be embedded in the structure of existing development programmes. It concludes that a synthesis between different approaches to development may bring a viable future for the world's poor people.
1992, Blackwell, Cambridge, MA, USA & Oxford, ISBNs 1-55786-299-0 (hbk) & 1-55786-300-8 (pbk) xi + 196 pp.

351 Friends of the Earth
How Green Is Britain? The Government's Environmental Record
This book reports on all aspects of the state of the environment in the UK. It strongly criticises the environmental policies of the British Government, which it considers to have one of the worst records of environmental care in Western Europe, despite declared 'Greenness'.
1990, Hutchinson Radius, London, ISBN 0-09-174598-5 (pbk) xiii + 242 pp.

352 Fukuyama, Francis
The End of History and the Last Man
This book attempts to set the post-Communist world in historical perspective, and offer a preview of its future evolution. Its author argues that the present time is probably the beginning of 'the end of history', with a 'permanent' dominance of 'liberal' market economics worldwide. He attempts to explain the collapse of Communism, not only in terms of the challenge of capitalist economics, but also in relation to historical factors. However, he comes nowhere near addressing the the radical rethinking of much of social and political theory, required by this collapse. He gives inadequate consideration to the irrational side of human nature, and to possible alternatives to 'capitalist liberal democracy', even though he admits that its world regime would be far from perfect and only slowly remove Third World poverty and inequality. The book has aroused a tremendous amount of interest and controversy, and has been reviewed by people with various opinions.
nd. Ed. 1993, Penguin Books, London, ISBN 0-14-013455-7 (pbk) 448 pp.; 1st. Ed. 1992, Hamish Hamilton, London & The Free Press, New York, ISBNs 0-241-13013-1 (HH) & 0-02-910975-2 (FP) (hbk).

353 Fulder, Stephen
The Handbook of Complementary Medicine
This comprehensive, fairly up-to-

date guide to alternative and complementary medicine surveys scientific, social, legal and other aspects. It describes and assesses each therapy in detail.
2nd. Ed. 1989, Coronet Books, Hodder & Stoughton, London, ISBN 0-340-49984-0 (pbk) xxiv + 411 pp; 1st. Ed. 1984.

354 Fuller, R. Buckminster
Operating Manual for Spaceship Earth
This book describes how we need to manage 'Spaceship Earth' if we are to survive. It identifies the key factors in the required new holistic thinking as general systems theory and the concept of synergy in which relationships between concepts can be explored.
1970, Simon & Schuster, New York, (hbk) 143 pp. & Pocket Books, New York, (pbk) 127 pp.

355 Fuller, R. Buckminster
Utopia or Oblivion: The Prospects for Humanity
This collection of essays illustrates the author's basic conviction that Utopia can be achieved, and ecological disaster prevented, by imaginative and fearless use of the latest technological discoveries.
3rd. Ed. 1972, Penguin Books, Harmondsworth, Middlesex, ISBN 0-14-021567-0 (pbk) 416 pp.; 1st. Ed. 1969, Bantam Books, New York.

356 Furnham, Adrian & Barrie Stacey
Young People's Understanding of Society
This book discusses the world of adults as perceived and understood by young people. It attempts to give a comprehensive and critical account of research on young people's attitudes. Where possible, it traces themes and issues common to all areas of research, and evaluates theories attempting to explain growth and development.
1991, Routledge, London & New York, ISBNs 0-415-01708-4 (hbk) & 0-415-01709-2 (pbk) xi + 215 pp.

357 Gablik, Suzi
The Reenchantment of Art
In this book, the author describes how her hope for the future of art depends on her hope for our culture's spiritual and ethical renewal, and presents her envisaged new cultural paradigm.
1991, Thames & Hudson, New York & London, ISBN 0-500-23619-4 (hbk) vii + 191 pp.

358 Gabor, D., U. Colombo, A. King & R. Galli (Eds.)
Beyond the Age of Waste: A Report to the Club of Rome
This book examines three major world problems: energy, diminishing material resources, and adequate food provision. It analyses the situation from humanitarian and technological

viewpoints, and assesses whether world resources will be sufficient to sustain a rapidly rising world population and enable basic needs to be satisfied.
1978, Pergamon Press, Oxford & New York, ISBNs 0-08-021825-0 (hbk) & 0-08-021824-2 (pbk) xviii + 237 pp.

359 Galbraith, John K.
The Culture of Contentment
This book discusses a relatively recent social problem, arising from the fact that about two-thirds of people in the USA and other affluent countries are wealthy or well-off, while, in most of them, up to a third belong to a poverty-stricken, under-privileged 'underclass'.
1992, Sinclair-Stevenson, London, ISBN 1-85619-147-8 (hbk) 256 pp.

360 Gale, Richard M.
On the Nature and Existence of God
This book critically evaluates the many defences of theism, together with pragmatic arguments to justify faith on the grounds of its prudential or moral benefits.
1991, Cambridge University Press, Cambridge & New York, ISBN 0-521-40300-6 (hbk) viii + 422 pp.

361 Gambling, Trevor & Rifaat Ahmed Abdel Karim
Business and Accounting Ethics in Islam

This book is about contemporary Islamic business and accounting, in theory and practice. It is intended for both Muslim and non-Muslim business people and accountants.
1991, Mansell, London & New York, ISBN 0-7201-2074-8 (hbk) vii + 152 pp.

362 Gamser, Matthew S. et al.
Mobilizing Appropriate Technology: Papers on Planning Aid Programmes
This book considers how to incorporate appropriate technology into a national aid programme, and discusses ecology, basic needs strategy, institutional activities, and what technologies are appropriate.
1988, Intermediate Technology Publications, London, ISBN 1-85339-045-3 (pbk) 112 pp.

363 Garrett, Martha J., Gerald O. Barney, Jennie M. Hommel & Kristen R. Barney (Eds.)
Studies for the 21st Century
This book contains the edited papers of the Second International Meeting on 21st Century Studies in 1989. Its introduction presents eight criteria that define a 21st century study, including a long-term perspective of at least 20 years from the present. A recurring theme is that humankind's advance will require better planning or technological advances, and also a deep change

in human systems, with a focus on sustainability as a key. Many nations are now planning 21st century projects, often to celebrate the year 2000. The book includes 37 national and regional studies about many parts of the world.
1991, UNESCO, Paris, ISBN ? (hbk) 642 pp.

364 Geddes, Sir Patrick
Cities in Evolution
This book shows how trends and tendencies can be made evident by surveying each city, and bringing its ordinary citizens to an understanding of its possible future.
2nd. Ed. 1949, Williams & Norgate, London, (hbk) 241 pp.; 1st. Ed. 1915.

365 Gee, David
Eco-Nomics for Children
This book aims to tell us what questions we should ask of our politicians and those who manage our economy. Its author argues that current ways of running the economy cannot deliver the goods that our children need: decent jobs for their parents, a caring community, and a healthy environment to sustain the economy in the future as well as today. He takes the key ideas from recent Green economic thinking to show how a new 'eco-nomics' can work for, not against our environment. Policies are available, whereby economic and environmental

recovery can support each other, but politicians have not yet adopted them.
1992, Earthscan, London, ISBN 1-85383-139-5 (pbk) 112 pp.

366 George, Susan
How the Other Half Dies: The Real Reasons for World Hunger
This book presents evidence that the main reason for the existence of world hunger on a large scale is the control of world food supplies by the rich and powerful for the wealthy consumer.
1976, latest reprint 1986, Penguin Books, London & Viking Penguin, New York, ISBN 0-14-022001-1 (pbk) 368 pp.

367 George, Susan
A Fate Worse than Debt: The World Financial Crisis and the Poor
This book shows how Third World debt has risen to a trillion dollars, bringing poverty to many millions of people, and causing great environmental damage to service debt. The author proposes an imaginative '3-D solution' ('Debt, Development, Democracy').
2nd. Ed. 1988, reprinted 1989, Penguin Books, London & Viking Penguin, New York, ISBN 0-14-022789-X (pbk) 304 pp.; 1st. Ed. 1987, Grove Weidenfeld, New York.

368 George, Susan
Ill Fares the Land: Essays on Food, Hunger and Power

This book is a collection of essays mostly about the role of power in perpetuating world hunger. The causes of poverty and hunger are not to be found primarily among the poor and the hungry, but rather in their relationships with the rich and powerful people who rule them locally, nationally and internationally.
2nd. Ed. 1990, Penguin Books, London & Viking Penguin, New York, ISBN 0-14-012790-9 (pbk) xvii + 261 pp.; 1st. Ed. 1984.

369 George, Susan
The Debt Boomerang: How Third World Debt Harms Us All
This book evaluates the relationship between the developed countries of the 'North' with the 'developing countries of the 'South' a decade after the onset of the Third World debt crisis. Its author advocates constructive measures for resolving the Third World debt crisi, and explains why solidarity with the South is not only ethically desirable, but also in the interests of the people of the North.
1992, Pluto Press, London, ISBNs 0-7453-0593-8 (hbk) & 0-7453-0594-6 (pbk) xxi + 202 pp.

370 Gershon, David & Robert Gilman
Household Ecoteam Workbook
This book presents a six-month programme to help bring a household into environmental balance in terms of six sets of actions.
1991, Global Action Plan for the Earth, Olivebridge, NY, USA, (pbk) iv + 145 pp.

371 Ghai, Dharam (Ed.)
The IMF and the South: The Social Impact of Crisis and Adjustment
This book examines how changes, occurring in the relative political and economic power within Third World societies, are being affected by International Monetary Fund Structural Adjustment Programmes. These changes, together with resistance to them by those whose living standards are most adversely affected, are shaping the future prospects of developing countries.
1991, Zed Books, London & Atlantic Highlands, NJ, USA, ISBNs 0-86232-950-7 (hbk) & ISBN 0-86232-951-5 (pbk) xii + 275 pp.

372 Ghai, Dharam & Jessica M. Vivian (Eds.)
Grassroots Environmental Action: People's Participation in Sustainable Development
This book provides some information and insights about how to achieve the goal of sustainable development in the Third World, and questions the viability of traditional management systems. Its case studies focus on areas where local people vigorously affect the determination of their own future

and that of their environment.
1992, Routledge, London & New York, ISBN 0-415-07762-1 (hbk) 304 pp.

373 Gilbert, Alan & Josef Gugler Cities, Poverty and Development: Urbanization in the Third World
This book offers a comprehensive analysis of the profound changes, which urbanisation is bringing to the countries of Africa, Asia & Latin America. It provides a critical synthesis of recent social science writing, seeking to show how people manage to survive in these cities.
2nd. Ed. 1992, Oxford University Press, Oxford & New York, ISBNs 0-19-874160-X (hbk) & 0-19-874161-8 (pbk) x + 331 pp.

374 Gilg, Andrew W. (Ed.) Restructuring the Countryside: Environmental Policy in Practice
This book discusses underlying issues and concepts relating to rural environmental policy, and presents nine contemporary case studies that attempt to link theory to practice; it reflects recent research.
1992, Ashgate Publishing, Aldershot & Brookfield, VY, USA, ISBN 1-85628-248-1 (hbk) xv + 238 pp.

375 Ginzburg, Lev R. (Ed.) Assessing Ecological Risks of Biotechnology
This book presents a comprehensive analysis of ecological risk assessment for biotechnology, from the viewpoint of researchers, research managers, regulators, and philosophers.
1990, Butterworth, London & Boston, ISBN 0-409-90199-7 (hbk) xix + 379 pp.

376 Girardet, Herbert Earthbase: How Can We Heal Our Injured Planet?
This book was published at the time of UNCED (United Nations Conference on Environment and Development) at Rio de Janeiro. It is a passionate polemic assessing the serious state of health of Earth's lands and seas, but it also has a message of hope. The tide is turning, and it must be encouraged by action as well as words. We *can* turn the tide against environmental catastrophes, but there are no quick and easy solutions. Many positive actions have already been taken, but the *only* real, all-encompassing solution for halting the destruction and healing Earth is for us to *change our attitudes* at work, at play, and in everything that we do.
1992, Paladin, Harper-Collins, London, ISBN 0-586-09252-8 (pbk) 239 pp.

377 Gleick, James Chaos: Making a New Science
This book is an excellent survey of the mathematical theory of chaos for those without mathematical knowledge. It gives a fascinating

account of how this theory was discovered during the 1960s and 1970s, and provides examples of its applications to science and technology and practical areas like weather forecasting and stock market prediction.
3rd. Ed. 1989, Cardinal, London, ISBN 0-7474-0415-5 (pbk) 368 pp.; 2nd. Ed. 1988, Heinemann, London, ISBN 0-434-29554-X (hbk).

378 Goldsmith, Edward
The Way: An Ecological World View
The author believes that only a clearly formulated ecological world view is likely to bring about a comprehensive strategy for assuring the preservation of the Earth, and thus the survival of humanity. In this book, he states and explains 66 principles which he considers fundamental for such a world view. He emphasises the need for people to realise that they are an integral part of nature, so that they cannot continue to exploit nature without damaging or destroying themselves. The book offers both profoundly argued theory and miscellaneous practical advice, and extensively discusses the philosophical, scientific, economic, and political foundations for an ecological world view. It throws much light on how the premises of deep ecology can best be realised in all spheres of thought and activity affecting our

world.
1992, Rider Books, London, ISBNs 0-7126-4646-9   (hbk)   & 0-7126-4663-9 (pbk) xxii + 442 pp.

379 Goldsmith, Edward et al.
A Blueprint for Survival
This book is one of the best-known statements of the problems facing humankind and the planet, the needs for change, and the goals required for establishing a stable society. It had enormous influence on the subsequent development of Green politics and Green philosophy.
3rd. Ed., 1972, Penguin Books, Harmondsworth, Middlesex, ISBN 0-14-052295-6 (pbk) 139 pp.; 1972, 2nd. Ed., Tom Stacy, London; 1st. Ed., January 1972, The Ecologist, Vol. 2, No. 1.

380 Goldsmith, Edward & Nicholas Hildyard (Eds.)
Earth Report 3: Monitoring the Battle for Our Environment
This book contains over 400 alphabetically arranged articles on a wide range of environmental issues and other world problems, together with over 40 diagrams displaying different aspects of the human and planetary situation. Its entries carry both internal cross-references to other relevant entries, and 'see also' references to entries containing specific types of additional information. Subjects covered include a wide

range of ecological and environ-mental topics, but also agricul-ture, development, economics, health, nutrition, population, Third World, and urban prob-lems. The book ends with a list of important international, British, North American, and Australa-sian organisations.
3rd. Ed. 1992, Mitchell Beazley, London, ISBN 0-85333-926-8 (pbk) 175 pp.

381 Goldsmith, Edward, Nich-olas Hildyard, Peter Bunyard & Patrick McCully
5000 Days to Save the Planet
This book is a plea by the Editors of *The Ecologist* (No. 1027) on behalf of our planet, explains what humanity is doing to the planet, and presents a manifesto of what needs to be done to save the planet. The authors present massive evidence about the damage being done to our environment, in land, air, fresh water and sea, and point out the resulting diminishing quality of life. In their view, we may have as little as 15 years, or even 5000 days, in which to save Earth. The book contains many beautiful col-our photographs and other illustrations.
1990, Hamlyn, London, & MIT Press, Cambridge, MA, USA (with the title "Imperiled Planet: Restoring Our Endangered Ecosystems"), ISBN 0-600-57156-4 (hbk) 288 pp.

382 Good, I. J.
Probability and the Weighing of Evidence
This book presents the theory of probability and its application to assessing the value of scientific evidence and legal testimony.
1950, Griffin, London, (hbk) viii + 119 pp.

383 Good, I. J., Alan J. Mayne & John Maynard Smith (Eds.)
The Scientist Speculates: An Anthology of Partly-Baked Ideas
"The intention of this anthology is to raise more questions than it answers". It is the first published collection of 'partly-baked ideas', namely scientific and other spec-ulations not yet 'baked' into a more rigorous form. It includes contributions on a wide range of scientific subjects from eminent scientists and others with power-ful imaginations.
3rd. Ed. 1965, Capricorn Books, New York, xvii + 413 pp.; 2nd. Ed. 1963, Basic Books, New York; 1st. Ed. 1962, Heinemann, London, (hbk).

384 Goodfield, June
The Planned Miracle
In 1983 was launched the Task Force for Child Survival, which aimed to give all the world's chil-dren a healthy start in life, includ-ing bringing protective immunisation to 80% of them by 1990. This book describes the enormous problems and dis-agreements surrounding this

international project, and the monumental humanitarian campaign that was eventually launched, involving over 100 countries worldwide.
1991, Cardinal, Sphere Books, London, 0-7474-0933-1 (pbk) xviii + 318 pp.

385 Goodman, David & Hall, Anthony (Eds.)
The Future of Amazonia: Destruction or Sustainable Development
This book aims to describe and analyse current strategies of economic development and frontier integration in Brazilian Amazonia, highlighting their major environmental, ecological and social impacts.
1990, Macmillan, London & St Martin's Press, New York, ISBN 0-333-46490-7 (hbk) xviii + 419 pp.

386 Goodman, David, Bernardo Sorj & John Wilkinson
From Farming to Biotechnology: A Theory of Agro-Industrial Development
This book offers an original interpretation of the industrialisation of agriculture and the evolution of the contemporary food system, and integrates previously diverse approaches.
1987, Blackwell, Oxford & New York, ISBN 0-631-14405-6 (hbk) 214 pp.

387 Gorbachev, Mikhail
Perestroika: New Thinking for Our Country and the World
In writing this book, the author, as General Secretary of the Communist Party of the USSR and subsequent President of the USSR, wishes to address directly the peoples of the USSR, the USA, indeed every country, about issues that concern them all. In his view, problems must be tackled in a spirit of cooperation rather than animosity, and the book is his contribution to the dialogue about them.
1987, Collins, London, ISBN 0-00-215660-1 (hbk) 254 pp.

388 Gordon, Anita & David Suzuki
It's a Matter of Survival
This book, about the shortsightedness of the human species and its failure to read the warning signs, is based on over 100 interviews with scientists and influential Green thinkers and activists.
1991, Harvard University Press, Cambridge, MA, USA, ISBN 0-674-469704 (hbk) 278 pp.

389 Gordon, Deborah
Steering a New Course: Transportation, Energy, and the Environment
This book discusses the deterioration of the USA's road, rail and air transport. Its infrastructure is becoming increasingly unable to handle the growing volume of

traffic, and a serious disintegration seems likely by the year 2000 unless corrective action is taken.
1991, Island Press, Washington, DC, USA, ISBNs 1-55963-135-X (hbk) & 1-55963-134-1 (pbk) 244 pp.

390 Gore, Senator Al
Earth in the Balance: Forging a New Common Purpose
The author of this book, who is now Vice-President of the USA, has been a passionate defender of the environment for over 20 years, and is now convinced that human civilisation has brought us to the edge of catastrophe. He argues that only a radical rethinking of our own relationship with nature can save the Earth's ecology for future generations. He presents many facts, some of them based on the latest research and not yet well-known, to show that the threat is real, that the quality of our air, water and soil is at grave risk. Global warming, the deteriorating ozone layer, and the rapid destruction of the world's rain forests must now be considered from a strategic viewpoint. The roots of the crisis reach into every aspect of society, and the author's analysis of where we have gone wrong covers politics, history, economics, science, psychology, and religion. In his view, a large part of the blame must be assigned to our political leaders, many of whom ignore the long-term consequences of timid policy choices. Similarly, most economists have failed to calculate the true cost of our greedy consumption of nonrenewable resources. Human civilisation must heal itself, psychologically and spiritually, before we can heal our ailing environment. But, at the same time, there needs to be a much broader awareness, a raising of consciousness among people in general, and the development of a new and inspiring common purpose, the rescue of our environment and the restoration of balance to our planet. This would be accompanied by a broad sense of justice and the promotion of true democracy and genuine freedom for all people. In the book's final part, the author argues that only a worldwide mobilisation can save us. He presents a comprehensive plan for action, which would be a 'global Marshall Plan' motivated by the new common purpose, and concentrating on strategic goals and actions likely to remove bottlenecks. The plan's objectives would be to: (1) stabilise world population; (2) develop and share appropriate technologies, encouraged by a Strategic Environment Initiative (SEI); (3) develop a new global eco-nomics; (4) bring about a new generation of international treaties and agreements; (5) achieve a new global environmental consensus

based on a broad cooperative programme for educating the world's
citizens. The plan takes account of
ecological and environmental
ideas, together with population
trends, appropriate technology,
and environmental education. It
would not all be immediately feasible, but it should be put into
effect as soon as the world climate
of opinion makes it practicable.
The author believes that the
severity of the world environmental crisis needs a bold and
visionary response, which
requires at least a new conception
of both the individual and the
environment. The book has a
bibliography.
1992, Houghton Mifflin, Boston,
ISBN 0-395-57821-3 (hbk), with sub-
title "Ecology and the Human
Spirit", & Earthscan, London, ISBN
1-85383-137-9 (pbk) viii + 408 pp.

391 Gosovic, Branislav
The Quest for World Environ-
mental Cooperation: The Case of
the UN Global Environment
Monitoring System
This book examines the origins
and effectiveness of the United
Nations' Global Environment
Monitoring System (GEMS), set
up in 1972 in response to an
acceptance of the world's global
interdependence   through
environmental issues. GEMS'
responsibilities, for providing
information on the environment,
present serious organisational

challenges.
1992, Routledge, London & New
York, ISBN 0-415-00458-6 (hbk) 368
pp.

392 Goudie, Andrew
Human Impact on the Natural
Environment
This book is concerned with the
ways in which people have
changed and are changing the
Earth, and with the human role in
natural processes and systems. Its
first   chapter   considers   the
development of ideas about interaction between humans and the
environment,   especially   the
effects of population increase and
technological change. Later chapters cover the human impact on
vegetation, animals, the soil,
water and water resources, land
forms, and the climate and
atmosphere. The final chapter
analyses the causes and consequences   of   environmental
change, and changing relations
between these processes and
phenomena.
3rd. Ed. 1990, Blackwell, Oxford,
ISBNs   0-631-17228-9   (hbk)   &
0-631-16164-3 (pbk) xi + 388 pp; 1st.
Ed. 1981.

393 Gould, Carol C.
Rethinking Democracy: Freedom
and Social Cooperation in Poli-
tics, Economy, and Society
This book proposes a fundamental rethinking of the theory of
democracy. Its author aims to lay

new philosophical foundations for democracy. She argues that democratic decision making should apply, not only to politics but also to economic and social life. As far as possible, it should be participative, but otherwise representative.
1988, Cambridge University Press, Cambridge & New York, ISBN 0-35048-4 (hbk) x + 363 pp.

394 Gourlay, K. A.
Poisoners of the Seas
This book discusses pollution of the seas and oceans, concluding that the situation should no longer be handled only by governments and their scientists, and that anti-pollution policies must be reversed.
1988, Zed Books, London, ISBNs 0-86232-685-0 (hbk) & 0-86232-686-9 (pbk) xvi + 256 pp.

395 Government of Canada
Canada's Green Plan for a Healthy Environment
This plan details *how* the Canadian Government proposes to achieve sustainable development. It was developed in a three-stage consultative process, with 58 sessions attended by nearly 10,000 Canadians, and a two-day final national sesion in August 1990.
1990, Environment Canada, Ottawa, Canada, free, (hbk) 174 pp.

396 Government Statistical Service
Government Statistics: A Brief Guide to Sources
This annual booklet provides a brief guide to the UK Government series of statistics. Its accompanying annual pocket-sized leaflet *United Kingdom in Figures* summarises key UK statistical series.
(updated yearly) Central Statistical Office, London, (pbk) 40 pp.

397 Goyder, George
The Just Enterprise
This book aims to develop the idea of the responsible company, and shows how the cause of freedom can be advanced through the creation of responsible industrial companies, which will require important reforms in company law. The author focusses on the social and human problems of industry, which have been ignored for far too long.
1987, Deutsch, London, ISBN 0-233-98157-8 (hbk) xii + 116 pp.

398 Gradwohl, Judith & Russell Greenberg
Saving the Tropical Forest
This book describes the concept of forest reserves, discusses the relationship between sustainable agriculture and forest conservation, and shows how to restore tropical forests at risk. It includes case studies of successful forest reclamation projects.
1988, Earthscan, London, ISBN 1-85383-014-3 (pbk) 288 pp.

**399 Graham-Brown, Sarah
Education in the Developing World: Conflict and Crisis**
This book traces the impact of international economic pressures and internal political conflicts on the provision of education in developing nations, and examines how these external factors are combined with in national struggles over educational policy. Case studies analyse issues round the setting of educational priorities, and explore how education can be made more relevant to societies' needs.
1991, Longman, London & New York, ISBN 0-582-06431-7 (pbk) xx + 332 pp.

400 Grahl, John & Paul Teague 1992 – The Big Market: The Future of the European Community
This book identifies the key issues of European Community (EC) policy, affected in the 1992 programme of European economic integration, and covers every major field of EC policy.
1990, Lawrence & Wishart, London, ISBN 0-85315-716-2 (pbk) 351 pp.

**401 Grainger, Alan
The Threatening Desert: Controlling Desertification**
This well-researched book provides one of the most comprehensive accounts of the enormous worldwide problem of dryland degradation. It shows where desertification is happening, and what effects it has. It explains its causes, together with the policies needed to prevent and reverse it. It describes the considerable successes already achieved.
1990, Earthscan, London, ISBN 1-83583-041-0 (pbk) 384 pp.

**402 Grainger, Alan
Controlling Tropical Deforestation**
This book provides a new analysis of the causes of deforestation, and presents an integrated strategy for controlling them, including appropriate actions in agriculture, forestry, and conservation. It outlines changes in government policies that could support sustainable forest management, and tackle causes of deforestation.
1992, Earthscan, London, ISBN 1-85383-142-5 (pbk) 208 pp.

**403 Grant, James P.
The State of the World's Children 1992**
This report by the Executive Director of UNICEF presents ten specific propositions for saving and improving the lives of children worldwide, and concludes by presenting UNICEF's goals for children.
1992, Oxford University Press, Oxford & New York, ISBN 0-19-262228-5 (pbk) 100 pp.

404 Grattan-Guinness, Ivor (Ed.)
Psychical Research: A Guide to Its History, Principles & Practices in Celebration of 100 Years of the Society for Psychical Research
This book provides an introduction to *all* the principal aspects of psychical research and parapsychology.
1982, The Aquarian Press, Wellingborough, Northants., ISBN 0-85030-316-8 (pbk) 424 pp.

405 Grayson, Lesley (Ed.)
Green Belt, Green Fields and the Urban Fringe: The Pressure on Land in the 1980s: A Guide to Sources
This bibliography surveys relevant literature since about 1980 on several aspects of topical environmental issues in southern England.
1990, The British Library, Boston Spa, Wetherby, West Yorkshire, ISBN 0-7123-0770-2 (pbk) 112 pp.

406 Grayson, Lesley (Ed.)
Recycling: New Materials from Community Waste: A Guide to Sources
This bibliography covers the legal, economic, political, technical, and social aspects of materials reclamation and recycling in the developed world over the last ten years.
1990, The British Library, Boston Spa, Wetherby, West Yorkshire, ISBN 0-7123-0778-8 (pbk) 154 pp.

407 Grayson, Lesley (Ed.)
Acid Rain and the Environment 1988-1991: A Select Bibliography
This bibliography has a broad coverage of the environmental and biological effects of acid rain and on acidification countermeasures.
1991, Technical Communications, Letchworth, Herts. & British Library, London, ISBN 0-946655-42-1 (hbk) iv + 217 pp.

408 Green, David
Green Living: Personal Action to Save Our Environment
This book argues that we must act more positively as individuals. Because the system that endangers us is driven by what we consume, resolute consumer action can speak very powerfully to governments. The author presents principles for Greener living, provides a checklist for a household audit, and clearly states what actions we can all take.
1990, Kogan Page, London, ISBN 0-7494-0266-0 (pbk) 144 pp.

409 Greenberg, Joseph
The Theory of Social Situations: An Alternative Game-Theoretic Approach
This book offers a new, integrative approach to the study of formal models in the social and behavioural sciences.
1990, Cambridge University Press, Cambridge & New York, ISBNs 0-521-37425-1 (hbk) & 0-521-37689-0 (pbk) ix + 187 pp.

410 Gribbin, John
The Hole in the Sky: Man's Threat to the Ozone Layer
This book shows how the ozone layer, Earth's shield against solar ultraviolet radiation, is being destroyed at an ever-increasing rate, mainly by human products containing CFCs and other halocarbons. The damage to the ozone layer harms human health, animals, food crops, life-support systems, and world climate. The author argues that, without *prompt* action, the ozone hole could widen disastrously, causing irreparable damage to our environment and ourselves.
1988, Corgi Books, London, ISBN 0-552-99329-8 (pbk) x + 160 pp.

411 Gribbin, John
Hothouse Earth: The Greenhouse Effect & Gaia
This comprehensive and readable book examines the greenhouse effect in relation to natural climatic processes. It explains why the Earth is becoming warmer, and what must be done to slow the pace of climatic change and prevent irreparable damage to the web of life on Earth.
1990, Black Swan, London, ISBN 0-552-99450-2 (pbk) x + 273 pp.

412 Griffin, David Ray (Ed.)
Sacred Interconnections: Postmodern Spirituality, Political Economy, and Art
This book presents a constructive postmodernism', that aims to revise modern premises and traditional concepts, and support the ecology, peace, feminist and other liberating movements of our time.
1990, State University of New York Press, Albany, NY, USA, ISBNs 0-7914-231-2 (hbk) & 0-7914-232-0 (pbk) xiii + 227 pp.

413 Griffin, Kate & John Knight (Eds.)
Human Development and the International Development Strategy for the 1990s
This book presents a strong case for a renewed emphasis on human development, both to promote economic growth and as an end in itself. Policies for human development enhance the capabilities of people to be and do things. The book advocates redirecting policies for the 1990s.
1990, Macmillan, London & New York, ISBNs 0-333-53512-X (hbk) & 0-333-53513-8 (pbk) viii + 258 pp.

414 Griffin, Kelley
Ralph Nader Presents More Action for a Change
This book presents a study of the Public Interest Research Groups (PIRGs) in the USA, which are taking action on various environmental issues. There is a description of PIRGs' achievements, success stories, history, structure, funding, and future goals.
1987, Dembner Books, New York,

ISBNs 0-934878-62-5 (hbk) & 0-934878-62-5 (pbk) xxix + 226 pp.

415 Group of Green Economists, The
Ecological Economics: A Practical Programme for Global Reform
This book assesses the thinking of various environmental, women's and human rights movements. The Group of Green Economists, associated with the German Greens, considers practical alternatives to the present vast inequalities and social and environmental dislocation. Its proposals are based on the principles of ecological balance, democracy, social equality, feminism, non-violence, and respect for cultural identity and diversity.
1992, Zed Books, London & Atlantic Highlands, NJ, USA, ISBNs 1-85649-069-6 (hbk) & 1-85649-070-X, xiv + 162 pp.

416 Grubb, Michael
The Greenhouse Effect: Negotiating Targets
This booklet reports a study of different countries' approaches to the emission of greenhouse gases.
1989, Royal Institute of International Affairs, London, ISBN 0-905031-30-X (pbk) viii + 56 pp.

417 Grubb, Michael
Energy Policies and the Greenhouse Effect: Volume One: Policy Appraisal
This book presents the results of two years' study by the Royal Institute of International Affairs' Energy and Environmental Programme. This first volume concentrates on policy issues, arising from attempts to reduce greenhouse gas emissions from the energy sector.
1990, Dartmouth Publishing, Aldershot, Hants. & Brookfield, VT, USA, ISBN 1-85521-175-0 (pbk) xvii + 294 pp.

418 Grubb, Michael
Energy Policies and the Greenhouse Effect: Volume Two: Country Studies and Technical Options
This book provides the detailed analysis, on which the conclusions of the first volume (No. 417) are based.
1992, Dartmouth Publishing, Aldershot, Hants. & Brookfield, VT, USA, ISBN 1-85521-198-X (hbk) xxi + 450 pp.

419 Gruen, Lori & Peter Singer
Animal Liberation: A Graphic Guide
This book describes the animal liberation movement, and then discusses the idea of animal liberation, what we should do to animals, and what we should do about animal rights. It lists useful organisations, direct action groups, and cruelty-free products.
1987, Camden Press, London, ISBN 0-948491-21-3 (pbk) 159 pp.

420 Gupta, Avijit
Ecology and Development in the Third World
This introduction to ecology and development in the Third World shows how the growing crisis of environmental degradation can be linked to the effects of global economic development and population pressure.
1988, Routledge, London & New York, ISBN 0-415-00673-2 (pbk) vii + 80 pp.

421 Haas, Michael
Polity and Society: Philosophical Underpinnings of Social Science Paradigms
This book is a critique of theoretical and conceptual literature on political science and the social sciences in general, and it has an extensive bibliography. Its author analyses the wide variety of paradigms used by their theorists, and exposes the metaphysical, epistemological and methodological bases of these paradigms. The most important chapters are the first chapter, whose basic discussion of paradigms sets the stage for the rest of the book, and the final chapter, which provides an excellent summing up of the author's arguments and conclusions. The intermediate chapters provide in-depth analyses of the leading paradigms proposed and formulated by political scientists and other social sciences in seven specific areas.

1992, Praeger, New York & London, ISBN 0-275-93558-2 (hbk) xiv + 306 pp.

422 Haila, Yrjo & Richard Levins
Humanity and Nature: Ecology, Science & Sociology
This book examines the different faces of ecology and their mutual connections. It discusses modern theory, disputes whether human beings ever existed in harmony with nature, and concludes that human activity should be included as part of nature.
1992, Pluto Press, London, ISBN 0-7453-0669-1 (pbk) xi + 270 pp.

423 Hall, Anthony L.
Developing Amazonia: Deforestation and Social Conflict in Brazil's Carajas Programme
This book describes in detail the Greater Carajas Project, Brazil's most recent and ambitious attempt to harness Amazonia's potential, assesses its social and environmental impacts, and argues that it has seriously worsened the region's crisis and damaged its environment.
1989, Manchester University Press, Manchester & New York, ISBN 0-7190-2494-3 (hbk) xxi + 295 pp.

424 Hall, Peter
Great Planning Disasters
This book discusses a series of great planning disasters, costing many millions of pounds. It analyses the planning processes

involved, isolates those which were disastrous, and distils valuable lessons.
2nd. Ed. 1981, Penguin Books, Harmondsworth, Midd. & New York, ISBN 0-14-081323-3 (pbk) x + 308 pp.; 1st. Ed. 1980.

425 Hall, Peter (Ed.)
Technology, Innovation and Economic Policy
This collection of essays shows the diversity of current research into the process of technological innovation, and the policies appropriate for its stimulation and direction.
1986, Philip Allen, Deddington, Oxon., ISBNs 0-86003-062-8 (hbk) & 0-86003-171-3 (pbk) viii + 248 pp.

426 Hall, Peter
Cities of Tomorrow: An Intellectual History of Urban Planning and Design in the Twentieth Century
This book provides a history of the ideas, events and personalities that shaped the cities of the world during the 20th century. It shows extravagant visions, grand ambitions, and often appalling failures.
1988, Blackwell, Oxford & New York, ISBN 0-631-13444-1 (hbk) xvi + 473 pp.

427 Hall, Peter
Urban and Regional Planning
This widely used textbook provides an up-to-date introduction to planning. It traces the evolution of urban and regional problems, planning philosophies, techniques, and legislation, and places modern planning in its historical context.
3rd. Ed. 1992, Routledge, London & New York, ISBNs 0-415-076234-4 (hbk) & 0-415-07624-2 (pbk) 304 pp.

428 Hall, Peter & Paschal Preston
The Carrier Wave: Information Technology and the Geography of Innovation 1846-2003
This book provides a comprehensive history of the new information technology industries, evolving across three economic long waves and into a fourth wave. It shows why the core of new information technology production has shifted from country to country and region to region.
1988, Unwin Hyman, London & Boston, MA, USA, ISBN 0-04445081-8 (hbk) xiv + 305 pp.

429 Hall, Stuart & Bram Gieben
Foundations of Modernity
This book provides an introduction to modern societies and sociological analyses. It describes the impact of historical processes, institutions, and ideas that have shaped modern societies. It traces the evolution of the distinctive forms of political, economic, social, and cultural life that are typical of contemporary society.
1992, Polity Press, Cambridge, ISBNs 0-7456-0959-7 (hbk) & 0-7456-0960-0 (pbk) viii + 342 pp.

430 Handy, Charles
The Future of Work: A Guide to a Changing Society
The author of this book argues for new forms of work and work pattern for the future, presenting both good and bad possibilities. He also considers the needs of the future workforce for education and training, and how to organise for these needs.
2nd. E. 1985, Blackwell, Oxford & New York, ISBN 0-631-14278-9 (pbk) xiii + 201 pp.; 1st. Ed. 1984, ISBN 0-631-14277-0 (hbk).

431 Handy, Charles
The Age of Unreason
This book presents a vision of an era of new discoveries, new enlightenments and new freedoms, and challenges readers' existing understanding of the world. Its three parts discuss changing, working and organisations, and living.
2nd. Ed. 1990, Arrow Books, London, ISBN 0-09-975740 (pbk) viii + 216 pp.; 1st. Ed. 1989, Business Books, Century Hutchinson, London.

432 Hardoy, Jorge E., Diana Mitlin & David Satterthwaite
The Environmental Problems of Third World Cities
This book analyses the severe problems faced by Third World cities, and shows how they affect human health, local ecosystems, and global cycles. It reveals political solutions by uncovering the political roots of many problems, and proposes new approaches.
1992, Earthscan, London, 1-85383-146-8 (pbk) 160 pp.

433 Hargreaves Heap, Shaun
The Theory of Choice: A Critical Guide
This book provides a comprehensive introduction to the fundamental principles of rational choice theory, game theory, and their implications and limitations. It includes surveys of social choice, democracy, power, and social justice.
1992, Blackwell, Oxford & Cambridge, MA, USA, ISBNs 0-631-17174-6 (hbk) & 0-631-18322-1 (pbk) xv + 398 pp.

434 Harman, Willis W.
An Incomplete Guide to the Future
This book explains and uses the approach of futures research to consider how we may be transforming ourselves. It examines the basic patterns of industrialised society, finds evidence for a massive change of direction, and discusses strategies for a viable future.
2nd. Ed. 1979, W. W. Norton & Co., New York & London; 1st. Ed. 1976 (pbk).

435 Harman, Willis W.
Global Mind Change: The Promise of the Last Years of the Twentieth Century

This book discusses the factors and forces that are already bringing about a global mind change, able to transform the present human and planetary situation. People can change the world by witholding legitimacy from its institutions and by modifying their internal image of reality. The fundamental change in 'Western' society is a shift of dominant metaphysic from materialistic monism to transcendental monism; this latter world-view sees the ultimate and primary substance of the universe as consciousness-mind, and implies a shift of moral and social values. The resulting transformation changes business and work, redefines global development, seeks global security, and brings the evolution of new politics and values.

1988, Knowledge Systems, Indianapolis, IN, USA, in cooperation with the Institute of Noetic Sciences, ISBN 0-941705-06 (pbk) 185 pp.

436 Harman, Willis W.
A Re-examination of the Metaphysical Foundations of Modern Science

This report criticises the foundations of modern science, then outlines some new possibilities. It describes some pointers to a philosophy of wholeness, addresses the issue of consciousness, and formulates the characteristics of a more complete 'wholeness'

science.

1991, Institute of Noetic Sciences, Sausalito, CA, USA, (pbk) vi + 110 pp.

437 Harman, Willis W. & John Hormann
Creative Work: The Constructive Role of Business in a Transforming Society

The authors of this book consider that the reassessment of the role of work seems to be central to the resolution of contemporary global problems. The problems of work may be solved through an emergence of new goals and values from "a vast, creative middle band of people who sense a new vision" and will not be satisfied until it is realised. This could lead to a new awakening, and to a society where everyone has the opportunity to be engaged in dignified and satisfying work.

1990, Knowledge Systems, Indianapolis, IN, USA, ISBNs 0-941705-11-0 (hbk) & 0-9041705-12-9 (pbk) 208 pp.

438 Harris, Nigel
The End of the Third World: Newly Industrialising Countries and the Decline of an Ideology

The author explores how some Third World countries have developed industrially, while others have not. Similarly, it considers why some have changed from peasant economies, and others have not.

2nd. Ed. 1987, Penguin Books, London & New York, ISBN 0-14-022563-3 (pbk) 240 pp.; 1st. Ed. 1986, Tauris, London & New York.

439 Harris, Nigel (Ed.)
Cities in the 1990s: The Challenge for Developing Countries
This book addresses some key issues facing urban areas in developing countries. It scrutinises the approaches emerging in a wave of change of aid policy, ranging from urban development to urban areas.
1992, UCL Press, London, ISBNs 1-85728-029-6 (hbk) & 1-85728-030-X (pbk) xxiii + 232 pp.

440 Harrison, Lawrence E.
Who Prospers: How Cultural Values Shape Economic and Political Success
In this book, the author asks why some nations and ethnic groups do better than others. Although climate, resources, geographical location and size, policy choices and luck are relevant factors, the author views culture, values and attitudes as mainly responsible.
1992, Basic Books, New York, (hbk) 280 pp.

441 Harrison, Paul
The Third Revolution: Environment, Population and a Sustainable World
This book covers a wide range of environmental and other global problems and issues. It provides pointers to a general theory, and presents options for action towards the coming 'third revolution' of sustainable balance with our natural environment. New values are spreading fast, and the coming new paradigm and philosophy involves an ethics of extended altruism, not only to people living today but also towards future generations and other species. Five Third World village case studies are included. The author's approach is empirical: to see the world as it is in all its complexity, to avoid preconceived ideas as far as possible, to look at the evidence and see where it leads. Although some simplification is necessary, oversimplifying or taking too polemical an attitude removes all hope of understanding the world properly or cooperating effectively in tackling its many problems.
1992, Tauris, London & New York, ISBN 1-85043-501-4 (hbk) xi + 359 pp.

442 Harte, John, Cheryl Holdren, Richard Schneider & Christine Shirley
Toxic A to Z: A Guide to Everyday Pollution Hazards
This book gives a general guide to pollution hazards, and discusses many specific aspects and issues. It provides detailed information on about 110 commonly encountered toxics, and has several appendices.

1991, University of California Press, Berkeley, CA, USA, (hbk) & (pbk) 479 pp.

443 Hartwick, John M. & Nancy D. Olewiler
The Economics of Natural Resource Use
This book discusses natural resource economics interms of land use and land value, non-renewable resources, energy economics, fisheries, forests, and pollution problems. Finally, it describes the role of government regulations and policies for natural resource use.
1986, Harper & Row, New York & London, ISBN 0-06-042695-0 (hbk) ix + 530 pp.

444 Harvey, Ronald
Our Fragmented World
This book looks at the present state of the world and our attitude to life and society. It explains why the world seems to be on the verge of destroying itself, and suggests how to contain the dangerous trends. It presents a plea for human unity and coherence, and seeks a resolution through balance, to which we can all contribute.
1988, Green Books, Hartland, Bideford, Devon, ISBN 1-87009-810-2 (pbk) 150 pp.

445 Harvey-Jones, Sir John
Making It Happen: Reflections on Leadership
The author reflects on his years as Chairman of ICI and on his experience of leadership. He offers a wealth of knowledge, know-how and common-sense advice about all aspects of business life and management. The principles and experience presented in this book represent some of the best aspects of practice in the contemporary mainstream world of business, and illustrate well how several leading companies and enterprises operate today. They can be applied not only to companies and businesses but also to many other types of organisation, ranging from governments to voluntary groups.
The author has repeatedly discovered that the people with whom he has worked have a fantastic capacity for achievement, that can be released by leadership tuned to their needs and aspirations. He has found that there is an art in motivating people to do their best. He believes that there are no limitations on what a company can achieve if the will is there. The key to achieving the required degree of coordination and mutual cooperation is effective team work throughout the organisation, with each team having a definite function and purpose and a definite set of objectives to which its members agree. The team will be more effective, the more its members are commited to their shared

goals.
3rd. Ed. 1989, Fontana Paperbacks, London, ISBN 0-00-637409-3 (pbk) 336 pp.; 1st. Ed. 1988, Collins, London.

446 Hawking, Stephen W.
A Natural History of Time: From the Big Bang to Black Holes
This book by one of the world's leading cosmologists and theoretical physicists explores the outer reaches of our knowledge of astrophysics and the nature of time and the universe, and introduces the most important contemporary theories on the physical universe.
1988, Bantam Press, Transworld Publishers, London & New York, ISBN 0-593-01518-5 (hbk) x + 198 pp.

447 Hayes, Peter & Kirk Smith (Eds.)
The Global Greenhouse Regime: Who Pays?
This book offers a remarkable interdisciplinary and multinational response to the challenge posed by the greenhouse effect. It presents the scientific, economic and political issues, and then describes the available policy options, in comprehensible terms.
1992, Earthscan, London, ISBN 1-85383-136-0 (pbk) 256 pp.

448 Hayter, Teresa
Exploited Earth: Britain's Aid and the Environment
This book examines the UK's aid policy, especially its effects on the world's forests. It describes the history and effects of different forms of aid, and shows how and why British aid needs to change.
1989, Earthscan, ISBN 1-85383-048-8 (pbk) xii + 276 pp.

449 Healey, John & Mark Robinson
Democracy, Governance and Economic Policy
This book explores the relation between democracy and development in developing countries, drawing on historical experience and the impact of economic development in the 1980s.
1992, Overseas Development Institute, London, ISBN 0-85003-171-0 (pbk) 188 pp.

450 Healy, Jane M.
Endangered Minds: Why Our Children Don't Think
Citing recent research on the neurophysiology of learning, the author of this book suggests that current educational practices and other aspects of modern life may influence not only how children think in affluent societies like the USA, but also the brain's physical structure. She suggests some possible remedies.
1990, Simon & Schuster, New York, ISBN 0-671-67349-1 (hbk) 382 pp.

451 Heatton, George R. & Robert Repetto
Back to the Future: US Government Policy Toward Environmentally Critical Technologies
New technologies are needed that increase the efficiency with which energy and raw materials are used, and eliminate or reduce pollution. This report provides an extensive list of 'environmentally critical' technologies that meet four specific criteria. It concludes that environmentally critical technologies are central to the creation of a sustainable economy, and urges much stronger public support for them.
1992, World Resources Institute, Washington, DC, USA, ISBN 0-915825-75-9 (pbk) 100 pp.

452 Heaword, Rose & Charmian Larke
The Directory of Alternative Technology
The 'alternative technologies' considered in this book are environmentally sound, provide technology in response to human need, are small scale and decentralised, promote self-reliance, enhance the whole human being, and foster human ingenuity. The ten sections cover: general aspects, land, food and agriculture, health, water and waste, buildings and structures, energy, goods and skills and services, work and employment, and transport.

1989, Routledge, London, ISBN 0-415-02761-6 (hbk) xiii + 317 pp.

453 Hedley, Rodney & Justin Davis Smith (Eds.)
Volunteering and Society: Principles and Practice
Volunteering is more prominent in the UK than ever before. This book examines important volunteering issues and future scenarios.
1992, NCVO Publications, London, ISBN 0-7199-9315-2 (pbk) x + 192 pp.

454 Heidenheimer, Arnold J. Hugh Heclo & Carolyn Teich Adams
Comparative Public Policy: The Politics of Social Choice in Europe and America
This book examines and compares the formulation, substance and implementation of public policy in the USA, UK, and Europe.
2nd. Ed. 1983, Macmillan, London & St. Martin's Press, New York, ISBNs 0-333-35359-5 (hbk) & 0-333-35360-9 (pbk) xvi + 367 pp.; 1st. Ed., 1975 (USA) & 1976 (UK).

455 Held, David
Models of Democracy
This book provides a coherent, analytical overview of the varieties of democratic theory, and examines changing responses to questions about the nature, application, and importance of democracy from ancient Athens

to modern times. It explores the context, nature, and limits of each of ten major models of democracy, and discusses the views of leading advocates of these approaches. The author defines a concept of democracy departing decisively from both liberal and Marxist positions.
1987, Polity Press, Cambridge, ISBN 0-7456-0043-3 (pbk) xii + 321 pp.

456 Heller, Frank (Ed.)
The Use and Abuse of Social Science
The contributors to this book assess the god and bad outcomes of social science research projects, and give some realistic guidelines. They argue, with supporting data, that social science research can lead to important developments in policy and changes of practice.
1986, Sage Publications, London & Newbury Park, CA, USA, ISBNs 0-8039-8016-7 (hbk) & 0-8039-8017-5 (pbk) x + 294 pp.

457 Helm, Dieter (Ed.)
Economic Policy Towards the Environment
This book examines the main economic concepts and policy instruments for the new environmental policy agenda of the 1990s. This agenda will require major changes in economic behaviour in power generation, transport, and use of natural resources. It introduces analytical issues on the

appropriate roles of markets and government, and pragmatic issues about the design of policies.
1991, Blackwell, Oxford & Cambridge, MA, USA, ISBNs 0-631-18201-2 (hbk) & 0-631-18202-0 (pbk) xvii + 326 pp.

458 Henderson, Hazel
Paradigms in Progress: Life Beyond Economics
This book summarises the author's own paradigms in progress, and her recent efforts to influence policy debates anywhere in the world and offers new directions and possibilities for 'win-win' solutions. In her view, such solutions are needed to create a saner and more equitable, gender-balanced and ecologically conscious future for our planet. She discusses many different aspects of this transition towards a new world order, and identifies elements needing further debate.
1991, Knowledge Systems, Indianapolis, IN, USA, ISBN 0-941705-21-8 (pbk) 293 pp.

459 Henderson, Paul (Ed.)
Signposts to Community Economic Development
This guide considers some of the key ingredients of community economic development. It explores how a community development approach can make a crucial difference to community

enterprises and businesses.
1991, Community Development Foundation, London, ISBN 0-902406-66-3 (pbk) xi + 51 pp.

460 Hepworth, Mark E.
Geography of the Information Economy
This book introduces the concept of the information economy to geography, and assesses the present and future role of information technology in the form and function of economic activity at all levels.
1992, Pinter, London, ISBN 1-85293-253-8 (pbk) 224 pp.

461 Hewitt, Tim, Hazel Johnson & Dave Wield (Eds.)
Industrialization and Development
Technological change, restructuring of industrial production, and the international division of labour all mean that developing countries are involved in a global process of industrialisation, This book aims to explain both positive and negative aspects of these processes in developing countries. Part 1 is about the international economy, and Part 2 explores specific aspects of industrialisation technology, gender relations, culture, and the environment.
1992, Oxford University Press, Oxford & New York, ISBNs 0-19-877332-3 (hbk) & 0-19-877333-1 (pbk) xi + 338 pp.

462 Hey, John D. (Ed.)
The Future of Economics
This book explores possible futures for economics as a discipline and a profession. Its contributors include leading economists and all consider the next 100 years of economics, covering many topics.
1992, Blackwell, Oxford & Cambridge, MA, USA, ISBNs 0-631-18493-7 (hbk) & 0-631-18487-2 (pbk) xi + 163 pp.

463 Hicks, David
Education for the Future: A Practical Classroom Guide
This practical handbook builds on much of the innovative work carried out in world studies during the past decade. It will enable teachers to help pupils look at the kind of future that they want for themselves, society, and our planet.
1993, World Wide Fund for Nature, Godalming, Surrey.

464 Higgins, Ronald
The Seventh Enemy: The Human Factor in the Global Crisis
This book describes the seven main threats to human survival as: the population explosion, food shortage, scarcity of natural resources, pollution and degradation of the environment, nuclear energy, uncontrolled technology, and human nature. The author considers that this 'seventh enemy' is so deeply entrenched

that there is a danger that we will do too little too late to meet the other six threats.
1978, Hodder and Stoughton, London (hbk).

465 Hildyard, Nicholas & Tracey Clunies-Ross
The Politics of Industrial Agriculture
This book assesses the revolution in agriculture in industrialised countries during the last 40 years, which has dramatically increased yields, but has had several undesirable economic, environmental and other side-effects. The authors ask several searching questions about 'sustainable agriculture', and report that groups throughout the world are working on a new approach that "depletes neither soil nor people".
1992, Earthscan, London, ISBN 1-85383-138-7 (pbk) 160 pp.

466 Hine, Thomas
Facing Tomorrow: What the Future Has been, What the Future Can Be
This book explores various aspects of the future, and enquires why past expectations of a good future have largely not been fulfilled, at least in the USA. For at least 20 years, "no compelling comprehensive vision of the future has captured the American imagination."
1991, Knopf, New York, ISBN 0-394-57785-X (hbk) xvi + 264 pp.

467 Hinrichsen, Don
Our Common Future: A Reader's Guide
This book summarises the core concepts of *Our Common Future* (No. 988), its discussions of the major world problems and issues, and its proposals for common action.
1987, Earthscan, London, & IIED/ Earthscan, Washington, DC, USA, ISBN 1-85383-010-0 (pbk) 40 pp.

468 Hirsch, Fred
Social Limits to Growth
This book argues that social, non-physical limits to growth, exist now, and that material affluence does not create an affluent society.
2nd. Ed. 1978, Routledge & Kegan Paul, London, ISBN 0-7100-8711-x (pbk) xiii + 208 pp.; 1st. Ed. 1977, ISBN 0-7100-8610-5 (hbk).

469 Hirshleifer, Jack & John G. Riley
The Analytics of Uncertainty and Information
Only recently has an accepted theory of information and uncertainty in economics and social sciences evolved. This book describes its practical applications, for example in analysing stock-market returns, evaluating accident prevention measures, and assessing patents and copyright laws. It unifies many important but partial results into one satisfying picture, clarifying

how the economics of uncertainty and information generalises and extends standard economic analysis.
1992, Cambridge University Press, Cambridge & New York, ISBNs 0-521-23956-7 (hbk) & 0-521-28369-8 (pbk) xi + 465 pp.

470 Hitchins, Derek K.
Putting Systems to Work
This book presents a radical approach to both systems thinking and systems practice. Building on general systems theory, the author introduces his own concept of 'Unified Systems Hypothesis', which provides a bridge between 'soft' and 'hard' systems thinking, and practical methods, tools, and techniques for best practice. The book is wide ranging, practical, and extensively illustrated with case studies.
1992, Wiley, Chichester & New York, ISBN 0-471-93426-7 (hbk) xv + 325 pp.

471 Hobbelink, Henk
Biotechnology and the Future of World Agriculture
This book connects genetic resources and biotechnology with the broader concern for sustainable agriculture. It adds genetic diversity to the three traditional foundations of agriculture: soil, water and sunlight. It surveys today's fastest growing technology biotechnology, and shows how giant corporations dominate

its research.
1991, Zed Books, London & Atlantic Highlands, NJ, USA, ISBNs 1-86232-836-5 (hbk) & 0-86232-837-3 (pbk) x + 159 pp.

472 Hoffman, W. Michael, Robert Frederick & Edward S. Petry, Jr
The Corporation, Ethics, and the Environment
This book surveys broad issues, such as the rights of future generations and conscientious corporate decision making, considers examples of both good and bad environmental impacts of large US corporations, and describes some examples of new corporate strategies. It discusses the future of business, ethics, and the environment.
1990, Quorum Books, Greenwood Press, Westport, CT, USA, ISBN 0-89930-603-9 (hbk) 319 pp.

473 Hogwood, Brian W.
Trends in British Public Policy: Do Governments Make Any Difference?
This book answers many questions about the policies and effects of British governments, and examines how British public policy and its effects have changed since 1945, especially over the last two decades.
1992, Open University Press, Buckingham & Philadelphia, ISBNs 0-35-15630-4 (hbk) & 0-335-15629-0 (pbk) viii + 260 pp.

474 Holdgate, Martin W.
A Perspective of Environmental Pollution
This book discusses a wide range of different aspects of environmental pollution, including possible future approaches.
1979, Cambridge University Press, Cambridge & New York, ISBNs 0-521-22197-8 (hbk) & 0-521-29972-1 (pbk) x + 278 pp.

475 Holman, Claire with Simon Festing (Fiona Weir (Ed.))
Air Quality and Health
This report shows that the UK continues to experience polluted air regularly, describes the effects on human health and the environment of five major pollutants, and compares their levels in the UK and other countries, examines how much air monitoring occurs in the UK, surveys selected air qualities worldwide, and comments on 'low' UK standards.
1991, Friends of the Earth, London, ISBN 0-905966-78-3 (pbk) ii + 72 pp.

476 Holmberg, Johan (Ed.)
Policies for a Small Planet: From the International Institute for Environment and Development
This book presents an integrated series of essays on policies for sustainable development from the IEED, a leading policy research institute on environment and development issues. It concentrates on the Third World and looks at specific sectors to which the policies have to be applied. It begins with a discussion of what sustainable development is, and considers what institutions are needed to mobilise human resources for change and economic policies for sustainable natural resource management. It then examines the policies needed in agriculture, urban development, industry, forests, dry lands, energy use, finance, population, and consumption. It shows how those directly involved are best placed to manage their costs and resources. Policies must support the experience and resourcefulness of local people. Sustainable development requires that they control their own futures.
1992, Earthscan, London, ISBN 1-85383-132-8 (pbk) 362 pp.

477 Holmberg, Johan, Koy Thomson & Lloyd Timberlake
Facing the Future: A Citizen's Guide to the Earth Summit
This concise, fully illustrated guide to the issues and results of the Earth Summit, held at Rio de Janeiro in June 1992, surveys its events and policy decisions. It describes clearly what was achieved, what might have been achieved, and what might emerge from it.
1993, Earthscan, London, ISBN 1-85383-154-9 (pbk) 40 pp.

478 Hopman, Conrad
The Book of Future Changes: Living in Balance in the Electronic Age
This book advocates a holistic approach, in which ideas from many fields of enquiry all have a part to play; this approach is needed to complement the old paradigms, that are not adequate in themselves. The book has provocative and at times paradoxical ideas, but its author views the universe as being full of paradoxes!
1988, Institute for Social Inventions, London, ISBN 0-948826-10-X (pbk) 154 pp.

479 Houghton, J. T., B. A. Callander & S. K. Varney
Climate Change 1992: The Supplementary Report to the IPCC Scientific Assessment
This report reviews the key conclusions of the 1990 report of the Intergovernmental Panel on Climate Change (IPCC) in the light of new evidence. Like its predecessor (No. 480), it represents the continuing efforts of the international scientific community to communicate to national and international policy makers the very latest scientific knowledge and understanding of the complex issues surrounding climate and climate change.
1992, Cambridge University Press, Cambridge & New York, ISBN 0-521-43829-2 (pbk) xii + 200 pp.

480 Houghton, J. T., G. J. Jenkins & J. J. Ephraims (Eds.)
Climate Change: The IPCC Scientific Assessment
This is the Report of Working Group 1 of the Intergovernmental Panel on Climate Change (IPCC). It provides the first comprehensive assessment and definitive report on global warming and other global climatic change under the influence of the greenhouse effect, prepared and reviewed by several hundred scientists from many countries.
1990, Cambridge University Press, Cambridge & New York, ISBN 0-521-40360-X (pbk) xxxix + 364 pp.

481 Howden, David (Ed.)
Energy Demand: Evidence and Expectations
This book reviews progress in energy demand modelling, and introduces new work. It has three types of contribution: (1) assessments of what has been learned from empirical work on energy demand; (2) new studies of energy demand in industry, transport and homes; (3) critical assessments of official demand forecasting methods, and developments in modelling that may affect future work. Caution is urged in interpreting and applying energy demand models.
1992, Academic Press, London & New York & Surrey University

Press,   Guildford,   ISBN
0-12-333310-5 (hbk) viii + 255 pp.

482 Howells, Gwyneth
Acid Rain and Acid Waters

This book provides a comprehensive description of acid rain and acid waters, resulting from various forms of air pollution. It covers man aspects and topics, and has a bibliography.
1990, Ellis Horwood, Chichester & New York, ISBN 0-13-004797-X (hbk) 215 pp.

483 Hoyle, Brian S. & Richard D. Knowles (Eds.)
Modern Transport Geography

This up-to-date book on transport is based on geographical theory and focussed on contemporary changes and issues. It examines the nature of transport geography and its place in transport studies, and covers many specific aspects and modes of transport, including its future.
1992, Belhaven Press, London & New York, ISBNs 1-85293-157-4 (hbk) & 1-85293-158-2 (pbk) x + 276 pp.

484 Hughes, Barry B.
World Futures: A Critical Analysis of Alternatives

This book supplies a framework for understanding and evaluating various widely divergent perspectives on the future by different forecasters, including global models. It considers how the forecasters' world views, theoretical assumptions, and interpretations of data all affect their predictions, and is sceptical about ability to forecast.
1985, Johns Hopkins University Press, Baltimore, MD, USA & London, ISBN 0-8018-3236-5 (pbk) xii + 243 pp.

485 Hunger Project, The
Ending Hunger: An Idea Whose Time Has Come

This book gives the basic facts about world hunger, food and agriculture, and presents arguments for and against many of the proposed approaches to the increase of food production and the alleviation of hunger. It also sets the problem of feeding the world within a wider context, and shows its relationships to the world economic order, development, foreign aid, population, and military security. It provides much important source material for those concerned with development. The Hunger Project aims to abolish hunger from the Earth by the year 2000.
1985, Praeger, New York & Eastbourne, Sussex, ISBN 0-03-006189-X (pbk) xi + 430 pp.

486 Hunt, Jenny & Heather Jackson (Eds.)
Vocational Education and the Adult Unwaged: Developing a Learning Culture

This book focusses on the education and training needs of adults

without employment. Its contributors argue strongly for developing a learning culture, which
enables individuals to realise their
potential, while supporting the
development of a competent and
capable work force.
*1992, Kogan Page, London, ISBN
0-7494-0493-0 (hbk) 188 pp.*

**487 Hurrell, Andrew & Benedict
Kingsbury**
**The International Politics of the
Environment: Actors, Interests,
and Institutions**
This book contains essays by leading specialists in various disciplines, who assess the strengths,
limitations, and potential of the
international system for global
environmental management, and
analyse relevant political, economic and moral issues.
*1992, Oxford University Press,
Oxford & New York, ISBNs
0-19-827365-7 (hbk) &
0-19-827778-4 (pbk) xiv + 492 pp.*

488 Huston, Aletha C. (Ed.)
**Children in Poverty: Child
Development and Public Policy**
This book examines important
questions about children in poverty in the USA, focussing on
children rather than on parents'
incomes or self-sufficiency. It
notes that the number of children,
living in poverty in the USA, rose
dramatically in the 1980s and
remains high.
1991, Cambridge University Press,

*Cambridge & New York, ISBN
0-521-39162-8 (hbk) x + 331 pp.*

489 Hutchinson, Colin
**Business and the Environmental
Challenge: A Guide for
Managers**
This guide to sound environmental performance of a business
presents a comprehensive, balanced statement of the corporate
environmental challenge. The
author provides helpful practical
advice, based on what companies
are actually doing. He suggests
simple but powerful techniques
for making change happen, provides a checklist for implementing an environmental
programme, discusses barriers to
progress, shows how to assess
environmental risk, states some
key principles for managing
change effectively, and indicates
the potential benefits of an
environmental programme.
*1991, The Conservation Trust, Read-
ing, (pbk) 52 pp.*

490 Hutchinson, Maxwell
**The Prince of Wales: Right or
Wrong? An Architect Replies**
This book is written in response to
HRH Prince Charles' book *A
Vision of Britain* (No. 764), and in
fact agrees with the Prince on
quite a lot of issues. The author
aims to reinstate the architect as
visionary and facilitator, and to
dispel the idea that he just adds to
the work of engineers and

developers. He points out that his true task is "to create the spaces which define our towns and cities."
1989, Faber, London, ISBN 0-571-14180-3 (pbk) 160 pp.

491 Huxley, Aldous
The Perennial Philosophy
The author of this book uses the 'perennial philosophy' as a metaphor representing the common ground of the world's religions. The book is an anthology of the perennial philosophy, containing many extracts from the writings of mystics, with accompanying commentary.
3rd. Ed. 1968, reprinted 1974, Chatto & Windus, London, ISBN 0-7011-0812-6 (hbk) viii + 358pp.; 1st. Ed. 1946.

492 Icke, David
It Doesn't Have to be Like This: Green Politics Explained
This book presents a Green perspective on: economic growth, Green economics and politics, food and agriculture, population, pesticides and pollution, energy, conservation, transport, war and peace, health, and housing. It is full of proposals for a safe and sustainable future.
1990, Green Print, London, ISBN 1-85425-033-7 (pbk) x + 212 pp.

493 Iivonen, Jyrki (Ed.)
The Changing Soviet Union in the New Europe
This book is a collection of papers, almost all presented at an international conference which discussed political, security, economic, and regional issues, relating to the USSR and Europe.
1991, Edward Elgar, Aldershot, Hants. & Brookfield, VT, USA, ISBN 1-85278-532-2 (hbk) xi + 250 pp.

494 Illich, Ivan D.
Celebration of Awareness: A Call for Institutional Revolution
In each chapter, the author questions some 'certainty', and discusses the 'deception' involved in some social institution.
1973, 3rd. Ed. Penguin Books, Harmondsworth, Midd., ISBN 0-14-080356-4 (pbk) 156 pp.; 2nd. Ed. 1971, Calder & Boyars, London; 1st. Ed. 1970, Doubleday, New York.

495 Illich, Ivan D.
Deschooling Society
This critique of schooling, schools and universities suggests the radical reform of education on a non-institutionalised basis.
3rd. Ed. 1973, Penguin Books, Harmondsworth, Midd., ISBN 0-14-080357-2 (pbk) 116 pp.; 2nd. Ed. 1971, Calder & Boyars, London; 1st. Ed. 1971, Harper & Row, New York.

496 Illich, Ivan D.
Energy and Equity
This book advocates containing the energy crisis by *limiting* traffic, too much of which will further

reduce our quality of life.
1974, Calder & Boyars, London, ISBNs 0-7145-1057-2 (hbk) & 0-7145-1058-0 (pbk) 95 pp.

497 Illich, Ivan D.
Limits to Medicine
This book argues that the medical 'establishment' and its ideology has become a threat to health in several respects.
2nd. Ed. 1977, Penguin Books, Harmondsworth, Midd., ISBN 0-14-022009-7 (pbk) 296 pp.; 1st. Ed. 1976, Calder & Boyars, London.

498 Ince, Martin
The Rising Seas
This book describes rises in sea level that seem likely to happen as a result of global warming, and indicates where the worst effects can be expected. He considers the different types of damage that higher seas could cause, and presents brief case studies from various parts of the world. He examines scientific and technological developments that could help, and states the policies on which governments need to agree.
1990, Earthscan, London, ISBN 1-85383-077-1 (pbk) viii + 152 pp.

499 Inglis, Mary & Sandra Kramer (Eds.)
The New Economic Agenda 1985
This book is based on a conference where practitioners of both conventional and alternative economics shared ideas. It brings together many of the elements that create the new economic theory and practice on local, national and international levels, and covers many topics.
1985, The Findhorn Press, Findhorn, Forres, Scotland, ISBN 0-905249-61-5 (pbk) 179 pp.

500 Irvine, Sandy & Alec Ponton
A Green Manifesto: Policies for a Green Future
This book offers a fresh look at contemporary world problems, from a Green political viewpoint. Its proposed new solutions appreciate our mutual interdependence and our relationship with our environment.
1988, Macdonald Optima, London, ISBN 0-356-15200-6 (pbk) xi + 178 pp.

501 Jackson, Ben
Poverty and the Planet: A Question of Survival
This book exposes the causes of Third World poverty. It describes how the poor are often hardest hit by environmental degradation, and how economic injustice and exploitation of the poor are major causes of environmental damage. It emphasises that the most important causes of Third World poverty and environmental destruction are hostile international economic and financial forces, and urges citizens of rich countries to use their influence to mitigate these forces. It rejects

short-term 'development' that destroys resources, and calls for ecological protection to match the desperate needs of the world's poor.

1990, Penguin Books, London & Viking Penguin, New York, ISBN 0-14-013149-3 (pbk) xvi + 226 pp.

502 Jackson, Wes
New Roots of Agriculture

This book offers a sound, well documented critique of the assumptions and effects of agribusiness, and goes beyond this to propose practical solutions. Its author, an agricultural researcher, claims that his new way of raising crops, by working with the soil's natural systems, would keep the world perpetually fed.

2nd. Ed. 1985, University of Nebraska Press, Lincoln, NB, USA & London, ISBN 0-8032-7562-5 (pbk) xv + 151 pp.; 1st. Ed. 19??.

503 Jacobs, Jane
Cities and the Wealth of Nations: Principles of Economic Life

The author of this book challenges how we think about wealth and poverty, and the rise and decline of nations or empires. She argues that almost all economic life depends on cities, which can make their surrounding rural economies prosperous or unbalanced and exploited.

2nd. Ed. 1985 Viking Penguin, New York & Penguin Books, Harmondsworth, Midd., ISBN 0-670-80045-7 (hbk) ix + 287 pp.

504 Jacobs, Michael
The Green Economy: Environment, Sustainable Development and the Politics of the Future

This book presents the central issues involved in creating an environmental economics, together with several practical proposals. It attempts to bring together and bridge the gap between two approaches that have historically remained apart, despite their common concerns: the academic discipline of environmental economics and the political ideology of the Green movement. It adopts an intermediate position that is probably 'unorthodox' from both environmental economic and Green viewpoints. It considers what economic policies could be implemented to achieve desirable environmental goals, if environmental concerns were already a government priority.

Part 1 attempts to explain the nature and causes of the environmental crisis. Part 2 describes the objectives of the Green economy, discusses its basic concepts, relates sustainability to other economic goals, and explores the environmental limits to economic activity. Part 3 considers how a sustainable economy would actually work in an industrialised

nation, sets out the broad framework of environmental economic policy making, looks at various available methods, and assesses the wider implications of these policies. Part 4 examines how the environment is measured for economic policy purposes, and discusses cost-benefit analysis and alternative economic indicators.
1991, Pluto Press, London & Concord, MA, USA, ISBN 0-7453-0412-5 (pbk) xxii + 312 pp.

505 James, P. D.
The Children of Men
In this novel, set in Oxford in the year 2021, human beings have lost their power to reproduce, and the UK is ruled by a dictator. 25 years since the last baby was born, the dependent old and anxious middle-aged have begun to outnumber the resentful and callous young. At the end, there is to be another birth, but readers are left in suspense as to whether it is a 'one-off' or the beginning of a new generation.
2nd. Ed. 1993, Faber, London, ISBN 0-871-16843-6 (pbk) 247 pp.; 1st. Ed. 1992, Faber, ISBN 0-571-16741-1 (hbk) 288 pp.

506 Jamison, Andrew, Ron Eyerman & Jacqueline Cramer
The Making of the New Environmental Consciousness: A Comparative Study of the Environmental Movements in Sweden, Denmark and the Netherlands
This book compares environmentalism in three nations, showing how each has developed its own Green approach. It links theoretical and empirical issues, and gives their cultural and historical background.
1990, Edinburgh University Press, Edinburgh, ISBN 0-7486-0180-5 (hbk) xiii + 216 pp.

507 Jantsch, Erich & Conrad H. Waddington
Evolution and Consciousness: Human Systems in Transition
This book brings together some emergent concepts and approaches, which could contribute to a scientific foundation for hope associated with life and further evolution. It helps to define a paradigm of self-realisation through self-transcendence.
1976, Addison-Wesley, Reading, MA, USA & London, ISBNs 0-201-03438-7 (hbk) & 0-201-03439-5 (pbk) xii + 259 pp.

508 Jaques, Elliott
Creativity and Work
This book is about the conditions under which people can work effectively and creatively, on their own and with others.
1990, International Universities Press, Madison, CT, USA, ISBN 0-8236-1088-8 (pbk) xii + 433 pp.

509 Jelinek, Marian & Claudia Bird Schoonhoven
The Innovation Marathon: Lessons from High Technology Firms
This book shows empirically that organisation theory does not sufficiently explain the world of high technology firms.
1990, Blackwell, Oxford & Cambridge, MA, USA, ISBN 0-631-15392-6 (hbk) viii + 469 pp.

510 Jennett, Christine & Rendal G. Stewart (Eds.)
Politics of the Future: The Role of Social Movements
This book is about the role of social movements in politics in Western Europe, North America, and the Third World. They may even represent the future form of politics, and their commitment to fundamental cultural transformation cannot be ignored.
1989, Macmillan Australia, ISBNs 0-333-50223-X (hbk) & 0-333-50222-1 (pbk) xii + 471 pp.

511 Joekes, Susan P. (Prepared by)
Women in the World Economy: An INSTRAW Study
This book presents the results of research on the role of women in developing nations, and the impact of the international economy on many types of women's work. It proposes innovative development policies.
2nd. Ed. 1989, Oxford University Press, New York & Oxford, ISBN 0-19-506315-5 (pbk) xi + 161 pp.; 1st. Ed., ISBN 0-19-504947-0 (hbk).

512 Johansson, Per-Olav
The Economic Theory and Measurement of Environmental Benefits
This advanced textbook surveys developments in measuring welfare, and its applications to environmental economics.
1987, Cambridge University Press, Cambridge & New York, ISBN 0-521-32877-2 (pbk).

513 Johansson, Per-Olav
An Introduction to Modern Welfare Economics
This book explores several concepts of welfare economics in detail. Topics include: market failures, public choice, government failures, taxation, intergenerational equity, and applied welfare economics.
1991, Cambridge University Press, Cambridge & New York, ISBNs 0-521-35616-6 (hbk) & 0-521-35695-4 (pbk) xii + 176 pp.

514 Johansson, Thomas B., Henry Kelly, Amulya K. Reddy & Robert H. Williams (Eds.)
Renewable Energy: Sources for Fuels and Electricity
This book is a state-of-the-art assessment, by 50 leading world energy specialists, of the technical and economic prospects for a wide range of options for making

fuels and electricity from renewable energy. Technical advances during the last decade suggest that, by the middle of the 21st century, renewable sources of energy could supply 60% of the world's electricity market and 40% of the market for fuels used directly. This could be done at competitive costs, and would bring many environmental and other benefits not usually costed into energy calculations. The book fully discusses these advances, the resulting options, and the policies needed to use renewable energy intensively.
1993, Earthscan, London, ISBNs 1-55963-139-2 (hbk) & 1-85383-155-7 (pbk) 1200 pp.

515 Johnson, Lawrence E.
A Morally Deep World: An Essay on Moral Significance and Environmental Ethics
In this book, the author, as a philosopher, presents his deep ecological viewpoint, and his re-evaluation of the natural world.
1991, Cambridge University Press, London & New York, ISBN 0-521-39310-8 (hbk) ix + 301 pp.

516 Johnson, Stanley P. & Guy Corcelle
The Environmental Policy of the European Communities
This reference book summarises the extensive legislation on environmental policy recently introduced by the European Community.
1989, Graham & Trotman, London & Boston, ISBN 1-85333-225-9 (hbk) xv + 349 pp.

517 Johnston, R. J.
Environmental Problems: Nature, Economy and State
This book considers the causes and solutions of many environmental problems and issues. It draws on a wide range of findings from the social sciences, linking them with environmental issues. It shows how continued and increased pressure on environmental resources is built into the operation of the modern capitalist world economy. It argues that the tackling of environmental problems will require collective state actions and illustrates the difficulty of achieving such actions.
1989, Belhaven Press, London & New York, ISBNs 1-85293-000-4 (hbk) & 1-85293-188-4 (pbk) xi + 211 pp.

518 Jones, Huw
Population Geography
This book incorporates the most modern perspectives on population geography, especially population dynamics, with worldwide case studies.
2nd. Ed. 1990, Paul Chapman Publishing, London, ISBN 1-85396-071-3 (pbk) x + 321 pp.

519 Jones, Tom & Soeren Wibe
Forests: Market and Intervention Failures: Five Case Studies

This book presents five European case studies, which examine failures of forest management and show how market and intervention policies can both increase the destruction of forests. The authors set out ways in which future policies can avoid the mistakes of the past.
1992, Earthscan, London, ISBN 1-85393-101-8 (pbk) 256 pp.

520 Jonsen, Albert R.
The New Medicine and the Old Ethics
This book shows how the precision and accuracy of new hi-tech medicine raise many ethical questions that have a blurred outline, such as responsibility, duties, rights, interests, beneficence, and justice.
1990, Harvard University Press, Cambridge, MA, USA & London, ISBN 0-674-61725-8 (pbk) xv + 171 pp.

521 Joranson, Philip N. & Ken Butigon (Eds.)
Cry of the Environment: Rebuilding the Christian Creation Tradition
This major theological study of the current environmental situation considers how and why 'Western' civilisation came to dominate the environment, and revisions creation consciousness so that we revere all life, including human beings, as part of the same creation.

1992, Bear & Co., Santa Fe, NM. USA & Element Books, Shaftesbury, Dorset, ISBN 0-939686-17-3 (pbk) xi + 476 pp.

522 Joseph, Lawrence E.
Gaia: The Growth of an Idea
This book describes the development and progress of the concept of Gaia, together with the reactions of both advocates and opponents.
1991, Arkana Penguin, London & Viking Penguin, New York, ISBN 0-14-019295-6 (pbk) pp.

523 Joseph, Martin
Sociology for Everyone
This introductory textbook on sociology emphasises 'doing' sociology and invites active participation rather than passive reading. Each chapter introduces a specific topic, explores several popular assumptions about it, and subjects them to searching criticisms in the light of theoretical and empirical findings.
1986, Polity Press, Cambridge, ISBNs 0-7456-0186-3 (hbk), 0-7456-0187-1 (pbk) & 0-7456-0301-7 (school pbk) xx + 310 pp.

524 Joste, Sten and Gillis Een
One Hundred Innovations for Development
This book describes 100 technical solutions that are simple, cheap, robust, easy to handle, maintain and repair, and can be manufactured in areas of demand. They

Keating, Giles et al. 115

cover most aspects of life in the Third World.
1988, IT Publications, London, ISBN 1-85339-045-X (pbk) 80 pp.

525 Jowell, Roger, Lindsay Brook & Bridget Taylor with Gillian Prior (Eds.)
British Social Attitudes : The 8th Report
This report is the eighth in a series seeking to chart changes in British social values, in relation to other changes in society. Since 1984, a report has been published each year, providing important information about the nature of British society and how it is changing.
1991, Dartmouth Publishing Co., Aldershot, Hants. & Ashgate Publishing Co., Brookfield, VT, USA, ISBN 1-85521-258-7 (hbk) (ISSN 0267-6869) xiv + 336 pp.

526 Jungk, Robert & Johan Galtung (Eds.)
Mankind 2000
This book reports the 1967 First International Future Research Conference, which pointed to new directions for futures research.
1969, 2nd. impression 1971, Universitetsforlaget, Oslo, Norway & Allen & Unwin, London, (hbk) 368 pp.

527 Jungk, Robert & Norbert Muellert
Future Workshops: How to Create Desirable Futures
This book is a manual of the

'futures workshop' technique, which helps small groups of people to dream up and implement creative ideas and projects for a saner society. The technique has had some notable successes, and is now widely used by groups throughout Europe.
2nd. Ed. 1987, Institute for Social Inventions, London, ISBN 0-948826-07-X (pbk) 123 pp.
528 Kahn, Herman, William Brown & Leon Martel
The Next 200 Years
This book outlines four different scenarios for the future of mankind and the planet, and details a variant of the most optimistic of these scenarios, with a strategy for both near and long-term future.
3rd. Ed. 1978, Abacus, Sphere Books, London, ISBN 0-349-12071-4 (pbk) xiv + 241 pp.; 1st. Ed. 1976.

529 Kast, Fremont E. & James E. Rosenzweig
Organization and Management: A Systems and Contingency Approach
This book uses systems concepts to understand organisations better, especially what makes them operate effectively and efficiently.
4th. Ed. 1985, McGraw-Hill, New York & London, ISBN 0-07-033443-9 (hbk) xv + 720 pp.; 1st. Ed. 1970.

530 Keating, Giles et al.
The State of the Economy 1992

This book considers the UK's economic outlook and economic policy, as it moves closer to monetary and political integration with Europe.
1992, Institute of Economic Affairs, London, ISBN 0-255-36304-4, ISSN 0305-814X (pbk) xv + 160 pp.

531 Kemball-Cook, David, Chris Mattingley & Malcolm Baker (Eds.)
The Green Budget: An Emergency Programme for the UK
This book proposes objectives, targets and indicators for a Green Budget, using various economic instruments.
1991, Green Print, London, ISBN 1-85425-055-8 (pbk) 120 pp.

532 Kemp, David D.
Global Environmental Issues: A Climatological Approach
This book considers various environmental problems in relation to atmospheric and climatic changes. The problems considered include: drought, desertification, famine, acid rain, the threat to the ozone layer, the greenhouse effect, and the possibility of a nuclear winter. The final chapter discusses future prospects in the light of present problems. There is an extensive bibliography.
1990, Routledge, London, ISBN 0-415-01109-4 (pbk) 240 pp.

533 Kemp, Penny & Derek Wall
A Green Manifesto for the 90s

This book is a fairly recent statement from members of the UK's Green Party, and presents a useful and wide ranging approach. It sets a political agenda for the changes required for the satisfactory resolution of world problems. It examines root causes besides symptoms.
1990, Penguin Books, London & Viking Penguin, ISBN 0-14-013272-4 (pbk) xii + 212 pp.

534 Kennedy, I. R.
Acid Soil and Acid Rain
This book examines the fundamental chemical processes involved in acidic deposition and soil acidification, to assess better their long-term influences on the health of soils life growing in them.
2nd. Ed. 1991, Research Studies Press, Taunton, Somerset & Wiley, New York, ISBNs 0-86380-124-2 (RSP) & 0-471-93404-6 (W) (hbk) xvii + 254 pp; 1st. Ed., 1986.

535 Kennedy, Ian
Treat Me Right: Essays in Medical Law and Ethics
This collection of papers covering a wide range of topics illustrating the relations between medicine, law and ethics.
1988, Oxford University Press, Oxford & New York, ISBN 0-19-825559-4 (hbk) 320 pp.

536 Kenton, Leslie & Susannah Kenton
Raw Energy: Eat Your Way to

Radiant Health
This book explains the merits of a diet with a high proportion of fresh uncooked foods and shows how it can improve health and bring extra vitality.
3rd. Ed. 1986, Arrow Books, London.

537 Keynes, John Maynard
The General Theory of Employment, Interest and Money (Vol. VII in *The Collected Writings of John Maynard Keynes*)
This is a complete version of the author's original classic, corrected in the light of later correspondence. It also includes some miscellaneous prefaces and articles by Keynes.
2nd. Ed. 1973, reprinted 1991, Macmillan, London & Cambridge University Press, New York, ISBNs 0-333-10729-2 (M) & 0-521-22099-8 (CUP) (hbk), 0-333-00942-8 (M) & 0-521-29382-0 (CUP) (pbk) xxxv + 428 pp.

538 Kidron, Michael & Ronald Segal
The New State of the World Atlas
This atlas has 57 maps, exploring many aspects of the state of the world. The maps are fully explained, and supplemented by tables.
2nd. Ed. 1984, Pan Books & Heinemann, London, ISBN 0-330-28432-0 (pbk) 176 pp.; 1st. Ed. 1981, Pluto Press, London, with the title "The State of the World Atlas".

539 Kidron, Michael & Dan Smith
The New State of War and Peace: An International Atlas
This atlas assesses the human, social and political costs of war and preparations for war, and examines the situation at the end of the Cold War: peace treaties, diplomacy, arms cuts and military withdrawals. Using powerful new maps and graphics, it presents and measures the world's arsenals, weapons, military spending, and arms trade. It looks at current conflicts and persistent pressure points.
2nd. Ed. 1991, Grafton Books, London, ISBNs 0-246-13867-X (hbk) & 0-246-13868-8 (pbk) 127 pp.; 1st. Ed. 1983, with the title "The War Atlas".

540 King, Anthony D.
Global Cities: Post-Imperialism and the Internationalization of London
This book explores the contemporary phenomenon of 'globalisation', and the spatial and cultural links between metropolitan core and global periphery centres. It presents the case for a global approach to understanding modern cities, and provides a useful guide for urban specialists seeking further understanding of global cities.
1991, Routledge, London & New York, ISBN 0-415-06241-1 (pbk) 208 pp.

541 King, Anthony D.
Urbanism, Colonialism and the World Economy: Cultural and Spatial Foundations of the World Urban System
This book explores in detail the historical, cultural and spatial links between the 'core' and the 'periphery' of cities and conurbations, and is a companion book to *Global Cities* (No. 540).
2nd. Ed. 1991, Routledge, London & New York, ISBN 0-415-06240-3 (pbk) xii + 185 pp.

542 King, Alexander & Bertrand Schneider
The First Global Revolution: A Report by the Council of the Club of Rome
In its thought-provoking analysis of contemporary world problems, this book brings important understanding of the present process of global development; we are in the early stages of forming a new type of world society. It also provides subtle insights into the interactions between these problems. It discusses 'the world problematique', the massive, complex and intricate mixture of global problems, and 'the world resolutique', the unified, coherent, comprehensive, and simultaneous attack that could be made on many of the problems. In this spirit, it rigorously tackles the key issues now facing Earth, warns of what could happen if appropriate actions are not taken, and presents a challenge to implement a sustainable solution like what it suggests.
2nd. Ed. 1991, Pantheon Books, ISBN 0-679-73825-8 (pbk) 259 pp.; 1991, Simon & Schuster, London & New York, ISBN 0-671-71094-X (hbk) xxi + 197 pp.

543 Kinsman, Francis
Millennium: Towards Tomorrow's Society
This book analyses the major social groupings in British society, each of which reflects different values, and shows how they are gradually changing and evolving. The changes in their proportions of the total population are influencing the development of patterns of public opinion, including the emergence of increasing support for Green attitudes and viewpoints. The book then considers three possible future scenarios for the UK and for the world as a whole, in the light of these changes. It points to the holistic vision of fulfilment of future potentialities, that could be fulfilled if the strengths and talents of the different social groupings and different people were to reinforce each other in mutual cooperation, and it discusses what needs to be done and is being done to move in this direction.
2nd. Ed. 1992, Penguin Books, London & Viking Penguin, New

York, ISBN 0-14-014721-7 (pbk) 304 pp.; 1st. Ed. 1989, W. H. Allen, London.

544 Kirdar, Uner (Ed.)
Change: Threat or Opportunity for Human Progress?
This book contains essays from the Round Table on Global Challenges meeting in Turkey in 1990. The titles of its five volumes are: (1) Political Change; (2) Economic Change; (3) Globalization of Markets; (4) Changes in the Human Dimensions of Development, Ethics and Values; (5) Ecological Change: Environment, Development and Policy Linkages.
1992, United Nations Publications, New York, (hbk) 5 volumes, 1209 pp.

545 Kline, Morris
Mathematics in Western Culture
This classic survey looks at the crucial role of mathematics in the natural sciences, applied sciences, engineering, the arts, and culture.
3rd. Ed. 1972, reprinted 1990, Penguin, London & Viking Penguin, New York, ISBN 0-14-013703-3 (pbk) 543 pp.; 1st. Ed. 1953, USA.

546 Klir, George J. (Ed.)
Trends in General Systems Theory
This book integrates the various branches of genera; systems theory into a unified study, and examines its state-of-the-art in the early 1970s. It discusses future research, education, and applications.
1972, Wiley, London & New York, ISBN 0-471-49190-X (pbk) ix + 462 pp.

547 Klir, George J. et al.
Facets of Systems Science
This book helps its readers to develop an adequate general impression of systems science, its main historical roots, its relationships with other areas of human affairs, its current status, and its likely future role. It also helps them to identify sources for further study of various aspects of systems science.
1991, Plenum Press, New York & London, ISBN 0-306-43959-X (hbk) xvi + 664 pp.

548 Kneller, George F.
Science as a Human Endeavour
This book provides examples from the history, philosophy, and sociology of science, to relate the concerns of scientific thought to human existence. The chapters discussing various modern theories of the philosophy of scientific method and on the complex interaction between science and technology are of special interest.
1978, Columbia University Press, New York & Guildford, Surrey, ISBN 0-231-04206-X (hbk) xi + 333 pp.

549 Koch, Tom
Journalism in the 21st Century:
Online Information, Electronic
Databases and the News
This book considers the relationship between the content of public information and the potential
effect of new technologies on the
degree and type of information
available in the public forum. The
author argues that remote access
by personal computers is poised
to change forever the ways in
which journalists write news and
people read it. The book's chapters consider: (1) myths and realities about news; (2) online news;
(3) the scale and focus of transformations; (4) information search
strategies; (5) electronic libraries.
2nd. Ed. 1991, Adamantine Press,
London, ISBN 0-7449-0032 (pbk)
(ISSN 0954-6103) 380 pp.; 1st. Ed.
1991, Greenwood, Westport, CT,
USA & Praeger, New York, ISBN
0-313-27750-8   (hbk)   (ISSN
0954-6103).

550 Koestler, Arthur
Janus: A Summing Up
This book summarises and continues 25 years of its author's
thought, during which he turned
attention   to   the   evolution,
creativity and pathology of the
human mind. His insights are
brought together into a coherent
and comprehensive synthesis. He
argues that materialism can no
longer claim to be a scientific
philosophy. He introduces the

concept of the 'holon', which, like
the two-faced Roman god Janus,
can behave both as an integrated
whole and as an integrated part of
a larger whole in the multi-level
hierarchies of existence.
1978, Hutchinson, London, ISBN
0-09-132100-X (hbk) xi + 354 pp.

551 Kohr, Leopold
The Breakdown of Nations
This stimulating and provocative
book presents the author's thesis
that the problems of the world are
due fundamentally to the bigness
of nations and other social organisations. He considers that the
idea and ideal of littleness is the
only antidote to the disease of
oversize.
2nd. Ed. 1974, Christopher Davies,
Swansea, SBN 7154-0107-6 (pbk) xii
+ 244 pp.; 1st. Ed. 1957, Routledge
& Kegan Paul, London.

552 Korten, David C.
Getting to the 21st Century: Vol-
untary Action and the Global
Agenda
This book examines the lessons of
the 1980s, giving us hope that the
1990s can achieve the transformation so essential to prepare our
entry into the 21st century. It
points out the importance of what
a society believes about itself in
determining how it behaves. It
focusses   on   the   voluntary
development sector, and indicates that it could become a people's movement. It outlines a

development agenda for the 1990s, focussing on a need for system transformation.
1990, Kumarian Press, West Hartford, CT, USA, ISBNs 0-931816-85-8 (hbk) & 0-931816-84-X (pbk) xvi + 254 pp.

553 Krause, Florentin, Wilfrid Bach & Jon Koomey
Energy Policy in the Greenhouse: From Warming Fate to Warming Limit
This book investigates global climate stabilisation targets for energy policy and planning, and presents the results of research by an international team of experts.
2nd. Ed. 1990, Earthscan, London, ISBNs 1-85383-080-1 (hbk) & 1-83583-081-X (pbk) 242 pp.

554 Kreisky Commission on Employment Issues in Europe
A Programme for Full Employment in the 1990s
This report maintains that the high levels of unemployment in Europe during the late 1980s are neither inevitable nore acceptable. It advocates and presents detailed recommendations for a strategy, which would increase economic growth, while adapting it to minimise damage to the environment and perhaps halve unemployment by about 1995.
1989, Pergamon Press, Oxford & New York, ISBNs 0-08-037761-0 (hbk) & 0-08-037760-2 (pbk) xxiii + 185 pp.

555 Kuhn, Thomas
The Structure of Scientific Revolutions
The author sees the history of a 'mature' science as essentially one of a succession of traditions, each with its own theory and research methods, each guiding a community of scientists for a number of years, and each eventually abandoned. He calls the ideas of a scientific tradition a 'paradigm', thus introducing a new definition for this word; although his definition is not explicit, it regards a paradigm approximately as a world view expressed in a theory. (From the 1960s onwards, something like this definition has gradually become widespread in intellectual discourse and discussion of ideas in general, not only in philosophy of science.)
In science, a paradigm determines what problems are investigated, what data are considered relevant, what research methods are used, and what types of solution are allowed. Taking examples from physics and astronomy, the author shows how science proceeds during periods when a give paradigm is dominant, and how it undergoes revolution during times of a paradigm shift; in the latter case, the revolution occurs when there are sufficient anomalous observations, which the old paradigm cannot explain, to lead to the challenge of that paradigm.

2nd. Ed., 1970, University of Chicago Press, Chicago, IL, USA, ISBNs 0-226-45803-2 (hbk) & 0-226-45804-0 (pbk) xii + 210 pp.

556 Kuik, O. J. et al.
Assessment of Benefits of Environmental Measures
This book presents possible guidelines for considering how the Single European Act will benefit the assessment of environmental measures in all European Community (EC) member states.
1992, Graham and Trotman, London, ISBN 1-85333-386-6 (hbk) vii + 125 pp.

557 Kule, E.
Economics of Natural Resources and the Environment
This book considers what economics says about the use of nature's scarce resources and associated environmental problems.
1992, Chapman & Hall, London & New York, ISBN 0-412-36330-5 (UK) & 0-442-31349-7 (USA) (hbk) xii + 287 pp.

558 Kung, Hans
Global Responsibility: In Search of a New World Ethic
This book discusses one of the most critical issues in the debate about the future of humankind on earth, the need for a common ethic. What is needed is at least a minimum of shared ethical principles on which we can all agree.

The author discusses how to achieve this.
1990, SCM Press, London, ISBN 0-334-0200-X (hbk) xix + 158 pp.

559 Lacey, Colin & Roy Williams (Eds.)
Education, Ecology & Development: The Case for an Education Network
This book, published for the World Wide Fund for Nature, explains the destructive trends in world systems, and the radical changes needed, so that young people can learn about environmental problems.
1987, Kogan Page, London, ISBN 1-85091-495-8 (pbk) viii + 163 pp.

560 Lal, Betty G. & John Tepper Martin
Building a Peace Economy: Opportunitie and Problems of Post-Cold War Defense Cuts
This book discusses the impact of defence cuts, resulting from the collapse of the Soviet Union, on US defence contractors and labour.
1992, Westview Press, Boulder, CO, ISBN 0-8133-8438-8 (hbk) 294 pp.

561 Lamb, David
Discovery, Creativity and Problem Solving
This book discusses several ways in which creative discovery can be investigated by rational enquiry, and presents many case studies.
1991, Avebury, Aldershot, Hants. & Gower Publishing Co., Brookfield,

VT, USA, ISBN 1-85628-043-8
(hbk) v + 183 pp.

562 Lancaster, Brian
Mind, Brain and Human Potential: The Quest for an Understanding of Self
This book explores the territory between brain science, psychology, and religion, and integrates ideas from contemporary brain research with insights from various spiritual traditions.
1991, Element Books, Shaftesbury, Dorest & Rockport, ME, USA, ISBN 1-85230-209-7 (pbk) xiv + 226 pp.

563 Lane, Andrew & Joyce Tait
Practical Conservation Woodlands
This book advocates the use of resources to link the profitable use of land with the enhancement of the conservation of wildlife and its species. This would be achieved by using a historical and cultural approach in terms of overall land management.
1990, Hodder and Stoughton, London, ISBN 0-340-53366-8 (pbk) 128 pp.

564 Lane, Jan-Erik, David McKay & Kenneth Newton
Political Data Handbook: OECD Countries
This book provides a reasonably comprehensive guide to the government and politics of 24 OECD countries, together with essential social and economic background information.
1991, Oxford University Press, Oxford & New York, ISBN 0-19-827718-0 (hbk) viii + 257 pp.

565 Lappe, Frances Moore & Joseph Collins
World Hunger: 12 Myths
In this book, the authors present evidence that the main cause of widespread world hunger is not natural disaster, overpopulation, or lack of fertile land. They examine directly the policies that prevent starving people from feeding themselves, and give illuminating examples of political and social attitudes that could and should be applied to help eliminate hunger. Twelve important misconceptions about causes of hunger are analysed and refuted, and the authors show which approaches are relatively ineffective. They discuss how to improve the situation, and list many of the organisations and projects concerned. There are extensive references to relevant literature.
2nd. Ed. 1988, Earthscan, London, ISBN 1-85383-012-7 (pbk) x + 182 pp.; 1st. Ed. 1986, Grove Press, New York.

566 Lappe, Frances Moore & Rachel Schurman
Taking Population Seriously: The Missing Piece in the Population Puzzle
This book analyses the reasons for the growth of population, and

examines the social and economic choices affecting fertility and people's choices about reproduction and number of children. Its authors advocate the removal of these causes as one requirement for the resolution of the world population problem; thus they argue that population policies need to address the underlying social and economic conditions, rather than implement crude population 'control'. They assess policies in several Third World countries that have already dramatically reduced their birth rates and population growth rates.
2nd. Ed. 1989, Earthscan, London, ISBN 1-83583-055-0 (pbk) vi + 96 pp.; 1st. Ed. 1988.

567 Lappe, Marc
Chemical Deception: The Toxic Threat to Public Health
This book gives details of ten misconceptions about chemical pollution. It concludes that we have overestimated the ability of the planet and people to cope with toxic substances, and must reduce emissions, eliminate certain chemicals, return to sustainable agriculture, and promote regenerative ecosystems.
1991, Sierra Club Books, San Francisco, CA, USA, ISBN 0-87156-603-6 (hbk) 420 pp.

568 Laszlo, Ervin (Ed.)
Goals for Mankind: A Report to the Club of Rome
This book presents the results of an international survey of national and international goals. Part 1 surveys current goals and aspirations worldwide. Part 2 considers the best long-term international policies whose achievement could make the world safer and more humane. Part 3 examines how people could work together towards these policies, and transform the values of international exchange. The book indicates how to change from self-centred, short-term goals to mankind-centred and long-term goals in the interest of *all* people.
1977, Hutchinson, London, ISBN 0-09-131301-5 (pbk) xxiii + 434 pp.

569 Laszlo, Ervin
The Inner Limits of Mankind: Heretical Reflections on Today's Values, Culture and Politics
The author of this book argues that we have been tackling the wrong problems and issues, and that humankind's truly decisive limits are inner, not outer. *We* are the cause of our problems, and can resolve them only by redesigning our thinking and acting, not the world around us. The book challenges contemporary values, beliefs, and practices, and explores ways how to contribute to their transformation.
1989, Oneworld Publications, London, ISBN 1-85168-009-8 (pbk) vi + 146 pp.

570 Laszlo, Ervin (Ed.)
The New Evolutionary Paradigm
This book examines the new evolutionary paradigm in science, for the elaboration of a general theory of evolution. The new paradigm presents a challenge to natural, human, and social scientists. The contributors explore its many, wide-ranging aspects through the creation, criticism, and elaboration of progressively more refined general evolution theories.
1991, Gordon and Breach Science Publishers, New York & Reading, ISBN 2-88124-375-4 (hbk) xxv + 204 pp.

571 Laszlo, Ervin
The Age of Bifurcation: Understanding the Changing World
This book reviews what seems likely to happen in the near future, and includes a model of holistic thinking for shaping a new age. It applies the emerging 'sciences of complexity' to the problems faced by humanity as it approaches the 21st century. It shows that society faces a major transformation, and applies new scientific concepts to the environment, economic systems, political systems and culture. It concludes with suggestions for changing the crisis-bound path of civilisation to create a stable, humanistic world for the future.
1991, Gordon and Breach Science Publishers, New York & Reading,

ISBN 2-88124-491-2 (hbk) xvii + 126 pp.

572 Laszlo, Ervin (Ed.)
The Destiny Choice: Survival Options for the 21st Century
This book describes the greatest challenge that humanity has ever faced: the challenge of choosing our future. Our generation finds itself called upon to decide the destiny of life on Earth. Making this decision can be neither ignored nor postponed. 'The Destiny Choice', that we now face, is a choice between evolution and extinction. The processes that we have initiated cannot continue in the lifetime of our children. What we do will either create the framework for continuing life and consciousness on this planet, or set the stage for its end. The book will help us to confront our first responsibility: to face the facts and become conscious of our condition. There is no reason for passivity and pessimism. As the author says, "The remarkable faculties of the conscious mind embrace the powers of reason and intelligence, of love and solidarity. If we grow conscious of our condition, if we recognize the choice facing us, we shall develop the insight and the will to opt for a life-enhancing path of survival and evolution rather than entering the life-destroying path toward extinction."
1993, Adamantine Press, London,

*ISBNs 0-7449-0079-4 (hbk) &
0-7449-0080-8 (pbk) (ISSN
0954-6103) 250 pp approx.*

573 Laszlo, Ervin (Ed.)
**The Evolution of Cognitive
Maps: New Paradigms for the
21st Century**
A 'cognitive map' is a representation of the environment in the
brain of an organism or organisation. This book traces the
development of cognitive maps,
and explores their dynamic interaction with the evolving world.
The dimensions of this process
and the resulting conclusions are
very important for everyone concerned with the match between
our changing world and our dominant views of it.
*1993, Gordon and Breach Science
Publishers, New York & Reading,
ISBN 2-88124-559-5 (hbk) about
288pp.*

574 Lavigne, Marie (Ed.)
**The Soviet Union and Eastern
Europe in the Global Economy**
This book focusses on recent
developments in the international
economic relations of the countries in the former Communist
bloc, and the international dimension of their domestic reforms and
revolutions. It discusses the
dilemmas in their transitions to
market economies.
*1992, Cambridge University Press,
Cambridge & New York, ISBN
0-521-414417-2 (hbk) xv + 220 pp.*

**575 Lean, Geoffrey, Don
Hinrichsen & Adam Markham
Atlas of the Environment**
This atlas contains over 200 full
color maps on what is happening
to our planet, and readable articles emphasising the issues, what
is being done, and what needs to
be done. It provides very extensive essential information for
understanding the crisis facing
our natural and human environments, together with the basic
facts about what is happening to
them. It presents possible solutions and problems impartially,
enabling readers to make up their
own minds.
*1990, Arrow Books, London, ISBN
0-09-984620-9 (pbk) 195 pp.*

576 Leebaert (Ed.)
**Technology 2001: The Future of
Computing   and
Communications**
This book convers trends in computing, information technology,
telecommunications, and public
policies for the information age.
*1991, MIT Press, Cambridge, MA,
USA   &   London,   ISBN
0-262-12150-6 (hbk) xvi + 392 pp.*

577 Le Grand, Julian
**Equity and Choice: An Essay in
Economics   and   Applied
Philosophy**
This book offers a new answer to
the old problem of the meaning of
a just or equitable pattern of

resources. It combines philosophical, economic, and policy analyses, and includes some simple mathematics.

1991, HarperCollins Academic, New York & London, ISBNs 0-04-350065-X (hbk) & 0-04-350066-8 (pbk) x + 190 pp.

578 Le Grand, Julian, Carol Propper & Roy Robinson
The Economics of Social Problems
Besides addressing recent and current social issues and concerns, this book explains key economic concepts, and analyses the economic mechanisms for allocating scarce resources in each problem area.

1992, Macmillan, London & New York, ISBNs 0-333-55257-1 (hbk) & 0-333-53258-X (pbk) xi + 262 pp.

579 Leggett, D. M. A.
Facing the Future: Towards Planetary Welfare
This book explores the spiritual implications of today's environmental threats, and reviews the more material aspects. It bridges the gap between the scientific and religious interpretations of life. The author believes that the planet's future is too important to be left to a limited approach; the solutions proposed by scientists must be balanced by informed opinion which understands the fundamental laws of life and recognises them as the 'ancient wisdom' lying at the foundation of the great religions. Part 1 covers important aspects of the current world situation. Part 2 gives examples of emerging fundamental changes in our understanding of the whole human being, and presents an outlook that is optimistic *provided that we act now*. The basic message is that the greatest mistake is to do nothing rather than do a little; in fact, the combined effect of our many small contributions to change will transform our total situation.

1990, Pilgrim Books, Tasburgh, Norwich, Norfolk, ISBN 0-946259-36-4 (pbk) x + 145 pp.

580 Leggett, Jeremy (Ed.)
Global Warming: The Greenpeace Report
This book presents a series of papers about the greenhouse effect, its effects (including global warming), and possible approaches (including energy conservation) to minimising these effects. The papers cover science, impacts, and policy responses; all are written by experts and well illustrated with diagrams and statistical information. The final paper formulates a plan of action from Greenpeace.

1990, Oxford University Press, Oxford & New York, ISBNs 0-19-217781-8 (hbk) & 0-19-286119-0 (pbk) xi + 554 pp.

581 Lemkow, Anna F.
The Wholeness Principle: Dynamics of Unity within Science, Religion & Society
This book states that we are now involved in a broad combination of mutually reinforcing developments, with far-reaching implications for all life on Earth. This illustrates a fundamental principle of the 'perennial philosophy', that the different dimensions of exstence – spiritual, moral, mental, emotional, and physical – are inseparable and inter-related dynamically. Thus we can achieve a more inclusive vision by integrating the consensual insights of the different modes of knowing and feeling – scientific, aesthetic, philosophical, religious, and mystical. However, the present educational environment hardly promotes such a holistic process.

The book attempts a remarkably broad synthesis of a wide range of different subjects and aspects, including: wholeness as a key to understanding, the perennial philosophy, the capacity of human beings to know, the rise of modern science, quantum reality, the biological view, the systems view, evolution, paranormal (psi) faculties, world religions, 'karma' as the ultimate law of wholeness, a planetary overview of human and social affairs, *Our Common Future* (No. 988), the need to redefine security, the need for a more holistic economy, grassroots movements for social change, prophetic thinkers, and the emergence of wholeness as a guiding principle in thought and action.

1990, Theosophical Publishing House, Wheaton, IL, USA, ISBN 0-8656-0655-4 (pbk) 456 pp.

582 Leonard, Ann (Ed.)
Seeds: Supporting Women's Work in the Third World
This book describes 'seeds', a remarkable development project organised by and for women in Africa, Asia, Latin America, and the Caribbean. The seeds projects were specifically designed to involve women in decision making and income generation, have been trainng leaders, and provide useful models for study and action.

1989, The Feminist Press, New York, ISBNs 0-935312-92-7 (hbk) & 0-935312-93-5 (pbk) xiii + 239 pp.

583 Leontief, Wassilly W., Anne R. Carter & Peter A. Petri
The Future of the World Economy: A United Nations Study
This report clearly describes the economic and policy measures needed to put the United Nations' 1974 vision of the New International Economic Order into practice over a 25-year period. After analysing the future in several critical areas, its authors reach rather optimistic conclusions

about the physical and environmental potential growth in the years ahead, and conclude that the limits to growth are not physical, but social, political and institutional. They conclude that appropriate reforms can bring improved living standards in developed and developing countries without inevitable environmental damage.
1977, Oxford University Press, Oxford & New York, ISBN 0-19-502233-5 (pbk) vii + 112 pp.

584 Lerner, Steve
Earth Summit: Conversations with Architects of an Ecologically Sustainable Future
This book attempts to raise the level of public dialogue focussing on the June 1992 Earth Summit. It presents interviews with 19 leading environmentalists and social activists on: (1) the global picture; (2) building and protecting sustainable communities; (3) reforming the institutions; (4) organising, lobbying, networking, and negotiating.
1991, Common Knowledge Press, Bolinas, CA, USA, (hbk) 263 pp.

585 Lessem, Ronnie
The Roots of Excellence
This book brings a fresh and lively approach to the analysis of business development in Britain's economy and culture, and identifies those features like individuality and creativity that make it unique.
1989, Fontana Collins, London, ISBN 0-00-636574-3 (pbk) 313 pp.

586 Lewenhak, Sheila
The Revaluation of Women's Work
This book surveys and analyses the different ways in which both paid and unpaid women's work is valued throughout the world.
2nd. Ed. 1992, Earthscan, London, ISBN 1-85383-115-8 (pbk) xii + 256 pp.; 1st. Ed. 1988, Croom Helm, London.

587 Lewis, H. W.
Technological Risk
This book is an introduction to the difficulties of risk assessment and management, attempting to correct the imbalance between trivial risks that scare the public and real risks that cause little concern.
1990, W. W. Norton & Co., New York & London, ISBN 0-393-02883-6 (hbk) xiii + 353 pp.

588 Lewis, Martin W.
Green Delusions: An Environmental Critique of Radical Environmentalism
The author of this book, who is a committed environmentalist, argues that many devoted 'Greens' propose a radical environmentalism that unwittingly has devastating implications for the global ecosystem. He distinguishes what he views as the main variants of 'eco-extremism', exposes its fallacies, and

indicates the catastrophic ecological effects that he considers its advocates would have if implemented. He argues that we must move forward into the 'solar age', invest more in our technological infrastructure, and retain a globally integrated economy. To promote the reforms needed to change our present course, capitalism must avoid divisive policies and try to create a broad consensus, to enable society to make desperately needed reforms.
1992, Duke University Press, Durham, NC, USA & London, ISBN 0-8223-1257-3 (hbk) ix + 289 pp.

589 Leytham, Geoffrey
Managing Creativity
This book attempts to describe the most creative people in their most creative moments. It has been written as a guide to help readers to encourage and manage creativity in themselves and others, and to know and understand the creative process. Its author believes that the analysis of creativity requires an interdisciplinary consortium.
1990, Peter Francis Publishers, Little Fransham, Dereham, Norfolk, ISBN 1-870167-23-6 (pbk) ix + 209 pp.

590 Linacre, Edward
Climate Data and Resources: A Reference and Guide
This book reviews the theory and practice underlying current climatic research. It is valuable for all interested in collecting and analysing climate data and assessing the global climate change impact.
1992, Routledge, London & New York, ISBNs 0-415-05702-7 (hbk) & 0-415-05703-5 (pbk) 384 pp.

591 Lipman, Matthew
Thinking in Education
In an increasingly complex world, thinking has become imperative, but evidence shows that our children are not learning how to think. This book makes practical suggestions for solving these problems.
1991, Cambridge University Press, Cambridge & New York, ISBNs 0-521-40032-5 (hbk) & 0-521-40911-X (pbk) ix + 280 pp.

592 Lipsey, Richard G.
An Introduction to Positive Economics
This introductory textbook starts at elementary level, and progressively increases to intermediate level. It presents empirical observations and policy issues as well as theories..
7th. Ed. 1989, Weidenfeld & Nicolson, London, ISBNs 0-297-79554-6 (hbk) & 0-297-79555-4 (pbk) xxii + 808 pp.; 1st. Ed. 1963.

593 Lipton, Michael & John Toye
Does Aid Work in India? A Case Study of the Impact of Official Development Assistance
The authors of this book examine the impact of foreign aid on

developing countries at macroeconomic and microeconomic levels. They focus on India as an example of an important recipient of aid, and evaluate the successes and some failures of aid in Indian development.

1990, Routledge, London & New York, ISBN 0-415-07160-7 (pbk) xii + 276 pp.

594 Litvinoff, Miles
The Earthscan Action Handbook for People & Planet
This book explains why so much damage has been done to our environment and what we can all do to help to clean up the mess. Each chapter discusses with one of the major world problems, indicates what actions can be taken to alleviate it, and shows how we can get involved. Part 1 is about human needs, including: food, wealth, population, health, the role of women, and human and civil rights. Part 2 is about our planet's resources: land, water, the atmosphere, habitats and species, and how to achieve a world without war. The book is full of suggestions for practical actions, and gives details of whom to contact, what to read, and where to go to do more. It is especially useful for individuals who wish to take action and influence people in power, many of whom respond inadequately to these growing problems.

1990, Earthscan, London, ISBN 1-85383-062-3 (pbk) xiv + 337 pp.

595 Loney, Martin et al. (Eds.)
The State of the Market: Politics and Welfare in Contemporary Britain
This book is a wide-ranging introduction to social welfare in Britain. It discusses the central questions about the relationship between the state and welfare, that have emerged from the break-up of the post-war political consensus about the welfare state.

2nd. Ed. 1991, Sage, London & Newbury Park, CA, USA, ; 1st. Ed. 1987.

596 Lovelock, James E.
Gaia: A New Look at Life on Earth
This is the first book about the Gaia Hypothesis, which can be used as one of the bases of holistic thinking about planet Earth. The basis of this philosophy is that Earth's life support system, both life and its immediate physical environment, is a self-regulating system. This system functions as a single organism, which actually defines and maintains the conditions necessary for its survival.

2nd. Ed., 1987, Oxford University Press, Oxford & New York, ISBN 0-19-286030-5 (pbk) xiii + 157 pp., 1st. Ed. 1979.

597 Lovelock, James E.
The Ages of Gaia: A Biography of

Our Living Earth
This book develops the basis of a new and unified view of Earth and the life sciences, based on the Gaia Hypothesis. It discusses in detail various recent human impacts on the global environment, and explores the interactions between the atmosphere, climate, oceans, and the Earth's crust in terms of the Gaia philosophy.
1988, Oxford University Press, Oxford & W. W. Norton & Co., New York, ISBN 0-19-217770-2 (hbk) xx + 252 pp.

598 Lovelock, James E.
Gaia: The Practical Science of Planetary Medicine
This book explores the Earth through the eyes of an imaginary physician. The concept of planetary medicine introduced here implies the existence of a planetary body that is in some way 'alive', and can experience health and disease. The book explores our planet with an imaginary planetary physician as guide, and subjects it to a thorough mid-life health check, with the diagnosis is Gaia is seriously sick. It remains to be seen whether Gaia will survive, and whether humans will still be part of 'her' living system.
1991, Gaia Books, London, ISBN 1-85675-040-X (hbk) 192 pp.

599 Lovins, Amory
Soft Energy Paths: Toward a Double Peace
This book proposes a placid, orderly transition to renewable and conserving 'soft' energy technologies based on 'energy income'.
1977, Penguin Books, Harmondsworth, Midd., ISBN 0-14-022029-1 (pbk) xx + 231pp.

600 Lutz, Mark A. & Kenneth Lux
Human Economics: The New Challenge
The authors of this book propose an alternative economic conceptual framework, based on a democratic and participatory human economy, that transcends old capitalist and socialist ideologies.
1988, The Bootstrap Press, New York, ISBNs 0-942850-10-6 (hbk) & 0-942850-06-8 (pbk) xii + 352 pp.

601 Lynch, James
Education for Citizenship in a Multicultural Society
This book is concerned with the educational implications of cultural diversity, and the role of schools and other educational institutions in combatting prejudices. It proposes a new multi-layered approach to education for democratic citizenship within a context of cultural pluralism and growing aspirations for democracy.
1992, Cassell, London & New York, ISBNs 0-304-31931-7 (hbk) & 0-304-31929-5 (pbk) v + 122 pp.

602 Macdonald, John J.
Primary Health Care: Medicine in Its Place
This book advocates and describes Primary Health Care, a radical new approach to health care systems, which is very different from the orthodox top-down medical model. It was launched in 1978, and is already applied in several Third World countries.
1992, Earthscan, London, ISBN 1-85383-112-3 (pbk) 192 pp.

603 Mackenzie, Dorothy
Green Design: Design for the Environment
This book attempts to provide guidance about Green design for those who need it. It defines clearly the issues and problems that designers may encounter, and provides many illustrative case studies.
1991, Lawrence King, London, ISBN 1-85669-001-6 (hbk) 176 pp.

604 Mackenzie, James J. & Mohamed T. El-Ashry (Eds.)
Air Pollution's Toll on Forests and Crops
This book presents evidence, with case histories, documenting the role of air pollution in causing damage to forest trees and crops.
1989, Yale University Press, New Haven, CT, USA & London, ISBNs 0-300-04569-7 (hbk) & 0-300-05232-4 (pbk) xi + 376 pp.

605 MacNeill, Jim, Pieter Winsemius & Taizo Yakushizi
Beyond Interdependence: The Meshing of the World's Economy and the Earth's Ecology
This book, written on behalf of The Trilateral Commission, shows that the interlocking of the world's economy and the Earth's ecology is the new reality of the late 20th century. Building on *Our Common Future*, it extends its analysis of the issues of global change and the changing international politics of the environment. It presents far-reaching recommendations for policy reform, and suggests how we can act urgently but intelligently to advance our common future.
The last chapter considers: (1) preparations for and the role of the Earth Summit, to be held in June 1992 in Rio de Janeiro; (2) the options for international conventions on global change, including seizing existing opportunities now; (3) strengthening institutional capacity and reforming our great international institutions. The book emphasises the 'growth of limits', through concentrated urgent human efforts, rather than the 'limits to growth', and argues that the only reasonable option is sustainable development.
1991, Oxford University Press, New York & Oxford, ISBNs 0-19-507125-5 (hbk) & 0-19-507126-3 (pbk) xx + 159 pp.

606 Macy, Mark (Ed.)
Solutions for a Troubled World
This book is the first volume in its publisher's Peace Series, and presents chapters contributed by leading thinkers from various countries, together with a general consensus of their views. As such, it provides a basis for agreement on the most sensible ways of solving the world's troubles and problems. Its contributions consider how to foster such basic ingredients of peace as clear community, fairness and equity, and sensible regulation and planet management; these components can be developed in our personal lives, our social groups, and our world. Other contributions discuss the transformation of people's values and belief systems towards greater mutual harmony, how to cope with the inequities which seem to be a leading cause of tension and conflict at all levels of society, and how to resolve or minimise the conflicts that actually occur. The final chapter presents the many points in the contributions on which there was substantial, sometimes extensive, agreement among the contributors.
2nd. Ed. 1989, Adamantine Press, London, ISBN 0-7449-0019-0 (pbk) 318 pp.; 1st. Ed. 1988, M. H. Macy & Co. (formerly Earthview Press), Boulder, CO, USA.

607 Maddison, Angus
The World Economy in the 20th Century
This book, based on a sample of 32 countries, analyses the magnitude, characteristics and causes of 20th century economic growth.
1989, OECD, Paris, distributed by HMSO,   London,   ISBN 92-64-13274-0 (pbk) 148 pp.

608 Malecki, Edward J.
Technology   and   Economic Development: The Dynamics of Local, Regional and National Change
This book shows how technology and the management decisions of global business enterprises affect the economies of nations, regions and localities. It views technology as   central   to   economic development.
1991, Longman Scientific and Technical, London & Wiley, New York, ISBNs   0-582-01758-0   (L)   & 0-470-21723-5 (W) (pbk) xvi + 495 pp.

609 Manley, Michael, Willy Brandt et al.
Global Challenge: From Crisis to Cooperation: Breaking   the North-South Stalemates
This report of the Socialist International Committee on Economic Policy extends the Brandt Commission's analyses (Nos. &), and maintains that, until the 'North' begins to resolve its crisis, there will be little progress for the

'South' or recovery in the 'North'. It is both visionary and pragmatic, and it presents proposals for what could have been radical action for the period 1985 to 1995.
1985, Pan Books, London, ISBN 0-330-29316-8 (pbk) 222 pp.

610 Mann, A. T.
Millennium Prophecies: Predictions for the Year 2000
This book reviews relevant and supposedly irrelevant historical and modern prophecies about the transition into the 21st century.
1992, Element Books, Shaftesbury, Dorset, 1-85230-323-9 (pbk) 160 pp.

611 Mannion, A. M.
Global Environmental Change
This book shows how natural and cultural agents have transformed the Earth's surface during the last 3 million years. It emphasises the need to understand past environmental changes to plan for the future.
1991, Longman, London & New York, ISBN 0-582-00351-2 (pbk) 404 pp.

612 Mannion, A. M. & S. R. Bowlby
Environmental Issues in the 1990s
This book examines key issues relating to debates on environmental change and policy in the 1990s. Section 1 presents various approaches to analysing the relationship betwen people and the environment, and explores the book's underlying theme of sustainability. Sections 2 and 3 discuss specific global and localised issues. Section 4 is an overview, exploring common themes and future possibilities.
1992, Wiley, Chichester & New York, ISBN 0-471-93326-0 (pbk) xv + 349 pp.

613 Manor, James (Ed.)
Rethinking Third World Politics
This book attempts to break out from the limitations of the 'political development' and 'dependency' paradigms, dominating the study of Third World politics, by asking more open-ended questions, and emphasising the need to analyse it in historical context. It does not prescribe policy, but aims to develop an understanding the nature of politics in the varied societies of the Third World.
1991, Longman, London & New York, ISBNs 0-582-07459-2 (hbk) & 0-582-07458-4 (pbk) x + 283 pp.

614 Mansbridge, Jane J. (Ed.)
Beyond Self-Interest
This book rejects the prevalent idea of human behaviour as being based mainly on self-interest, narrowly conceived, and argues for a more complex view of individual behaviour and social organisation.
1990, University of Chicago Press, Chicago, ISBNs 0-226-50359-3 (hbk) & 0-226-50360-7 (pbk) xiii + 402 pp.

615 Mansfield, Peter
The Good Health Handbook: Help Yourself Get Better
This book by a general practitioner shows how to solve a wide range of health problems. It enables its readers to understand what has gone wrong with their bodies, and how to help themselves to get better.
1988, Grafton Books, London, ISBN 0-246-13169-1 (pbk) ix + 322 pp.

616 Marco, Gino J., Robert M. Hollingworth & William Durham (Eds.)
Silent Spring Revisited
This book is a collection of essays on Rachel Carson's book *Silent Spring* (No. 168) and subsequent developments on its theme. It outlines the book, and discusses the regulation of pesticides.
1987, American Chemical Society, Washington, DC, USA, ISBNs 0-8412-20980-4 (hbk) & 0-8412-20981-2 (pbk) xviii + 214 pp.

617 Marien, Michael & Lane Jennings (Eds.)
What I Have Learned: Thinking About the Future Then and Now
This book of essays updates and revises previous thinking about the future. It then summarises the lessons learned from it, which can be classified into three groups: (1) the wide variety of futurists and possible futures; (2) the agreed value of thinking about the future; (3) the value of a mid-career assessment for future studies.
1987, Greenwood Press, New York & London, ISBN 0-313-25071-5 (hbk) xv + 204 pp.

618 Markandya, Anil & Julie Richardson (Eds.)
The Earthscan Reader in Environmental Economics
This book brings together many important contributions to the development of environmental economics. Its sections cover: the theoretical issues, the different ways of valuing the environment, economic instruments of environmental policy, environment and development, and global environmental problems. The introduction outlines the area and its principal arguments and viewpoints.
1992, Earthscan, London, ISBN 1-85383-106-9 (pbk) 352 pp.

619 Martyn, John, Peter Vickers & Mary Feeney (Eds.)
Information UK 2000
This book, commissioned by the British Library, projects and analyses the potentially vast future changes in technologies, working practices, and market opportunities within the library and information community. It examines how information will be generated, handled, stored and used during the 1990s and into the 21st century.
1990, Bowker-Saur, London & New

York, ISBN 0-86291-620-8 (hbk) vii + 293 pp.

620 Maruyama, Magoroh (Ed.)
Context and Complexity: Cultivating Contextual Understanding
This book offers a reformulation of interdisciplinarity and its practical applications. It illustrates throuh examples several ways to understand and handle today's complex problems contextually.
1992, Springer, London, New York, etc., ISBNs 0-387-47542-X & 3-540-47542-X (hbk) xii + 145 pp.

621 Maslow, Abraham
Motivation and Personality
This book is a highly readable, systematic presentation of the author's general theory of human motivation, based primarily on a synthesis of holistic and dynamic principles. It explains the psychology of health, describes self-actualisation, and introduces the author's well-known formulation of the hierarchy of human needs. Applications are made to aspects of psychology and personal growth.
3rd. Ed. 1987, Harper & Row, New York & London, ISBN 0-06-041987-3 (pbk) 336 pp.

622 Masulli, Ignazio
Nature and History: The Evolutionary Approach for Social Scientists
This book presents a morphological analysis of the historical and social sciences, which have traditionaly been viewed as too random in their progressions to conform to a model. By empirical evaluation, it constructs a case for an evolutionary paradigm for the natural and social sciences. It presents working historical case studies.
1990, Gordon and Breach Science Publishers, New York & Reading, ISBN 2-88124-376-2 (hbk) 184 pp.

623 Mathews, Jessica Tuchman (Ed.)
Preparing the Global Environment: The Challenge of Shared Leadership
This book aims to lay the foundations for scholarship in a new area of policy analysis at the intersection of four fields: foreign policy, environmental science, international law, and economics.
1990, W. W. Norton & Co., New York, ISBN 0-393-02911-3 (hbk) 362 pp.

624 Mathews, John
Tools of Change: New Technology and the Demoralisation of Work
This book takes a comprehensive look at how work is changing in relation to technology. Work could be alienated, or become more open, participative, democratic, productivity, efficient, and successful.
1989, Pluto Press Australia, Sydney, Australia & London, ISBN 0-949138-22-3 (pbk) xvi + 234 pp.

625 Maxwell, Nicholas
From Knowledge to Wisdom: A Revolution in the Aims & Methods of Science
This book advocates a comprehensive intellectual revolution, affecting all branches of scientific research, technology, scholarship, and education. This new philosophical approach requires a radical change in fundamental intellectual aims and methods of enquiry. No longer would philosophy and science be confined to the pursuit of abstract knowledge, isolated from practical affairs. Instead, one of their central concerns would be the problems experienced by the people of the world in their daily lives. The author presents an original proposal to help humankind to resolve such problems rationally and in a spirit of cooperation. He discusses the detailed implications of putting this proposal into practice in various areas of thinking.
2nd. Ed. 1987, Blackwell, Oxford & New York, ISBN 0-631-15641-0 (pbk) vi + 298 pp.; 1st. Ed. 1984, ISBN 0-631-13602-9 (hbk).

626 Maybury-Lewis, David
Millenium: Tribal Wisdom and the Modern World
This beautifully illustrated book attempts to capture the wisdom of tribal people before it is lost. It introduces tribal people and explains tribal approaches from many parts of the world, in a way that may inform, inspire, and change the way we think.
1991, Viking, New York & Penguin Books, London, ISBNs 0-670-84606-6 (US Body Shop Ed.) & 0-670-82935-8 (hbk) xvi + 398 pp.

627 Mayo, Deborah G. & Rachelle D. Hollander
Acceptable Evidence: Science and Values in Risk Management
This book concentrates on the entry of values in collecting, interpreting, communicating, and evaluating the evidence of risk, thus on issues of the acceptability of evidence of risk.
1991, Oxford University Press, New York & Oxford, ISBN 0-19-506372-4 (hbk) xii + 202 pp.

628 Mazur, Allan
Global Social Problems
This book presents the study of global problems, as a complement to the study of American social problems, over a long historical timespan.
1991, Prentice-Hall, Englewood Cliffs, NJ & London, ISBN 0-13-35013-4 (pbk) xi + 207 pp.

629 McBurney, Stuart
Ecology into Economics Won't Go: Or Life Is not a Concept
The author of this book argues that 'sustainable economic growth' is a delusion, and that the current attempts to solve environmental problems through market-

based incentives are ineffective.
1990, Green Books, Hartland, Bideford, Devon, 1-870098-28-5 (pbk) xii + 196 pp.

630 McCormick, John
The Global Environmental Movement: Reclaiming Paradise
This book traces the history of the global environmental movement from its roots to the mid-1980s, assesses the prospects for greater international cooperation in addressing environmental problems, and looks at its prospects of environmentalism into the 21st century.
1989, Belhaven Press, London & Indiana University Press, Bloomington, IN, USA, ISBN 1-85293-086-1 (hbk) xvii + 250 pp.

631 McCormick, John
British Politics and the Environment
This book describes the growth of the immense range of environmental law in the UK, and the pressures, compromises, and parliamentary and civil service opportunity to allow this structure to develop over most of the 20th century.
1991, Earthscan, London, ISBN 1-85383-090-9 (pbk) 201 pp.

632 McGregor, Alan & Andrew A. McArthur (Eds.)
Community Enterprise in the Local Economy
This report discusses different forms of community enterprise in the UK. It describes a range of different models of community enterprise in very varied places, with great diversity of experience.
1990, Training and Employment Research Unit, Glasgow, TERU Research Paper, No. 2, xii + 149 pp (ISSN 0956-5566).

633 McKenzie, George & Stephen Thomas
Financial Instability and the International Debt Problem
This book is based on a research report, prepared for the United Nations Conference on Trade and Development. It details many aspects of the dynamic nature of the international banking problem.
1992, Macmillan, London & New York, ISBN 0-333-46419-2 (hbk) ix + 211 pp.

634 McKibben, Bill
The End of Nature
This book presents the views of an American writer on how we are destroying our natural environment *now*. By our actions, we may *already* have irredeemably altered nature and stepped over the threshold of recovery which will result in the end of nature and put us all at risk. It includes extensive discussions of the causes and consequences of global warming and ozone layer depletion. It emphasises the interdependence of all life and the frailty of the

human species.
1990, Penguin Books, London & Viking Penguin, New York, ISBN 0-14-012306-7 (pbk) xi + 212 pp.

635 McLeish, Ewan
The Spread of Deserts: Conserving Our World
This book describes the increase of deserts and its major implications. It surveys the growing problems and issues of desertification, examines its causes, and considers its prevention.
1990, Wayland, London, ISBN 1-85210-694-9 (hbk) 48 pp.

636 McMillan, Della E. (Ed., with Jeanne Harlow)
Anthropology and Food Policy: Human Dimensions of Food Policy
This collection of seven essays covers a wide variety of food and agriculture topics in Africa and Latin America, and suggests many ways by which anthropologists can help in the struggle against world hunger.
1991, University of Georgia Press, Athens, GA, USA & London, ISBNs 0-8203-1287-8 (hbk) & 0-8203-1288-6 (pbk) vii + 191 pp.

637 McNay, Ian (Ed.)
Visions of Post-Compulsory Education
This book reviews significant changes in post-compulsory education during recent decades, and sets a new framework for the next decade.
1992, Open University Press, Buckingham & Philadelphia, ISBNs 0-335-09779-0 (hbk) & 0-335-09778-2 (pbk) xi + 180 pp.

638 McNealy, Jeffrey A.
Economics and Biological Diversity: Developing and Using Economic Incentives to Conserve Biological Resources
This book presents an attempt to design economic and other mechanisms to promote the conservation of biological diversity.
1988, IUCN, Gland, Switzerland, ISBN 2-88032-964-7 (pbk) xvi + 236 pp.

639 McRobie, George
Small Is Possible
This sequel to *Small Is Beautiful* (No. 818) optimistically assesses the potential of alternative technology, and shows that there is an alternative future for us all. It presents the work of the Intermediate Technology Development Group (No. 1497), and describes the alternative technology movements in developed and developing countries.
2nd. Ed. 1982, Abacus Books, Sphere, London, 0-349-12307-1 (pbk) xvi + 336 pp.; 1st. Ed. 1981, Cape, London.

640 Meadows, Donella H.
The Global Citizen
This sample of the author's weekly column of *The Global Citizen* presents a system-based,

globally oriented, long-term viewpoint of future planetary prospects, covers a wide range of global issues, and reassesses *The Limits to Growth* (No. 642).
1991, Island Press, Washington, ISBN 1-55963-058-2 (pbk) xiv + 300 pp.

641 Meadows, Donella H., Dennis L. Meadows & Jorgen Randers
Beyond the Limits: Collapse or a Sustainable Future
This book is a sequel to *The Limits to Growth* (No. 642), first published in 1972, whose conclusions were rejected by many people. It presents global evidence, collected since then, that is claimed by its authors to confirm their earlier findings. The authors' use of their previous World 3 global model is rather surprising, in view of the extensive criticisms made of that model and the many alternative, more detailed, models introduced since then. They conclude, from their runs of the new, personal computer version of World 3, that our planet has *already* overshot some of its limits. *If present trends continue*, a global collapse seems almost certain, perhaps within a few decades, through a rapid and uncontrolled decline in food production, industrial capacity, population, and life expectancy.

Having presented this warning, the authors argue that a sustainable society *is* technically and economically feasible, if growth in material consumption and population are reduced, and if materials and energy are used much more efficiently. To make that transition, long and short term goals must be balanced with much greater care, to enhance equity and quality of life. That will require wisdom, far-sightedness, and magnanimity. Despite the limitations of its model, the book gives useful insights into understanding the vital issues of our time and the achievement of a transition to a sustainable future.
1992, Earthscan, London, ISBNs 1-85383-130-1 (hbk) & 1-85383-131-X (pbk) xix + 300 pp.

642 Meadows, Donella H., Dennis L. Meadows, Jorgen Randers & William W. Behrens II
The Limits to Growth: A Report for the Club of Rome's Project on the Predicament of Mankind
This book outlines the World 3 global model, developed by a team led by Dennis Meadows and sponsored by the Club of Rome. This model recognised the existence of limits to economic growth, determined by the finite amount of resources in the planet, and attempted to find the extent of these limits. It forecast several possible major crises to mankind during the 21st century, which

could occur if current trends were to continue. The authors present alternative policies that, in their view, could increase the stability of the world situation and achieve a sustainable equilibrium. These policies could be applied successfully only if some prevailing economic and political assumptions are abandoned. For critiques of the book, see Nos. 189 & 208.
2nd. Ed. 1974, Universe Books, New York, ISBNs 0-87663-165-0 (hbk) & 0-87663-901-5 (pbk) 205 pp.; 1st. Ed. Universe Books & Earth Island, London.

643 Meek, Jennifer
Sick Earth Syndrome and How to Survive It
This book considers many of the environmental problems that we face today as a result of widespread pollution. They include: office machinery, air conditioning, noise pollution, domestic chemicals, cosmetics, and food additives. The author gives specific advice on what we can do to improve our health and quality of our lives.
1992, Macdonald Optima, London, ISBN 0-356-19672-0 (pbk) x + 198 pp.

644 Meier, Gerald M.
Leading Issues in Economic Development
This book focusses on the fundamental analytical principles and empirical relationships of development economics. It presents a variety of viewpoints and perspectives, includes up-to-date case studies, and covers a broad range of theory, applications, and policies.
5th. Ed. 1989, Oxford University Press, New York & Oxford, ISBN 0-19-505572-1 (pbk) xvi + 560 pp.; 1st. Ed., 1964.

645 Miller, E. J. & P. K. Rice
Systems of Organization: The Control of Task and Sentient Boundaries
This book develops a theory of organisation that reconciles tasks, human activities, and organisation within one general framework.
1967, Tavistock Publications, London & New York, (hbk) xviii + 286 pp.

646 Miller, Eric & the Editors of *Research Alert* **(Eds.)**
Future Vision: The 189 Most Important Trends of the 1990s
This book considers five primary major groups of trends that are reshaping America, discussing 189 trends that belong to them.
1990, Sourcebooks, Naperville, IL, USA, ISBNs 0-942061-17-9 (hbk) & 0-942061-16-0 (pbk) 329 pp.

647 Miller, Eric & the Editors of *Research Alert* **(Eds.)**
The Lifestyle Odyssey: 2001 Ways American Lives Are Changing
This book contains abstracts from

various periodicals and business reports focussing on changing lifestyles, considering six overall principles and factors that will shape 1990s American lifestyles.
1992, Sourcebooks, Naperville, IL, USA, ISBNs 0-942061-31-4 (hbk) & 0-940261-36-5 (pbk) 304 pp.

648 Miller, G. Tyler
Living in the Environment: An Introduction to Environmental Science
This students' textbook on the environment has six parts: (1) humans and nature: an overview; (2) basic concepts; (3) population; (4) resources; (5) pollution; (6) environment and society.
6th. Ed. 1992, Wadsworth Publishing Co., Belmont, CA, USA, ISBN 0-534-0852-9 (pbk) xxiv + 668 pp.; 1st. Ed. 1975.

649 Minsky, Marvin
The Society of Mind
This book proposes a revolutionary answer to the question of how the human mind works, and explores human intelligence.
2nd. Ed. 1987, Heinemann, London, ISBN 0-434-46758-8 (hbk) 339 pp.; 1st. Ed. 1985, USA.

650 Mintzer, Irving M. (Ed.)
Confronting Climate Change: Risk, Implications and Responses
This book presents the risks, differences, and opportunities of the emerging political era, where the

impact of prospective global warming could affect all regional, public and even individual decisions.
1992, Cambridge University Press, Cambridge & New York, ISBNs 0-521-42091-1 (hbk) & 0-521-42109-8 (pbk) xiv + 382 pp.

651 Mitchell, Senator George J.
World on Fire: Saving an Endangered Earth
In this book, the US Senate Majority Leader warns of major environmental problems (global warming, ozone layer destruction, acid rain, and tropical rain forest destruction, population growth in developing countries, threats to human health, and extinction of species). He argues that the USA has been a major contributor to global environmental problems, and must take the lead in an essential global crusade. He gives high priority to the global climate change problem and a new US energy strategy. He lists ten major actions that the nations of the world should take together. His findings are based largely on the work of the Worldwatch Institute (No. 1651).
1991, Scribners, New York, ISBN 0-684-19231-4 (hbk) viii + 247 pp.

652 Mitsch, William J. & Sven Erik Jorgensen (Eds.)
Ecological Engineering: An Introduction to Ecotechnology

This book introduces the basic principles of ecological enginerring and ecological modelling, with case studies of ecological engineering.
1989, Wiley, New York & Chichester, ISBN 0-471-62559-0 (hbk) xv + 472 pp.

653 Moder, Joseph J. & Salah E. Elmaghraby (Eds.)
Handbook   of   Operations Research.
This is a comprehensive two-volume survey of the state of the art in the late 1970s of most aspects of 'operational research' (known in the USA as 'operations research'), which is essentially an application of scientific method to problems of organisation and decision making.
1978, Van Nostrand Reinhold, New York   &   London,   ISBNs 0-442-24595-5   (Vol.   1)   & 0-442-24596-3 (Vol. 2) (hbk) l + 1281 pp.

654 Moll, Peter
From Scarcity to Sustainability: Future Studies and the Environment: The Role of the Club of Rome
This study of the Club of Rome (No. 1635) is based on interviews with about 100 members, as well as with leading futurists, UN officials, and environmentalists. It emphasises the impact of the Club's first report, *The Limits to Growth* (No. 642), and its interactions with the World Future

Society (No. 1647) and the World Future Studies Federation (No. 1648). It concludes that there are many potentials for the Club.
1991, Peter Lang, Frankfurt, Germany & New York, (pbk) 328 pp.

655 Møller, Bjorn
The Dictionary of Alternative Defence
Now that the Cold War has ended and alternative defence concepts are more topical than ever before, this important reference source is a key tool in transforming the present low awareness of defence alternatives. Covering the whole field of alternative defence, it is intended for both specialists and interested general readers. It has many annotations referring to its comprehensive bibliography.
1993, Adamantine Press, London & Lynne Rienner, Boulder, CO, USA, ISBNs   0-7449-0048-4   (AP)   & 1-55587-386-3 (LR) (hbk) 448 pp.

656 Mollison, Bill
Permaculture: A Practical Guide for a Sustainable Future
Permculture is the conscious design and maintenance of economical, agriculturally productive ecosystems, having the diversity, stability, and resilience of natural ecosystems. This valuable reference book brings together the latest thinking on permaculture, and is a detailed working manual. Topics covered include: design methods, climatic

factors, trees, water, soils, earth resources, agriculture, and strategies for an alternative global nation. The book explains how to grow food, fix broken land, and devise a better society.

1990, Island Press, Washington, ISBN 1-55963-048-5 (hbk) xii + 579 pp.

657 Monbiot, George
Amazon Watershed: The New Environmental Investigation

This book explores the underlying reasons for deforestation in the Amazon basin, and shows why efforts to prevent it have not succeeded.

1991, Michael Joseph, London & New York, ISBN 0-7181-3428-1 (hbk) ix + 276 pp.

658 Moorcroft, Sheila (Ed.)
Visions for the 21st Century

This book is a challenging and stimulating collection of 21 essays by invited international contributors. Their topics and insights are as varied as the cultures and disciplines that they represent. Some of them look far into the future, and others address urgent global issues of the 1990s into the 21st century. They illustrate several ways whereby Earth and humankind *could* have a good future. These visions resonate with the interconnectedness and interdependence of the issues that face us, and also show that the solutions that we seek can be achieved, as long as they reflect that interconnectedness. If we regard the future only with trepidation, we become mesmerised by the problems and lose sight of the possibilities. Although offering no easy solutions, these visions provide a refreshing remedy for the growing tide of pessimism. The richness of their ideas is highly stimulating.

U.S. Ed. 1993, Praeger, Westport, CT, USA, 0-275-94572-3 (pbk); U.K. Ed. 1992, Adamantine Press, London, ISBNs 0-7449-0052-2 (hbk) & 0-7449-0053-0 (pbk) (ISSN 0954-6103) x + 178 pp.

659 Morgan, David R. (Ed.)
The BMA Guide to Pesticides, Chemicals and Health

This guide addresses the growing public concern over the levels of pesticide use and their adverse effects in the environment and, in the long run, on human health. It presents extensive evidence about the hazards and benefits of pesticide use, and looks forward to a future safer role and improved regulation of pesticides.

2nd. Ed. 1992, Arnold, London, ISBN 0-340-54924-6 (pbk) viii + 215 pp.; 1st. Ed. 1990.

660 Morone, Joseph G. & Edward J. Woodhouse
The Demise of Nuclear Energy? Lessons for Democratic Control of Technology

This book considers the economic viability of the American nuclear energy industry, assesses the lessons learned from it, impartially diagnoses of the decision making processes, and argues that a radically altered form of nuclear power could be much more acceptable.
1989, Yale University Press, New Haven, CT, USA & London, ISBNs 0-300-04448-8 (hbk) & 0-300-04449-6 (pbk) xii + 172 pp.

661 Mounfield, Peter
World Nuclear Power
This book provides a comprehensive, detailed, and well illustrated survey of the nuclear power industry worldwide.
1991, Routledge, London & New York, ISBN 0-415-00463-2 (hbk) 416 pp.

662 Mullineux, Andy W.
Business Cycles and Financial Crises
This book argues that the time has come for a major research programme, designed to throw light on the many unanswered questions about business cycles. Seven of these questions are discussed.
1990, Harvester Wheatsheaf, London & University of Michigan Press, Ann Arbor, MI, USA, ISBNs 0-7450-0545-4 (HW) & 0-472-10181-1 (UM) (hbk) 200 pp.

663 Mumford, Lewis
Technics and Civilization
This book discusses the rise of the machine and mechanisation from earliest times and their good and bad influence on civilisation.
1934, Routledge, London, (hbk) 495 pp.

664 Mumford, Lewis
The Culture of Cities
This book discusses the rise of the city from earliest times, and the need for a new urban order based on a wide range of human values, and greatly influenced the garden city and new town movement.
1938, Harcourt Brace, New York, (hbk) 586 pp.

665 Mumford, Lewis
The Condition of Man
This book considers the importance of human values from a historical and current perspective, and sees little hope that humanity will meet the challenge of its crisis during World War 2.
1944, Secker & Warburg, London, (hbk) 467 pp.

666 Mumford, Lewis
The Conduct of Life
This book explains how we should change our attitudes, beliefs, and actions if we are to survive. Its references to the importance and complexity of the environment are especially interesting.
1952, Secker & Warburg, London, (hbk) 341pp.

667 Mumford, Lewis
The Transformations of Man
This book is a collection of essays on the successive stages and transformations of man, and on world culture and human prospects.
1957, Allen & Unwin, London, (hbk) 192 pp.

668 Mumford, Lewis
The City in History
This book is a rewrite of *The Culture of Cities* (No. 664), with much new material, but omitting much of the original material.
2nd. Ed. 1966, Penguin Books, Harmondsworth, Middlesex, (pbk) 696 pp.; 1st. Ed. 1961, Secker & Warburg, London.

669 Mumford, Lewis
The Myth of the Machine, Vol. 1: Technics and Human Development
This book explores the history of the rise of the use of technology and its social impact.
1967, Secker & Warburg, London, (hbk) 342 pp.

670 Mumford, Lewis
The Myth of the Machine, Vol. 2: The Pentagon of Power
This book discusses 'mega-machine attitudes', and criticises the rise of automation and the use of computers. Since it was written, the rise of personal computing and computer-aided manufacturing has enabled us to use them for our needs, but Mumford's warning should be heeded.
1971, Secker & Warburg, London, (hbk) 496 pp.

671 Mumford, Lewis
Interpretations and Forecasts, 1922-1972
This collection of useful extracts from Mumford's major books, also includes some useful articles; Mumford made all the selections.
1973, Secker & Warburg, London, ISBN 1-15-167680-1 (hbk) 522 pp.

672 Munday, Brian (Ed.)
The Crisis in Welfare: An International Perspective on Social Services and Social Work
This book presents the results of a major international comparative study of the crisis in the welfare state on social work and personal social services. It covers social and welfare services from different countries with a variety of types of society and political system, and shows that economic and ideological pressures often result in the adoption of 'welfare pluralism', a 'mixed economy of welfare'.
1989, Harvester Wheatsheaf, London & New York, ISBNs 0-7450-0163-7 (hbk) & 0-7450-0849-6 (pbk) x + 230 pp.

673 Munro, David A. & Martin W. Holdgate (Eds.)
Caring for the Earth: A Strategy for Sustainable Living

In this book, the world's three most powerful environmental organisations (the IUCN, UNEP & WWF) provide a new strategy for sustainable development, updating their 1979 World Conservation Strategy. They build on all that has been learned during recent years about the complexity of environmental problems and the radical and far-reaching objectives and actions needed to handle them. Nine principles for defining a sustainable society are stated. Part 1 covers principles for sustainable living, Part 2 indicates additional actions for sustainable living, and Part 3 considers implementation and follow-up. Eight annexes review the global situation of people and the areas where they live, with respect to the environmental crisis.
1991, Earthscan, London, distributed in the USA by Island Press, Washington, ISBN 1-85383-126-3 (pbk) 232 pp.

674 Murray, Michael & Joseph Pizzorno
Encyclopaedia of Natural Medicine
This book is perhaps the most comprehensive guide and reference to the use of natural measures in maintaining good health and preventing and treating disease. It explores the principles of natural medicine, and outlines their application through the safe and effective use of herbs, vitamins, minerals, diet, and nutritional supplements.
1990, Macdonald Optima, London, ISBN 0-356-17218-X (hbk) xi + 622 pp.

675 Muses, Charles & Arthur M. Young (Eds.)
Consciousness and Reality: The Human Pivot Point
This book is a multidimensional, kaleidoscopic exploration of the nature of consciousness, covering many mathematical, scientific, and parascientific topics and human evolutionary development.
1969, Outerbridge & Lazard, New York, ISBN 0-87690-028-7 (hbk) xii + 472 pp.

676 Myers, Isabel Briggs with Peter B. Myers
Gifts Differing
This book is about human personality and its roles in life. It presents and assesses the Myers-Briggs Type Indicator (MBTI) of psychological types, whose understanding and use can lead to personal fulfilment. It outlines MTBI's applications to education, training, and the effective running of organisations, business, and industry.
1980, reprinted 1991, Consulting Psychologists Press, Palo Alto, CA, USA, ISBNs 0-89106-015-4 (hbk) & 0-89106-011-1 (pbk) xiii + 217 pp.

677 Myers, Norman (Ed.)
The Gaia Atlas of Planet Management for Today's Caretakers of Tomorrow's World
This atlas presents the facts of the present human and planetary situation, together with its future options. There are many coloured maps and illustrations, with extensive accompanying text. It shows the increasing divisions caused by human conflict, and proposes that we redirect our course and become caretakers for our future. Each of its seven sections – Land, Ocean, Elements, Evolution, Humankind, Civilisation, and Management – considers its subject area from three perspectives: potential resources, crises, and management alternatives.
2nd. Ed., 1986, Pan Books, London, ISBN 0-330-28491-6 (pbk) 272 pp.; 1st. Ed. 1985.

678 Myers, Norman
The Gaia Atlas of Future Worlds: Challenge and Opportunity in an Age of Change
This atlas aims to present approaches leading to a better future for ourselves and our planet. Part 1 discusses the pressures and processes leading to planetary change, including world problems, ideas and beliefs, and advances in science and technology. Part 2 evaluates the resulting global, regional, sectoral, and personal impacts and outcomes. Part 3 considers several positive options for planetary change and creating the future. Part 4 indicates a wide range of future possibilities and prospective developments.
1990, Robertson McCarta, London & Archer Books, Doubleday, New York, ISBN 1-85365-123-0 (pbk) 191 pp., US ISBNs 0-385-26605-7 (hbk) & 0-385-26606-5 (pbk).

679 Myers, Norman
Population, Resources and the Environment: The Critical Challenges
The author of this book presents the latest UN population projections. He argues that these figures, and the resulting increased consumption of food and other products that they imply, combine to reduce drastically the capacity of environmental resources to sustain human communities on the Earth. The problems of land and water shortages are especially severe. The author presents case studies of population carrying capacity in seven developing countries and several proposed policies for sustainability. He concludes that we know what to do, but at present lack the will to do it.
1991, United Nations Population Fund, New York, (pbk) 154 pp.

680 Naisbitt, John
Megatrends: Ten New Directions in Transforming Our Lives
This book analyses ten observable

trends of the 1980s, pointing to the future. It forecasts some consequences of sophisticated technology, how we will be governed, and how social structures will change.
2nd. Ed. 1984, Macdonald Futura, London, ISBN 0-7088-2058-7 (pbk) xvi + 127 pp.; 1st. Ed. 1982, Warner Books, New York.

681 Naisbitt, John & Patricia Aburdene
Megatrends 2000
This book analyses what its authors view as the most important trends of the 1990s. Above all, the world of the 1990s will be a world of possibilities; the direction of these megatrends will strengthen society's ability to handle its worst problems and social difficulties.
2nd. Ed. 1991, Pan Books, London, ISBN 0-330-32139-0 (pbk) 288 pp.; 1st. Ed. 1990, Sidgwick & Jackson, London.

682 National Academy of Sciences, National Academy of Engineering & Institute of Medicine
Policy Implications of Greenhouse Warming
This report describes what is needed to make informed decisions about global warming, advises US policy makers, and addresses the need for an international response to potential global warming.

1992, National Academy Press, Washington, DC, USA, ISBN 0-309-04440-5 (pbk) 144 pp.

683 National Research Council
Global Environmental Changes: Understanding the Human Dimensions
This report offers a strategy for combining the efforts of natural and social scientists, to understand better how our actions influence the global environment and how changes in our environment influence us.
1992, National Academy Press, Washington, DC, USA, ISBN 0-309-04494-4 (hbk) 320 pp.

684 Naya, Seji, Miguel Urrutia, Shelley Mark & Alberto Fuentes
Lessons in Development: A Comparative Study of Asia and Latin America
This book directs special attention to differential growth rates, as shown by the relatively rapid growth of countries in East and South East Asia, and the slower growth trends of some Latin American countries, richly endowed with resources.
1989, International Centre for Economic Growth, San Francisco, CA, USA, ISBN 1-55815-052-8 (pbk) xvii + 361 pp.

685 Newell, John
Playing God? Engineering with Genes
This book discusses developments in genetics and genetic

engineering, and how they can benefit present and future generations.
1991, Broadside Books, London, ISBN 0-951629-5-9 (pbk) 111 pp.

686 Newsom, Malcolm (Ed.)
Managing the Human Impact on the Natural Environment: Patterns and Processes
This is a textbook about the physical processes of the natural environment and its human management, use, and abuse.
1992, Belhaven Press, London & New York, ISBNs 1-85293-184-1 (hbk) & 1-85293-185-X (pbk) xxi + 282 pp.

687 Nicholson, Max
The New Environmental Age
This book shows that we have now entered the environmental age, and that society now recognises and accepts environmental management as a significant factor. It encourages us to think about the urgent environmental issues that currently face us, and raise awareness of the relevance of environmentalism in our lives today.
2nd. Ed. 1989, Cambridge University Press, Cambridge & New York, ISBN 0-521-37992-X (pbk) xvii + 232 pp.; 1st. Ed. 1987, ISBN 0-521-33522-1 (hbk).

688 Nilsson, Sten, Ola Sallnaes & Peter Duinker
Future Forest Resources of Western and Eastern Europe
This book presents the results of the IIASA (International Institute for Applied Systems Analysis) Forest Study, whose first aim was to develop a consistent, formal dynamic model for European forests.
1992, Parthenon Publishing Group, Carnforth, Lancs. & Park Ridge, NJ, USA, ISBN 1-85070-424-4 (hbk) xiii + 496 pp.

689 Nisbet, E. G.
Leaving Eden: To Protect and Manage the Earth
This book shows how humanity has become the regulator and manager of the Earth, even if largely unaware of this role. Its author concludes that global economics must now be based on good stewardship.
1991, Cambridge University Press, New York & Cambridge, ISBNs 0-521-39311-6 (hbk) & 0-521-412579-4 (pbk) xviii + 358 pp.

690 Norgaard, Richard B.
Development Betrayed
This book thoroughly assesses the ideas behind development, and explains how they have contributed to its current perilous state. Its author proposes to replace the 'sustainable development' paradigm by a co-evolutionary paradigm, where people develop in relation to their environment; they neither control it nor are they at its mercy.
1992, Routledge, London & New

*York, ISBNs 0-415-06861-4 (hbk) &
0-415-06862-2 (pbk) 224 pp.*

691 Northcott, Jim (Ed.)
**Britain in 2010: The Psi Report (I
Report, No. 704)**
This report highlights and examines the options for change in
Britain in the next 20 years,
reporting the findings of a multidisciplinary team of experts at the
Policy Studies Institute (PSI) and
elsewhere. Its important findings
are presented in a clear, concise,
readable form. It uses economic
projections   prepared   by
Cambridge Econometrics, to present
three possible scenarios of possible economic strategies for the
UK: (1) the market-oriented scenario; (2) the interventionist scenario;   (3)   the   environment-
oriented scenario. Surprisingly,
the   projections   of   economic
growth were found to be similar
for all three scenarios.
*1991,   Policy   Studies   Institute,
London, ISBNs 0-85374-492-0 (hbk)
& 0-85374-439-9 (pbk) xx + 364 pp.*

**692 Nye, Joseph S., Jr., Kurt
Biedenkopf & Motoo Shinaa**
**Global Cooperation After the
Cold War: A Reassessment of
Trilateralism (** *The Triangle Papers,*
No. 41)
This booklet presents an agenda
of ten global challenges, on which
to concentrate the efforts of the
Trilateral   Commission;   they
include: sustainable global eco-
nomic growth and development,

peaceful change, better international security, further European
integration,   and   international
cooperation. It urges the Commission to have ongoing study
groups in these areas; groups
have already been set up for common security, economic interdependence,   and   transnational
global issues.
*1991, The Trilateral Commission,
New York, (pbk) 64 pp.*
693 Oberai, A. S.
**Assessing   the   Demographic
Impact of Development Projects:
Conceptual, Methodological and
Policy Issues**
This book reviews Third World
development projects, assesses
their demographic impact, and
uses case studies from developing
countries.
*1991, Routledge, London & New
York, ISBN 0-415-06841-X (hbk) 128
pp.*

694 O' Biso Soaha, Laura
Endangered Species of the World
This book describes some rare
threatened species of wild animals, includes stunning colour
photographs of them, and discusses conservationists' efforts to
save them from extinction.
*1991, Friedman Group, New York,
ISBN 0-7924-5305-0 (hbk) 127 pp.*

695 O'Connor, Anthony
**Poverty in Africa: A Geograph-
ical Approach**
This geography text examines the

extent, nature, and distribution of material poverty in Africa, and identifies contributing factors.
1991, Belhaven Press, London, ISBNs 1-85293-087-X (hbk) & 1-85293-088-8 (pbk) x + 184 pp.

696 OECD
INTERFUTURES: Facing the Future: Mastering the Probable and Managing the Unpredictable
This report presents the main results of the INTERFUTURES project, which aimed to assess OECD member governments' alternative patterns of longer-term world economic development. It considers six scenarios, reflecting various growth rates and trade patterns, and makes many policy recommendations for developed and developing countries.
1979, OECD, Paris, distributed by HMSO, London, ISBN 92-64-11967-1 (pbk) vi + 425 pp.

697 OECD
Responding to Climate Change: Selected Economic Issues
This report examines three specific aspects of economic efficiency in relation to climate change resulting from global warming. It is one of several OECD studies on the economic aspects of climate change.
1991, OECD, Paris, distributed by HMSO, London, ISBN 92-64-13565-0 (pbk) 149 pp.

698 OECD
Environmental Policy: How to Apply Economic Instruments
This report presents OECD guidelines for the practical application of economic instruments to deal with air, water and noise pollution. It targets economic sectors which have a major impact on the environment: energy, transport, agriculture, and industry.
1991, OECD, Paris, distributed by HMSO, London, ISBN 92-64-13568-5 (pbk) 130 pp.

699 OECD
Global Warming: The Benefits of Emission Abatement
This report describes a framework for evaluating and assessing the benefits of potential strategies for abating global warming.
1992, OECD, Paris, distributed by HMSO, London, ISBN 92-64-13639-8 (pbk) 69 pp.

700 Olson, William C. & A. J. R. Groom
International Relations Then and Now: Origins and Trends in Interpretation
This book reviews the development of international relations theory, and assesses its contemporary schools of thought.
1991, HarperCollins Academic, London, ISBN 0-04-495101-6 (pbk) xvii + 358 pp.

701 Oman, Charles P. & Ganeshan Vignaraja
The Postwar Evolution of Development Thinking
This book provides a critical summary of the different schools of development thought, and covers both orthodox and heterodox approaches to development thinking in an up-to-date, non-technical way.
1991, Macmillan, London & New York, ISBNs 0-333-54620-2 (hbk) & 0-333-54621-0 (pbk) xv + 272 pp.

702 Omara-Ojungu, Peter H.
Resource Management in Developing Countries
This highly practical book, with many case studies, focusses on problems facing effective resource management in the Third World, and on strategies for handling the issues arising. It outlines the basic ecological, economic, technological, and ethnological aspects of resource management, requiring the immediate attention of governments. It stresses the importance of adopting an integrated approach to resource management and addressing poverty as a critical issue. It discusses in detail the specific problems and issues of resource management in agriculture, and also draws attention to issues facing mountainous, coastal and marine areas.
1992, Longman, London & Wiley,
New York, ISBN 0-582-30102-5 (pbk) xvi + 213 pp.

703 o Murchu, Diarmuid
Coping with Change in the Modern World
This book begins with the assumption that change happens and will continue to happen for the benefit of humanity. It shows how we can use it to our advantage at the global, social and personal levels, when we surrender to its momentum, even though we cannot halt, control or modify it. This book has an unusual interpretation of change, assuming both an evolutionary and a spiritual basis to the world that we experience. It is very much concerned with new paradigms.
1987, The Mercier Press, Cork & Dublin, Ireland, & Fowler Wright Books, London, ISBNs 0-85342-827-1 (MP) & 0-85244-144-4 (FW) (pbk) 231 pp.

704 Onimode, Bade
A Future for Africa: Beyond the Politics of Adjustment
The author of this book argues against the 'adjustment programmes', imposed on Africa by the World Bank and International Monetary Fund, and assesses their damage to Africa's economic and political life. He believes that real development will be achieved only by returning control of Africa's abundant resources to its own people, and ensuring greater

democracy and accountability in African political structures.

1992, Earthscan, London, ISBN 1-85383-100-X (pbk) 208 pp.

705 Ornstein, Robert & Paul Ehrlich
New World, New Mind: Moving Toward Conscious Evolution

The authors of this book argue that the human mental system is failing to understand the increasingly changing and dangerous modern world. The only way to resolve this problem is by conscious change, where we take our evolution into our own hands and create a new evolutionary process, based on a different kind of education and training, to help us understand long-term threats.

2nd. Ed. 1990, Touchstone Books, New York, ISBN 0-621-69606-8 (pbk) 302 pp.; 1st. Ed. 1989, Double-day (hbk).

706 Orr, Bill
The Global Economy in the 90s: A User's Guide

This book describes the global economy as it moves into the 1990s. It presents and comments on facts and figures about different nations.

1992, Macmillan, London & New York, ISBN 0-333-57977-1 (hbk) xxiii + 330 pp.

707 Osborne, David
Laboratories of Democracy

This book describes the innovative economic activities of many US States during the 1980s, and mostly describes the activities of six States that serve as models of a new, 'neoprogressive' political paradigm, contrasting with US liberalism and Reaganite conservatism.

1988, Harvard Business School Press, Boston, ISBN 0-87584-233-X (pbk) 369 pp.

708
Reinventing Government: How the Entrepreneurial Spirit Is Transforming the Public Sector from Schoolhouse, City Hall to Pentagon

This book, written from a non-partisan viewpoint, focusses not on *what* government should do, but on *how* it should work. As such, its ideas have been endorsed in the USA by both Democrats and Republicans. It is about pioneers of a new form of government, and a new paradigm for government, an 'American perestroika'. Its case histories point to a *third* way of government, different from both bureaucratic and laissez-faire approaches, which aims to combine positive and caring government with high efficiency and productivity.

1992, Addison-Wesley, Reading, MA, USA & Wokingham, Berks., ISBN 0-201-52394-9 (hbk) xxii + 405 pp.

709 Owen, Richard & Michael Dynes
The Times Guide to the Single European Market: A Comprehensive Handbook
This guide is an authoritative, comprehensive, and up-to-date handbook about the Single European Market, due to come into effect after the end of 1992, and covers many of its aspects.
1992, Times Books, London, ISBN 0-7230-0470-6 (pbk) 340 pp.

710 Paehlke, Robert C.
Environmentalism and the Future of Progressive Politics
This book analytically treats both environmentalism as it is today, and its potential as an effective set of ideas. Its author considers that environmentalism will be able to attract a mass following, only if environmentalists develop clear and consistent positions on the whole range of political and social issues. He indicates eight priorities for a contemporary progressive environmental approach.
1989, Yale University Press, New Haven CT, USA & London, ISBNs 0-300-04021-0 (hbk) & 0-300-04826-2 (pbk) viii + 325 pp.

711 Park, Chris C.
Acid Rain: Rhetoric and Reality
This book examines the full implications of scientific studies to date, and sets the political debate in the USA and UK about acid rain in its proper context. It considers evidence from around the world.
2nd. Ed. 1990, Routledge, London & New York, ISBNs 0-416-92190-6 (hbk) & 0-416-05082-0 (pbk) xiv + 272 pp.; 1st. Ed. 1987, Methuen, London.

712 Parkin, Sara (Ed.)
Green Light on Europe
In this book, leading Green activists from all over Europe discuss some key issues facing Europe. They concluded that any realistic planning for Europe's future must have a Green perspective.
1991, Heretic Books, London, ISBN 0-946097-29-1 (pbk) 367 pp.

713 Parry, Martin
Climate Change and World Agriculture
This book assesses the substantial uncertainties about issues of climate change and its global effects. It then analyses the sensitivity of the world food system, and analyses various ways in which it will be affected if scientifically predicted climatic changes occur. Some countries may be better off, but many will not. The book describes the possible effects on agriculture, estimates the impacts on plant and animal growth, and looks at the geographical limits to different types of farming. It considers a range of possible ways to adapt agriculture, ato mitigate otherwise disastrous effects. The author considers that world food

production could probably be maintained at about the same level as without climate changes, but perhaps only at a very large cost.

1990, Earthscan, London, ISBN 1-85383-065-8 (pbk) xv + 157pp.

714 Pask, Gordon
An Approach to Cybernetics
This book by a leading pioneer of cybernetics is a useful non-specialist introduction to cybernetics in its fairly early days.
1961, Hutchinson, London, (hbk) 128 pp.

715 Pasztor, Janos & Lars A. Kristoferson (Eds.)
Bioenergy and the Environment
Biomass is already a very important source of energy, especially in the Third World, and this book explores its possible future. The author considers that biomass, especially wood fuels, will probably become more important, shows how to use and manage it more efficiently, and discusses the impacts of various biomass fuels are discussed.
1990, Westview Press, Boulder, CO., USA, ISBN 0-8133-8062-6 (pbk) 410 pp.

716 Patterson, Walter C.
The Energy Alternative: Changing the Way the World Works
This book draws on the thought of energy experts and analysts from all parts of the world, and discusses most aspects of energy. Topics covered include: the present energy dilemma, the nature of energy, the history of energy, the onset of the global energy crisis, energy supply and demand, the use of energy, energy efficiency, fossil fuels, renewable energy, and future alternatives for energy.
1990, Boxtree, London, ISBN 1-85283-284-3 (hbk) viii + 186 pp.

717 Payer, Cheryl
Lent and Lost: Foreign Credit and Third World Development
The author of this book is one of North America's most prominent critics of the International Monetary Fund and World Bank. She offers a strong critique of the widely held assumption that poor countries are natural importers of capital, and argues that capital inflows in the form of debt must ultimately reverse themselves, to become net capital outflows that have to be paid in hard currency. This argument has profound implications. The Third World debt crisis cannot be solved until the debt is extinguished or written down to a fraction of its present total. Future debt crises can be avoided only if the developing countries abandon their simplistic approach to financial problems.
1991, Zed Books, London & Atlantic Highlands, NJ, USA, ISBNs 0-86232-952-3 (hbk) & 0-86232-953-1 (pbk) xii + 154 pp.

718 Pearce, David (Ed.)
Blueprint 2: Greening the World Economy
This book is a sequel to *Blueprint for a Green Economy* (No. 721), and applies its economic analyses to global environmental problems. It begins by describing the reasons for using economic approaches to common resources like climate and biodiversity. It then examines in detail the economic approaches to tackling the issues involved in: global warming, ozone depletion, environmental degradation in the Third World, population growth, rain forests, environmentally sensitive aid, equity between generations, international environmental cooperation, and Green international policies. It shows not only how economic theory can consider all these problems, but also the economic price of failing to consider them in economic policies. It argues that Western nations must pay twice to combat global warming and deforestation, by cleaning up the pollution that they have caused and by paying the Third World not to add to environmental destruction. It proposes a range of economic measures to achieve these objectives, including national and international carbon taxes on polluting fossil fuels. It provides an agenda for international and governmental action.

1991, Earthscan, London, ISBN 1-85383-076-3 (pbk) ix + 232 pp.

719 Pearce, David
Economic Values and the Natural World
This book considers how, in economics, we attribute values to the environmental goods and services from which we choose. It presents the reasons for valuing aspects of the natural world, the different ways in which this can be done, case studies showing how it has been done, and its consequences for economic development and global economic problems.
1993, Earthscan, London, ISBN 1-85383-152-2 (pbk) 144 pp.

720 Pearce, David, Edward Barbier & Anil Markandya
Sustainable Development: Economics and Environment in the Third World
This book attempts to structure the concept of sustainable development, and uses case studies to illustrate ways in which environmental economics can be applied to the developing countries.
2nd. Ed. 1991, Earthscan, London, ISBN 1-85283-088-7 (pbk) 228 pp.; 1st. Ed. 1990, Edward Elgar, Aldershot, Hants.

721 Pearce, David, Anil Markandya & Edward B. Barbier
Blueprint for a Green Economy
This book, also known as *The*

Pearce Report, was originally commissioned by the UK Department of the Environment (No. 1463), and is published in association with the International Institute for Environment and Development (No. 1501). It presents a series of practical proposals for achieving and financing sustainable economic development. It has set a new agenda for environmental protection, and adds a vital new dimension to the economic management of environmental concerns. It assesses the meaning of sustainable development, considers how to value and account for the environment, and discusses project appraisal, discounting the future, and prices and incentives for environmental improvement.
1989, Earthscan, London, ISBN 1-85383-066-6 (pbk) xvi + 192 pp.

722 Pearce, David & Kerry Turner
Economics of Natural Resources and Environment
This book provides a thorough grounding in the principles of economics, including sustainable economics, required to understand national, international, and environmental problems.
1990, Harvester Wheatsheaf, London & Johns Hopkins University Press, Baltimore, MD, USA, ISBNs 0-7450-0202-1 (HW), 0-8018-3986-6 (JH) (hbk) & 0-8018-3987-4 (JH) (pbk).

723 Pearce, Fred
Green Warriors: The People and Politics behind the Environmental Revolution
This book studies the changes in the Green movement during the last 20 years, and its current thinking and conflicts. It suggests a possible agenda for the movement for the 1990s into the 21st century.
1990, The Bodley Head, London, ISBN 0-370-31401-8 (hbk) xvi + 331 pp.

724 Pearson, Carol S.
Awakening the Heroes Within: Twelve Archetypes to Help Us Find Ourselves and Transform Our World
This book describes twelve archetypal patterns in the human personality, each having both good and bad qualities. It shows how these 'heroes within' can be developed in a balanced way to help fulfil our inner lives, our potentialities, and our quest for wholeness.
1991, HarperSanFrancisco, Harper-Collins, New York & London, ISBN 0-06-250678-1 (pbk) xiii + 319 pp.

725 Pearson, David
The Natural House Book
This book concentrates on removing the load of pollutants from our homes and replacing the offending materials with safe ones. It ha lovely illustrations of nearly perfect houses, mostly new.

1989, Conran Octopus, London, 1-85029-175-6 (hbk) 287 pp.

726 Pearson, Mark & Stephen Smith
The European Carbon Tax: An Assessment of the European Commission's Proposals
This report describes the main features of the European Commission's proposals for introducing a carbon tax throughout the European Community, as part of its response to global warming.
1991, The Institute for Fiscal Studies, London, ISBN 1-873357-12-5 (pbk) v + 58 pp.

727 Pearson, Peter (Ed.)
Energy Policies in an Uncertain World
This book discusses various aspects of energy policies and how best to regulate energy.
1989, Macmillan, London & New York, ISBN 0-333-47076-1 (hbk) xxxi + 113 pp.

728 Peccei, Aurelio
One Hundred Years for the Future: Reflections of the President of the Club of Rome
This book is a strong personal statement by the Founder of The Club of Rome (No. 1635), which expresses his belief that humankind can build an acceptable future by intelligent use of our natural and human resources. He points out both the factors leading to humanity's present decline,

and the fundamental changes of human thought and behaviour that are essential to bring about a new age of renaissance.
1981, Pergamon Press, New York & Oxford, ISBN 0-08-028110-9 (hbk) xvi + 192 pp.

729 Peck, M. Scott
The Different Drum: The Creation of True Community: The First Step to World Peace
The author of this book points out that we have failed to build a true, global community. If we are to prevent present civilisation from destroying itself, we must urgently rebuild community at all levels – local, national, and international – as the first step to spiritual survival. He describes how true communities work, how to develop group actions on the principles of love, and how we can start to transform society into a true community.
2nd. Ed. 1988, Rider, London, ISBN 0-7126-1862-7 (pbk) 334 pp.; 1st. Ed. 1987, Simon & Schuster, New York.

730 Pereira, Winin & Jeremy Seabrook
Asking the Earth: Farms, Forestry and Survival in Africa
This book presents the lessons to be learned from local, indigenous knowledge in developing modern techniques of sustainable farming.
1991, Earthscan/WWF, London, ISBN 1-85383-045-3 (pbk) 240 pp.

731 Peters, B. Guy
The Politics of Bureaucracy
This book compares the political
and policy making roles of public
bureaucracies in various nations
around the world.
*3rd. Ed. 1989, Longman, New York
& London, ISBN 0-8013-0066-5
(pbk) xiii + 303 pp.*

732 Peterson, Robyn
**Managing Successful Learning:
A Practical Guide for Teachers
and Trainers**
This book sets out the essentials
of the process, whereby attitudes,
experience, training, and objectives combine to provide excellence and quality performance in
an organisation.
*1992, Kogan Page, London, ISBN
0-7494-0547-3 (pbk) 258 pp.*

**733 Pfaller, Alfred, Lan Gough &
Goeran Therborn (Eds.)**
**Can the Welfare State Compete?
A Comparative Study of Five
Advanced Capitalist Countries**
In this book, the authors address
the real problems of the welfare
state, although they sympathise
with the now less popular idea of
social and economic citizenship
for all. The authors argue that the
virtues of discipline and solidarity
must be institutionalised more
firmly, if the ideas and practices of
a welfare state are to regain their
appeal and effectiveness by the
beginning of the 21st century.
*1991, Macmillan, London & New
York, ISBN 0-333-48755-9 (hbk) xi
+ 354 pp.*

734 Philips, Derek L.
Towards a Just Social Order
This book develops in great depth
the concept of a 'just social order',
where the available resources
within a stable society are shared
out fairly, to meet the basic needs
of all its citizens as far as possible.
The author indicates why he considers that this concept needs
serious attention in the present
human situation, and he urges
social scientists, as well as thinking people in general, to tackle
questions of this sort that involve
fundamental moral values. He
examines at considerable length
some of the previous attempts to
discuss these questions.
He develops in detail his own
approach to a just social order, in
which the rights to freedom and
well-being constitute the moral
principles that should regulate
the just social order. He presents
a list of policies and programmes
that he considers feasible and
capable of realisation in all of
today's 'advanced' societies; no
vast transformations would be
required, only the active involvement of governments in meeting
the required goals.
*1986, Princeton University Press,
Princeton, NJ, USA & Guildford,
Surrey, ISBNs 0-691-09422-5 (hbk)
& 0-691-02834-6 (pbk) x + 460 pp.*

735 Piel, Gerard
Only One World
This book contains many facts of importance to concerned citizens, and was written with the June 1992 UNCED Conference in mind. Although its information is not new, it reports it in a novel way, and reinterprets familiar data. It is strictly rational, with every assertion based on a verifiable fact. The theme of the book is how to turn possibilities into actualities. Theoretically, there should be enough food, energy, and consumer goods to give everyone a reasonable standard of living, but the political tasks of the future, to ensure this and other desirable aims, are formidable.
1992, W. H. Freeman, New York & Reading, (pbk) 367 pp.

736 Piel, Jonathan (Ed.)
Managing Planet Earth
This special issue contains a series of articles on managing planet Earth, exploring such problem areas as climatic change, population growth, pollution, and energy conservation, and providing information about planet management, manufacturing, and sustainable growth.
September 1989, Scientific American, Vol. 261, No. 3, ISSN 0036-8753.

737 Piel, Jonathan (Ed.)
Energy for Planet Earth
This special issue examines our ability to meet the world's energy needs without destroying our planet; it shows that, with the right incentives, much can be done. It considers most aspects of energy conservation and generation, including energy efficiency, solar energy, and the transition to sustainable energy provision worldwide.
September 1990, Scientific American, Vol. 263, No. 3, ISSN 0036-8753.

738 Piel, Jonathan (Ed.)
Communications, Computers and Networks
This special issue considers various aspects of communications, computers and computer networking, for example in relation to products and services, work, business, education, and public policy. It also considers future possibilities in the 1990s and into the 21st century.
September 1991, Scientific American, Vol. 265, No. 3, ISSN 0036-8753.

739 Piel, Jonathan (Ed.)
Mind and Brain
This special issue considers various aspects of mind and brain, including: the developing brain, visual imagery, learning and individuality, language, memory, sex differences, major disorders, ageing, artificial neural networks, and consciousness.
September 1992, Scientific American, Vol. 267, No. 3, ISSN 0036-8753.

740 Pierce, John T.
The Food Resource
This book analyses the impact of human-induced physical factors on the supply and adequacy of food produced, and the implications for future global agriculture. It views the most important factor as our willingness to modify our approaches to the use of the environment.
1990, Longman, London & Wiley, New York, ISBN 0-470-21512-7 (pbk) xx +344 pp.

741 Pietroni, Patrick C. (Ed.)
Reader's Digest Family Guide to Alternative Medicine
The entries in this guide are arranged in alphabetical subject order and written by distinguished medical contributors. There is also a chapter on first aid and medical action in emergencies.
1987, Reader's Digest Association, London, ISBN 0-276-42010-1 (hbk) 400 pp.

742 Pimentel, David & Marcia Pimentel
Food, Energy and Society
This book explores the interdependence between food and energy and their impacts on society. It includes tables giving specific energy figures and other statistics. Its important last chapter discusses: future food needs, strategies for meeting food needs, energy needs in food production, land and water constraints, climate, environmental pollution, and future prospects. It has an extensive bibliography.
1979, Arnold, London, ISBN 0-7131-2761-9 (pbk).

743 Plant, Judith (Ed.)
Healing the Wounds: The Promise of Ecofeminism
This book develops a new departure in feminist thinking, which states that we must rediscover and act out the wholeness at the core of the feminist and Green perspectives. The challenge is to bring body, mind and spirit together, to unite the personal, the political and the spiritual in terms of theory, practice and reflection.
1989, Green Print, London, ISBN 1-85425-016-7 (pbk) xiii + 262 pp.

744 Plant, Judith & Christopher Plant (Eds.)
Putting Power in Its Place: Create Community Control!
The contributors to this book advocate bringing control over our lives back to the community, where it belongs. Only when it has power can a local community learn from the consequences of its decisions. As government bureaucrats and company executives make decisions on our behalf, our communities and environments suffer from pollution, resource mining, and social neglect. The book states which

vested interests have incentives to abuse our communities and planet, by holding to power and applying it far from the communities where most of us live. Creating community power will require a tremendous struggle to transform the current power structure, together with continued attention to our own homes and communities. The authors identify some of the obstacles, and tell the inspiring stories of several communities that have reclaimed their power to determine their own future.

1992, New Society Publishers, Philadelphia, ISBNs 0-86571-217-4 (hbk) & 0-86571-217-4 (pbk) vi + 137 pp.

745 Pomfret, Richard
Diverse Paths of Economic Development

This book explores the varied experiences of developing countries, and describes the full range of economic development issues, including: industrialisation, agriculture, labour, and trade policy. It shows the very divergent circumstances and prospects of Third World countries, and presents many case studies, from a wide variety of countries, to illustrate these different patterns of economic development.

1992, Harvester Wheatsheaf, London & New York, ISBNs 0-7450-0972-7 (hbk) 0-7450-09733-5 (pbk) xiii + 231 pp.

746 Ponting, Clive
A Green History of the World: The Environment and Collapse of Great Civilizations

This book interprets world history from a global Green perspective, relating how today's urgent global and environmental problems. It presents a new approach to history from early hunter-gatherers to today. Instead of relating political, diplomatic and military events, it considers the fundamental forces shaping human history, how and why humans have changed the world, and the consequences of their actions. The author's historical overview, illustrated by many detailed examples, tells how humans have destroyed much of the natural world, and how great societies have collapsed by abusing their environment. He shows the deep-seated nature of global problems, and emphasises the importance of our relationship with the environment.

2nd. Ed., 1992, Penguin Books, London & St. Martin's Press, New York, ISBN 0-14-016642-4 (pbk) 448 pp.; 1st. Ed. 1991, Sinclair-Stevenson, London.

747 Poore, Duncan
No Timber without Trees: Sustainability in the Tropical Forest

This book reports an extensive study of the impact of the tropical timber trade, and its current

failure to follow sustainable practices. Proposed remedies for the situation are presented.
1990, Earthscan, London, ISBN 1-85383-060-7 (pbk) 412 pp.

748 Popcorn, Faith
The Popcorn Report: Faith Popcorn on the Future of Your Company, Your World, Your Life
Using information from a whole range of popular media, this book first distils and discusses ten trends of American life. Then it has brief chapters on tracking the future, trend analysis, and possible business strategies to cope with the future.
1991, Doubleday, New York, ISBN 0-385-40000-4 (hbk) 226 pp.

749 Pope, Stephen, Mike Pope & Elizabeth Anne Wheal
The Green Book: The Essential A-Z Guide to the Environment
This book provides information on over 500 key environmental topics, and suggests positive values that will help readers to play an active part in improving the environment. It includes a list of environmental organisations in various parts of the world.
1991, Hodder and Stoughton, London, ISBN 0-340-53298-X (pbk) viii + 337 pp.

750 Popper, Karl R.
The Open Society and Its Enemies: Vol. I: The Spell of Plato: Vol. II: The High Tide of Prophecy: Hegel, Marx, and the Aftermath
This critical introduction to the philosophy of politics and history examines some of the principles of social reconstruction. It tries to contribute to our understanding of totalitarianism, and of the significance of the perennial fight against it.
5th. Ed. 1966, Routledge and Kegan Paul, London, 2 vols., xvii + 781 pp.; 1st. Ed. 1945.

751 Porritt, Jonathon
Seeing Green: The Politics of Ecology Explained
This book outlines the nature and significance of the recent upsurge in Green politics, presents an alternative, Green approach to our current problems that goes far beyond conventional political solutions, and challenges *all* politicians to change their ways.
1984, Blackwell, Oxford, ISBNs 0-631-13892-7 (hbk) & 0-631-13893-5 (pbk) xvi + 249 pp.

752 Porritt, Jonathon (Ed.)
Friends of the Earth Handbook
This handbook provides much practical advice on how each of us can limit the damage that we cause to the environment in our daily lives. Its essential message is that only by protecting Earth can

we protect ourselves. It addresses the whole range of environmental problems and related questions of economics, health, and education. It states some of the available options, selects important issues, and shows how a few simple steps can lead towards a saner, more responsible world.

1987, Macdonald Optima, London, ISBN 0-356-12560-2 (pbk) 160 pp.

753 Porritt, Jonathon
Where on Earth Are We Going?

This book is based on the author's television series with this title. It reviews a wide range of Green and other issues that have recently become known to large numbers of people, and considers how far that public awareness has been translated into action. It is not so much about the environment as about a Green approach to mainstream political and economic concerns. Its author considers that we can protect the environment only by: finding better ways of meeting our food and energy needs, organising our education and health services, and providing environment-friendly industry and fulfilling work. Just as we have largely created the problems, so can we achieve their solution, if we have the right kind of vision, political will, and spiritual values. Viable options *do* exist, and some successful Green policies, already operating around the world, are evaluated.

1990, BBC Books, London, ISBN 0-563-20847-3 (pbk) 241 pp.

754 Porritt, Jonathon (Ed.)
Save the Earth

This book has a Foreword by HRH The Prince of Wales and an Introduction by Sir David Attenborough. It provides a dramatic statement of the damage already being done to planet Earth, and an urgent appeal to the people of all nations to change their ways. It states that the answers to world problems lie in *our* hands, as voters, citizens, parents, and consumers. Over 100 eminent people and committed activists have contributed articles or signed statements, uniting to condemn the destruction already caused to Earth.

To highlight the immensity of the task of ensuring a healthy, sustainable future, the book visualises several 'worst-case' crises, each of which is an 'environmental time bomb', now being set in place, that could cause extreme harm to future generations. In every case, a distinction is made between irreversible damage and lesser damage that could still be restored if we act soon enough. The book encourages us with realistic hope for our future, based on the actions of those who support change *now*, before it really is too late. It is intended to stimulate action, and is accompanied by an

Action Pack, originally designed to encourage readers' pledges and contributions of ideas in support of the process of the June 1992 Earth Summit.
1991, Dorling Kindersley, London & New York, ISBN 0-86318-642-4 (hbk) 208 pp.

755 Porritt, Jonathon & David Winner
The Coming of the Greens
This book examines how Green thinking began to influence all aspects of 'Western' society during the 1980s, and explores the increasing concern expressed about the environment. It assesses how deep the Green revolution is going, and puts the British experience in the context of the growth of the international Green movement. It includes interviews with politicians, activists, writers, and artists.
1988, Fontana Paperbacks, Fontana Collins, London, ISBN 0-00-637244-9 (pbk) 287 pp.

756 Porteous, Andrew
Dictionary of Environmental Science and Technology
This dictionary provides students and general readers with a working knowledge of the scientific and technical terminology, associated with environmental studies and appraisals of current issues.
2nd. Ed. 1992, Wiley, Chichester & New York, ISBN 0-471-93544-1

(pbk) xiv + 439 pp.; 1st. Ed.1991, Open University Press.

757 Porter, Alan L. et al.
Forecasting and Management of Technology
This is a basic textbook on technology forecasting and assessment for business and engineering students, with many case studies.
1991, Wiley-Interscience, New York & Chichester, ISBN 0-471-51223-0 (hbk) 448 pp.

758 Porter, Gareth & Janet Welsh Brown
Global Environmental Politics
This book discusses the whole range of global environmental problems, whose new agenda of these problems will demand international attention in the 1990s and into the 21st century. The authors consider the three broad strategies of incremental change, global partnership, and global governance, that have been suggested. They conclude that something beyond traditional power politics is at work, and that the issue is not whether nations will move fast enough towards more effective cooperation on global environmental threats.
1991, Westview Press, Boulder, CO, USA, (hbk) & (pbk) 208 pp.

759 Postel, Sandra
The Last Oasis: Facing Water Scarcity

This book shows how scarcity of fresh water may limit growth more severely than shortage of land. Investing in water efficiency, recycling, and conservation can do much to meet rising demands and prevent disaster. The priorities are a common recognition of the water situation's seriousness, and a widespread drive to develop proper institutions and frameworks to manage the sustainable use of water.
1992, Earthscan, London, ISBN 1-85383-148-4 (pbk) 224 pp.

760 Prescott-Allen, Robert & Christine Prescott-Allen
Genes from the Wild: Using Wild Genetic Resources for Food and Raw Materials
This book describes the growing contribution made by wild genetic resources to food and other production in the Third World, and examines their sustainable management.
1988, Earthscan, London, ISBN 1-85283-026-7 (pbk) 112 pp.

761 Preston, David (Ed.)
Latin American Development
This book considers the major aspects of development problems and policies in Latin America today, and shows the significant contributions made by geographers during recent decades. It is a companion book to *South East Asian Development* (No. 291).

1987, Longman, London & Wiley, New York, ISBN 0-470-20783-3 (pbk).

762 Preston, Ronald H.
Religion and the Ambiguities of Capitalism
This book discusses how far Christians understand modern economic realities, and throws much light on economics in relation to religion.. It examines and criticises various Christian statements on current economic issues, surveys the use and misuse of the Bible and Christian doctrine in economic questions, discusses the economic problems arising from the gulf between the developed and developing countries, and suggests the basis for a more adequate social theology.
1991, SCM Press, London, ISBN 0-334-02305-X (pbk) x + 182 pp.

763 Primavesi, Anne
From Apocalypse to Genesis: Ecology, Feminism and Christianity
This book re-evaluates Christianity in the light of a growing realisation that traditional interpretations of important biblical texts support a damaging concept of human domination over the rest of creation. The author proposes an ecological paradigm of humankind and nature, which expresses the essential wholeness, interrelatedness and interdependence of all creation.

1991, Burns & Oates, Tunbridge Wells, Kent, ISBN 0-86012-174-7 (pbk) xii + 324 pp.

764 Prince of Wales, HRH The
A Vision of Britain: A Personal View of Architecture
In this outspoken book, the Prince of Wales develops the arguments first presented in his 1988 TV programme. He indicts, examines, and convicts the post-war architectural 'establishment' of 'crimes' against Britain, that range all the way from arrogance and failing to consult the people to architectural monstrosities and thoughtless sacrilege. In contrast, the traditional buildings of the past, and just *a few* of those being built now are harmonious and full of character. The key part of the book is his prescription for a more humane and harmonious future, expressed as ten principles of architectural propriety. For a response by a leading architect, see No. 490.
1989, Doubleday, London & New York, ISBN 0-385-26903-X (hbk) 169 pp.

765 Prins, Gwyn & Robbie Stamp
Top Guns and Toxic Whales: The Environment and Global Security
This book explains clearly and comprehensively the implications of environmental degradation for global, national and individual security, and the policies available in the light of the changing world order.
1991, Earthscan, London, ISBN 1-85383-094-1 (pbk) 168 pp.

766 Przeworski, Adam
Democracy and the Market: Political and Economic Reforms in Eastern Europe and Latin America
This analysis of recent events in Eastern Europe and Latin America, focussing on transitions to democracy and market-oriented reforms.
1991, Cambridge University Press, Cambridge & New York, ISBNs 0-521-41225-0 (hbk) & 0-521-42335-X (pbk) xii + 210 pp.

767 Pylkkanen, Paavo
The Search for Meaning: The New Spirit in Science and Philosophy
This book is a highly original and stimulating collection of essays, based on the work of the distinguished theoretical physicist David Bohm. Its contributors, from various fields – including mathematics, physics, biophysics, biology, medicine, and psychology – discuss and explore concepts of meaning suitable for our own time.
1989, Aquarian Press, Wellingborough, Northants., ISBN 1-85274-061-2 (pbk) 318 pp.

768 Ralston, Kathleen & Chris Church
Working Greener: Sustainable Work Strategies for Organisations, Industry and Business

This book is an invaluable guide to employers and employees who want to improve and 'Green' their workplaces. It includes sections on: (1) moving people to action; (2) working in sustainable ways; (3) ways for future action. It looks at how to change both working environments and the work that is done. It is a very practical guide, full of examples of the changes that can be and are being, made throughout the world, and it provides ideas applicable to *any* workplace. The authors show how to use management techniques successfully to achieve the desired changes. The book is based on their practical experience.
1991, Green Print, London, ISBN 1-85425-064-7 (pbk) vi + 152 pp.

769 Rambler, Mitchell, Lynn Margulis & Rene Fester (Eds.)
Global Ecology: Towards a Science of the Biosphere
This interdisciplinary book presents various aspects of the global system and biosphere within a planetary perspective, using an expanded science of ecology. It shows that scientific tools may now be available for understanding the Earth and viewing it as an integrated system.
1989, Academic Press, London & New York, ISBN 0-12-576890-7 (?bk) xii + 204 pp.

770 Ramphal, Sridath
Our Country, The Planet
This book was produced for the June 1992 Earth Summit. Its objectives go beyond changing the climate of opinion to changing behaviour at individual levels and policies at government levels. He presents three scenarios: (1) muddling through along the current unsustainable path; (2) adopting a new 'imperialism', ordered by the powerful economic insitutions; (3) choosing a path of enlightened change and shared responsibility for our future, which would be a new commitment to the ideals which created the United Nations.
1992, Limetree Books, London, ISBN 0-413-45581-5 (hbk) xvii + 293 pp.

771 Redclift, Michael
Sustainable Development: Exploring the Contradictions
This book discusses the transformation of the environment in the course of development in both developing and developed countries. Its author argues that environmental problems need to be viewed in terms of the global economic system, and that the degradation of the environment is linked to economic and political structures. At the same time, biotechnology offers possibilities for recreating nature.
1987, Methuen, London & New York, ISBNs 0-416-90240-5 (hbk) & 0-416-90250-2 (pbk) vii + 221 pp.

772 Reed, David
Structural Adjustment and the Environment
This book details the environmental effects of 'structural adjustment programmes' (SAPs), imposed on the Third World by the World Bank and the International Monetary Fund. The results are far from encouraging, and it is an urgent priority to understand why. The author concludes that failure to appreciate environmental impacts will perpetuate trends that have already generated environmental crises.
1993, Earthscan, London, ISBN 1-85383-153-0 (pbk) 230 pp.

773 Rees, Arthur
The Pocket Green Book
This pocket reference guide to Green issues provides a useful introduction to environmental problems and environmental destruction.
1991, Zed Books, London, ISBNs 0-86232-998-1 (hbk) & 0-86232-999-X (pbk) vi + 154 pp.

774 Rees, Judith
Natural Resources: Allocation, Economics and Policy
This textbook on economic geography provides a fundamental perspective on all aspects of resource management in the modern world.
2nd. Ed. 1990, Routledge, London & New York, ISBNs 0-415-05104-5 & 0-415-05103-7 (pbk) 528 pp.; 1st. Ed. 1985.

775 Reich, Robert B.
The Work of Nations: Preparing Ourselves for 21st Century Capitalism
The author of this book argues that we are in an economic transformation, where there will no longer be national economies, products, technologies, corporations, or industries as we know them. Nations' principal assets will be their people's skills and insights.
1991, Knopf, New York, ISBN 0-394-58352-3 (hbk) xii + 331 pp.

776 Reiser, Oliver L.
Cosmic Humanism and World Unity
This book by a scientific philosopher develops his concepts of cosmic humanism, "a complete world view, a theory of knowledge, a cosmology, and a possible universal religion. It aims to discover and formulate some of the main principles that may help mankind to integrate the world ethically, aesthetically, and spiritually."
1975, An Interface Book, Gordon and Breach Science Publishers, New York & London, ISBNs 0-677-03870-4 (hbk) & 0-677-03875-5 (pbk) xii + 274 pp.

777 Renner, Michael
Economic Adjustments after the Cold War: Strategies for Conversion
This book presents the results of

research on the winding down of the military-industrial sectors in many countries, and the redeployment of human and material resources hitherto directed to military purposes.
1992, Dartmouth Publishing Co., Aldershot, Hants. & Brookfield, VT, USA, ISBN 1-85521-259-5 (hbk) xiii + 269 pp.

778 Rhys-Thomas, Deirdre
The Future is Now
This book reports interviews with a wide range of people on their concerns about the environment and the world for future generations.
1989, Macdonald Optima, London.

779 Ricklefs, Robert E.
Ecology
This book reflects the broadening interests of biological ecologists. It places equal emphasis on phenomena and theory, and includes a balanced set of examples from various habitats.
3rd. Ed. 1989, W. H. Freeman & Co., New York, ISBN 0-7167-2077-9 (hbk) 832 pp.

780 Riddell, Carol
The Findhorn Community: Creating a Human Identity for the 21st Century
This book describes the Findhorn Community in Scotland as it is today, and shows how it gradually evolved from its small beginning. It is one of several successful communities providing a living example of a new, sustainable way of living together, and continues to be an important part of a worldwide movement for spiritual transformation.
1990, Findhorn Press, Findhorn, Forres, Scotland, ISBN 0-905249-77-1 (pbk) xiv + 286 pp.

781 Rietbergen, Simon (Ed.)
The Earthscan Reader in Tropical Forestry
This collection of important papers on tropical forestry presents the issues, diagnoses, and solutions, together with the different approaches advocated for the preservation and management of forests.
1992, Earthscan, London, ISBN 1-85383-127-1 (pbk) 352 pp.

782 Rifkin, Jeremy
Biosphere Politics: A New Consciousness for a New Century
This book traces the history of the ecological crisis, predicts the end of the nation state and the multinational corporation, and suggests new ways of living in a global community.
2nd. Ed. 1992, HarperCollins, London & New York, ISBN 0-00-593164-? (pbk) 338 pp.; 1st. Ed. 1991, Crown Publishers, New York.

783 Rivers, Patrick
The Stolen Future: How to Rescue the Earth for Our Children

The author of this book argues the future that our children deserve is being stolen from them; humankind must recreate societies in which we can again get in touch with human roots. Only then can we replace our competitive, ruthless, selfish self-image with one of cooperation, trust, compassion, and tolerance. Exploitation must be replaced by renewal, to reconcile human and planetary needs. This awareness of Earth's needs will be the next vital leap in human evolution. The new values cannot come from existing power structures, and will have to emerge from the critical mass of ideas at the grassroots, as people everywhere unite against deepening crisis.
1988, Green Print, Basingstoke, Hants, ISBN 1-85425-005-1 (pbk) viii + 248 pp.

784 Robbins, Christopher
Poisoned Harvest: A Consumer's Guide to Pesticide Use and Abuse
This reference book details the main pesticides used in the UK, shows the foods and other products likely to contain them, and assesses their level of risk.
1991, Gollancz, London, ISBN 0-575-04797-6 (pbk) xv + 320 pp.

785 Roberts, Peter C.
Modelling Large Systems: *Limits to Growth* **Revisited**
The author of this book approaches global modelling through the eyes of a scientist. He describes and compares several global models and trends in global modelling. The last chapter discusses the scope for global models and the possible next generation of global models.
1978, Taylor & Francis, London, ISBN 0-85066-170-6 (pbk) x + 120 pp.

786 Robertson, James
The Sane Alternative: A Choice of Futures
This book takes a holistic look at the human situation and its future prospects. The first chapter presents five scenarios of possible alternative futures, of which the author prefers the Sane Humane Ecological (SHE) scenario. The book then develops the new economic approach and indicates the paradigm shift, both of which are implicit in the SHE scenario. Towards the end of the book, the process of transformation, needed to achieve something like the SHE scenario, is described, and an appropriate strategy for change is outlined. The book is remarkable, not only for its breadth of vision, but also for its wealth of annotated references to literature and projects about many different aspects of the human and planetary situation. It is clearly written, and combines important new theory with practical guidance.

2nd. Ed., 1983, self-published, ISBN 0-9505962-1-3 (pbk) 156 pp.; 1st. Ed. 1978

787 Robertson, James
The Future of Work
This book looks at possible scenarios for the future of work in our society. Its author concludes that the only one offering any real hope is the 'SHE' scenario (see No. 786), with the development of a whole new approach to work, 'ownwork', where an increasing number of people will be self-employed or work in small groups, often to meet the needs of comparatively localised communities. He explores new patterns of work will mean, and suggests that a new work ethic is already emerging.
1985, Gower Publishing Co., Aldershot, Hants., ISBNs 0-85117-259-8 (hbk) & 0-85117-260-1 (pbk) xiv + 220 pp.

788 Robertson, James
Future Wealth: A New Economics for the 21st Century
This book emphasises the need for a more fundamental revolution in the way in which the world thinks about economics and organises economic life. It advocates a new worldwide economic order for the 21st century, adapted and applied to real human and planetary needs; the key aspects of this new economy must be to enable people and conserve the Earth. It is necessary to make people fully aware that human beings can replace competitive greed between individuals and nations by a much better main motivation for economic life. Systematic wastefulness, destruction and pollution must be replaced by systematic conservation and monetary reform. Taxation should be shifted away from what people contribute to society towards what they take out of it. This book is important for those concerned with the present state of the world, and interested in economical and social reforms and a better future.
1990, Cassell, London & Bootstrap Press, New York, ISBNs 0-304-31930-9 (hbk) & 0-304-31933-3 (pbk) xii + 178 pp.

789 Rodda, Annabel (Ed.)
Women and the Environment
This book looks at the role of women in relation to the environment, the present degradation of the environment, and the importance of women as change agents on the Third World.
1991, Zed Books, London, ISBNs 0-86232-984-1 & 0-86232-985-X (pbk) viii + 180 pp.

790 Roe, M. F. H.
Ethical Issues
This book addresses the current state of society from an ethical perspective, and explores contemporary society and social

problems through a holistic framework.
1989, The Britannia Press, London, ISBN 1-872571-00-X (pbk) vii + 304 pp.

791 Rogers, Paul & Malcolm Dando
A Violent Peace: Global Security after the Cold War
This book attempts to think through and answer some questions about the new kind of world being constructed out of the old East-West confrontation. Its authors believe that a stable order can arise only through an intensive effort to address the problems of world poverty, environmental destruction, and the continuing militarisation of international relations. They foresee a new North-South axis of conflict, aggravated by divisions between rich and poor, and think that conflict will become much less, if the North finds imaginative and radical new means of cooperation to structure a new world order, rather than exerting economic and military coercion over the Third World.
1992, Brassey's (UK), London, New York & Washington, ISBN 0-08-036694-5 (hbk) v + 209 pp.

792 Rolt, L. T. C.
High Horse Riderless
This book is a remarkable early statement of many of the essential features of Green philosophy, including the needs for a self-sufficient society based on spiritual principles and values, and for a new sustainable economics promoting decentralisation and conservation of resources. Its approach is well summarised in its twenty conclusions.
2nd. Ed. 1988, Green Books, Bideford, Devon; 1st. Ed. 1947, Allen & Unwin, London.

793 Roszak, Theodore
Person/Planet: The Creative Disintegration of Industrial Society
This book is concerned with the meeting point between psychology and natural ecology. It suggests that the the needs of the person and of the planet have become one, and have begun to transform the central institutions of society, with eventual prospects of cultural renewal. *2nd. Ed. 1979, Gollancz, London, ISBN 0-575-02595-6 (hbk) xxx + 349 pp.; 1st. Ed. 1977, USA.*

794 Rotfeld, Adam Daniel (Ed.)
Global Security and the Rule of Law
This book addresses both traditional and new items on the post-cold-war security agenda. Topics include: arms control, the future security order in Europe, peaceful settlement of disputes, the human dimension of security, the production and proliferation of weapons, military expenditure, debt, disarmament, and international security.

1992, Oxford University Press, Oxford & New York, ISBN 0-19-829163-9 (hbk) 320 pp.

795 Rothkrug, Paul & Robert L. Olson (Eds.)
Mending the Earth: A World for Our Grandchildren
The book presents a wide variety of leading-edge ideas for a sustainable society, and suggests at least five areas needing change.
1990, North Atlantic Books, Berkeley, CA, USA, ISBN 1-55643-091-4 (pbk) 250 pp.

796 Rowlands, Ian H. & Malory Greene (Eds.)
Global Environmental Change and International Relations
This book investigates the implications for international relations of global environmental change, exploring both theoretical and policy problems. The variety of perspectives used shows the complexity of the relevant issues.
1992, Macmillan, London, ISBNs 0-333-55440-6 (hbk) & 0-333-55441-4 (pbk) xiii + 202 pp.

797 Russell, Peter
The Awakening Earth: Our Next Evolutionary Leap
This book explores human potential from a planetary viewpoint, outlines a vision of humanity evolved into a single social superorganism, with individual people as its cells, and presents evidence that this radical transformation might occur within a few decades. We now face a critical challenge in a period of accelerating change, to meet which we need a change of heart, a new model of ourselves, and a new sense of who we are. It is not yet known whether this trend to inner awakening is occurring fast enough.
2nd. Ed., 1991, Arkana Penguin, London & Penguin Books, New York, ISBN 0-14-019304-9 (pbk) xvii + 242 pp., 1st. Ed. 1982, Routledge & Kegan Paul, London.

798 Russell, Peter
The White Hole in Time: Our Future Evolution and the Meaning of Now
In this book, the author argues that something very significant is happening on our planet. We could be standing on the threshold of a movement towards which the universe has been moving since the beginning of time, an evolutionary climax of tremendous depth and unimaginably rapid transformation. The book attempts to portray our possible destiny in the near future, based on currently accepted theories and ideas. It aims to explore unconsidered possibilities and see where they lead, rather than prove a theory, to raise questions rather than giving answers. Its ideas are drawn from many subjects.
1992, The Aquarian Press/Thorsons,

London, & HarperCollins, New York, ISBN 1-85538-188-5 (pbk) ii + 214 pp.

799 Ryding, Sven-Olof
Environmental Management Handbook: The Holistic Approach from Problems to Strategies

This reference book offers an overall framework for developing effective strategies for the very wide range of problems that we face today. It aims to assess ongoing environmental activities from scientific, technical, and political viewpoints, by reviewing significant past and present knowledge and experience in a logical, structured way. It considers the assessment and control of the major environmental problems in both developed and developing countries. Special attention is paid to applying science and technology to prepare a global master plan on environment and development.
1992, IOS Press, Amsterdam, Oxford & Burke, VA, USA, ISBN 90-5199-062-6 (hbk) 797 pp.

800 Sachs, Ignacy & Dana Silk
Food and Energy: Strategies for Sustainable Development

This booklet summarises the research findings and their policy implications, derived from the UN University Food-Energy Nexus Programme (FEN). It has a bibliography of about 100 FEN publications.

1990, United Nations University Press, Tokyo, (pbk) 83 pp.

801 Sahtouris, Elisabet
Gaia: The Human Journey from Chaos to Cosmos

This book provides a fascinating vision of planet Earth and the place of humanity on the planet. It explains the Gaia Hypothesis and its wider implications, tracing the history of Earth from its formation, through the origin of life and its evolution through several successive stages, to the emergence and development of humankind. Today, the environment has become one planetwide economic empire, dominated and in effect ruled by the rich industrial countries, yet split into competing blocs. Therefore, the human species has reached a critical limit, but it has the potential to reorganise itself with incredible speed, using its modern information and communications networks. The book shows how, by understanding Gaia and the true roles of competition and cooperation in evolution, we may be able to solve our greatest problems: economic, ecological, political, and spiritual.
1989, Simon & Schuster, New York & London, ISBN 0-671-68002-1 (pbk) 252 pp.

802 Sakaiya, Taichi
The Knowledge-Value Revolution or a History of the Future

The author of this book argues that civilisation is disrupted by technology, the changing natural resource situation, and the global population explosion. In the new society, the main source of economic growth will be the creation of knowledge-value; the new economy will have a tremendous variety of diverse products, and an exponential growth in the number of people employed in creating knowledge-value.
2nd. Ed. 1991, Kodansha International, Tokyo & New York, ISBN 0-87011-948-7 (hbk) 256 pp.; 1st. (Japanese) Ed. 1985.

803 Saks, Mike (Ed.)
Alternative Medicine in Britain
This book is a collection of readings on alternative medicine, that fills a major gap in the social science literature on this subject. It brings together writings, relevant to social scientists, on the past, present and future of alternative medicine in Britain.
1992, Oxford University Press, Oxford & New York, ISBNs 0-19-827278-2 (hbk) & 0-19-827277 (pbk) x + 271 pp.

804 Salkie, Ralph
The Chomsky Update: Linguistics and Politics
This book aims to produce an accurate, clear, responsible guide to Noam Chomsky's work, which deals with crucial issues rather than details. It introduces his revolutionary work in linguistics, his sharp critique of US foreign policy, and his keen analysis of the role of intellectuals in modern society,
2nd. Ed. 1992, Routledge, London & New York, ISBN 0-415-08398-2 (pbk) xv + 238 pp.; 1st. Ed. 1990, Unwin Hyman, London.

805 Samuelson, Paul A. & William D. Nordhaus
Economics
This is the latest edition of a classic text-book of economics, covering basic concepts, macroeconomics, microeconomics, and other topics including: incomes, efficiency, equity, government, economic growth, and international trade. Each chapter or appendix to a chapter ends with a summary, concepts for review, and questions for discussion.
9th. Ed. 1989, McGraw-Hill, New York & London, ISBN 0-07-054786-6 (hbk) xii + 1013 pp.; 1st. Ed. 1948.

806 Sands, Philippe (Ed.)
Greening the Law: The Politics of International Environmental Law
This book of essays shows how law is only beginning to address environmental problems, and states the role of law in international environmental politics. It discusses how law prevents or brings about change, and the

legal system's inadequacies in handling environmental issues and current developments. It describes the changes needed to make international law really effective.
1993, Earthscan, London, ISBN 1-85383-151-4 (pbk) 240 pp.

807 Santos, Miguel A.
Managing Planet Earth: Perspectives on Population, Ecology, and the Law
This book provides a realistic and easily understood analysis of the scientific and legal dimensions of environmental stability.
1990, Greenwood Press, Westport, CT, USA, 0-89789-216-X (hbk) 192 pp.

808 Sargant, William
Battle for the Mind: A Physiology of Conversion and Brainwashing
This book describes the basic principles of brainwashing and other forms of psychological control of the mind.
3rd. Ed. 1963, Pan Books, London, (pbk) 218 pp.; 1st. Ed. 1957, Heinemann, London.

809 Sargent, Caroline & Stephen Bass (Eds.)
Plantation Politics: Forest Plantations in Development
This book examines the purpose of industrial plantations in the developing world, and explains how they can promote sustainable development. It describes how both simple and complex plantations can best contribute to sustainable development, and how their ownership, management, and cultivation present new challenges that will be much easier to meet by aplying the book's carefully researched guidelines.
1992, Earthscan, London, ISBN 1-05383-113-1 (pbk) 192 pp.

810 Sarre, Philippe, Paul Smith & Eleanor Morris
One World for One Earth: Saving the Environment
This general course book on the environmental problems facing the world advocates a combination of local action and international cooperation to achieve sustainable development.
1991, Earthscan, London, ISBN 1-85383-119-0 (pbk) iv + 188 pp.

811 Satin, Mark
New Age Politics: Healing Self and Society
This book describes the new form of politics that has been emerging since the early 1970s from feminist, human potential, spiritual, simple living, nonviolent action, environmental, and appropriate technology movements, and out of the ideas of people sharing these concerns.
1978, Whitecap Books, West Vancouver, B.C., Canada, ISBN 0-920422-01-2 (pbk) 240 pp.

180 Satin, Mark

812 Satin, Mark
New Options for America: The Second American Experiment Has Begun
This book seeks a decentralist, ecological, and globally responsible society beyond economic growth, the welfare state, and policing international conflicts, and covers a wide variety of topics.
1991, California State University Press, Fresno, CA, USA, ISBN 0-8093-1794-X (pbk) 251 pp.

813 Schatter, Andre
Free Market Economics: A Critical Appraisal
This book provides a clear, accessible account and critique of the free market system's potential to solve social and economic problems.
2nd. Ed. 1990, Blackwell, Oxford & Cambridge, MA, USA, ISBNs 1-55876-074-2 (hbk) & 1-55876-066-1 (pbk) xiv + 171 pp.; 1st. Ed. 1985, St. Martin's Press, New York.

814 Scherer, Donald (Ed.)
Upstream/Downstream: Issues in Environmental Ethics
This book is a stimulating and satisfying collection of essays by philosophers on environmental ethics.
1990, Temple University Press, Philadelphia, PA, USA, ISBN 0-87722-747-0 (hbk) 228 pp.

815 Schmidheiny, Stephan with the Business Council for Sustainable Development
Changing Course: A Global Business Perspective on Development and the Environment
This book provides an extensive analysis of how the world business community can adapt and contribute to the crucial goal of sustainable development worldwide, which combines the objectives of environmental protection and economic growth. Its chief author is a Swiss industrialist, who is Chairman of the Business Council for Sustainable Development (No. 1453), and was Principal Adviser for Business and Industry to the June 1992 UNCED Conference. The book starts with the Council's declaration, expressing its commitment to sustainable development, to meet the needs of the present without compromising the welfare of future generations. It affirms that the quality of present and future life rests on meeting basic human needs without destroying the environment on which all life depends. To achieve this goal, new forms of cooperation between government, business, and society are required, and the Council's members commit themselves to promoting this new partnership in changing course towards our common future.

The expertise of more than 50 leaders of multinational corporations, based in various countries, was gathered to help prepare analyses and case studies, showing how companies and governments can make ecological imperatives part of the market forces that govern production, investment, and trade. Issues addressed include: energy efficiency and sustainable energy, reduction of pollution, renewable agriculture and forestry, innovation and technology cooperation, managing corporate change, trade, and financing sustainable development. There are 38 case studies, describing how firms, including some very well-known multinationals, have modified production and products, and entered new markets. These examples of successful steps towards sustainable development, from many parts of the world, show existing best practices that reinforce the case presented here that industry need not be incompatible with the environment. They provide evidence that an increasing number of business and government leaders are finding ways to produce *simultaneously* both economic development and a cleaner, safer environment. The book presents what was perhaps the most hopeful initiative during 1992 for the improvement of the human and planetary situation, although tremendous efforts will still be needed.
1992, MIT Press, Cambridge, MA, USA & London, ISBNs 0-262-19318 (hbk) & 0-262-69153-1 (pbk) xxv + 374pp.

816 Schneider, Bertrand
The Barefoot Revolution: A Report to the Club of Rome
This book calls for a redirection of the economic strategy of the last 20 years. It identifies and measures the impact of an alternative approach, using small development projects run by non-governmental organisations in Latin America, Africa and Asia. It aims to offer a valid development alternative for the Third World and provide some practical answers to the urgent needs of its rural people.
1989, Intermediate Technology Publications, London, ISBN 1-85339-037-2 (pbk) 296 pp.

817 Schneider, Stephen H.
Global Warming: Are We Entering the Greenhouse Century?
This book, by a leading scientific researcher on climate change, discusses our common future, and how we might deal with the prospect of rapid climate change caused by human economic and population growth.
2nd. Ed. 1990, Lutterworth Press, Cambridge, ISBN 0-7188-2185-6 (pbk) xv + 343 pp.; 1st. Ed. 1989, Sierra Club Books, San Francisco, CA, USA.

818 Schumacher, E. F.
Small Is Beautiful: A Study of Economics as if People Mattered
This book looks at the world's economic structure in a revolutionary way. It argues that man's current pursuit of profit and progress has resulted in gross economic inefficiency, environmental pollution, and inhumane working conditions. It challenges excessive specialisation, and proposes a system of 'intermediate technology', based on smaller working units, communal ownership, and regional workplaces using local labour and resources. Emphasising the person, not the product, it points the way to a world where capital serves man instead of man remaining a slave to capital.
2nd. Ed. 1974, with later reprints , Abacus, Sphere Books, London, ISBN 0-349-13172-5 (pbk) 255 pp.; 1st. Ed. 1973, Blond and Briggs, London.

819 Schumacher, E. F.
Guide for the Perplexed
The author of this book states that philosphy's task is to provide a map of knowledge devoting proper attention to the most important aspects of life, restores science to its true territory, and presents his own Christian philosophy of life.
2nd. Ed. 1978, Cape, London, ISBN 0-224-01496-X (hbk) 166 pp.; 1st. Ed. 1977.

820 Schumacher, E. F.
Good Work
The author of this book threee basic purposes of human work is threefold, and offers important and thought provoking ideas about work, which point the way to humanity's physical and mental liberation. He shows that small is not only beautiful but essential, if our working lives are to be satisfactory, meaningful and creative.
2nd. Ed. 1980, Abacus, Sphere Books, London, ISBN 0-349-13133-3 (pbk) xi + 148 pp.; 1st. Ed. 1979, Cape, London.

821 Schurr, Sam H.
Energy, Economic Growth, and the Environment
This book covers the conflict between two of the USA's objectives: (1) providing energy to meet the needs of future economic growth; (2) protecting the quality of the environment.
1992, Johns Hopkins University Press, Baltimore, MD, USA & London, ISBNs 0-8018-1414-6 (hbk) & 0-1553-3 (pbk) viii + 232 pp.

822 Schwartz, Peter
The Art of the Long View
In this book, the author discusses the lessons of his varied and valuable experience, including starting the Global Business Network (No. 1392) in 1987. He describes the writing of scenarios to take a long view in a world of great

uncertainty, gives details of three scenarios of the world of 2005, and lists eight steps in developing a scenario.

1991, Doubleday Currency, New York & London, ISBN 0-385-26731-2 (hbk) xi + 258 pp.

823 Schwarz, Walter & Dorothy Schwarz
Breaking Through: The Theory and Practice of Wholistic Living
This book discusses how we can break away from negative, destructive attitudes that ruin our environment and threaten our planet, and break through to positive, constructive approaches that could transform our lives and society. It presents a 'wholistic' philosophy and approach to life that makes this new way of living possible, and looks at many aspects of the 'wholisitc movement'.

1987, Green Books, Hartland, Bideford, Devon, ISBN 1-870098-05-6 (pbk) xvi + 272 pp.

824 Seabrook, Jeremy
The Race for Riches: The Human Cost of Wealth
In this book, the author exposes what he considers to be the myths of modern economics that underpin our society, including the myth of progress which offers people the illusion of relief from poverty.

1988, Green Print, London, ISBN 1-85425-003-5 (pbk) vi + 182 pp.

825 Seabrook, Jeremy
The Myth of the Market: Promises and Illusions
The author of this book argues that no more efficient mechanism than the market economy could have been devised to bring our planet to the edge of ruin, yet most people in the West believe that it holds promise for the future. If a sustainable economy and society are to be achieved, he considers that Greens in the West must passionately and energetically resist the growing dictatorship of the market.

1990, Green Books, Hartland, Bideford, Devon, ISBN 1-870098-36-6 (pbk) 189 pp.

826 Seager, Joni (Ed.)
The State of the Earth: An Atlas of Environmental Concern
This atlas examines the quality of human life and the health of the planet in 37 coloured maps of the world, for each of which a textual commentary is provided. It presents a wide range of world problems and environmental issues, reflects worldwide public concern for the environment, and considers the need for more effective international organisations and agreements. It is concise, up-to-date, and full of facts on a broad range of topics. Its visual approach provides a valuable basis for formulating an overall global action plan.

1990, Unwin Hyman, London,

ISBNs 0-04-440693-2 (hbk) & 2nd. Ed. 1988, Dorling Kindersley,
0-04-440692-4 (pbk) 127 pp.   *London, ISBN 0-86318-364-6 (pbk)*
                                      192 pp, 1st. Ed. 1987.

827 Sen, Amartya
Poverty and Famines: An Essay   **829 Seymour, John & Herbert**
on Entitlement and Deprivation   **Girardet**
This book is analyses the causes of   **Far from Paradise: The Story of**
hunger and starvation in general,   **Human   Impact   on   the**
and famines in particular. After a   **Environment**
very general discussion of the   This book discusses the impor-
problem of starvation, it examines   tance of the land as the vital
in detail four specific famines,   resource to supply us with food. It
two in Africa and two on the   sees the key to sustainable agri-
Indian subcontinent. Finally, it   culture as mixed farming, in
shows how people's lack of   which animals are kept on the
entitlement leads to their depriva-   land in rotation with the growing
tion of food.   of crops, so that chemical fertil-
1991, Oxford University Press,   isers are no longer needed. It
Oxford & New York, ISBN   views the continuing use of chem-
0-19-828463-2 (hbk) xi + 257 pp.   icals on the land as leading to a
                                       major crisis. Technological fixes
                                       will not solve any of the problems
828 Seymour, John & Herbert   of the land; instead, we should be
Girardet   adopting biologicaly sound ways
Blueprint for a Green Planet: of using the land, which could
How You Can Take Practical   sustain us all indefinitely.
Action Today to Fight Pollution   *2nd. Ed. 1988, Green Print,*
This book is a well-presented,   *Basingstoke,   Hants.,   ISBN*
well-illustrated guide for positive   *1-85425-006-X (pbk) 216 pp.; 1st.*
and practical action in various   *Ed. 1986, BBC Books, London.*
aspects of everyday life. It   **830 Shahidullah, Shahid M.**
attempts to provide a handbook   **Capacity-Building in Science and**
of positive measures and real   **Technology in the Third World**
alternatives, both in the home   The author of this book discusses
and beyond. It covers in detail   four major aspects of science and
many specific aspects of environ-   tecnology in the Third World. He
mental problems and sustainable   considers that their expansion has
living on the land and in the cities. not been very productive or
It discusses various practical   socially relevant in most develop-
approaches to general environ-   ing countries, even though their
mental action, including positive   political and scientific leaders
actions that *we* can take.

have generally strongly recognised their key role in development progress.

1991, Westview Press, Boulder, CO, USA, ISBN 0-8133-7948-2 (pbk) 258 pp.

831 Sheldrake, Rupert
A New Science of Life: The Hypothesis of Formative Causation
In this book the author presents his view of the possible outlines of a broader science of life.
2nd. Ed. 1985, Anthony Blond, London, ISBN 0-85634-198-3 (pbk) 277 pp.; 1st. Ed. 1981, Blond & Briggs, London.

832 Sheldrake, Rupert
The Presence of the Past: Morphic Resonance and the Habits of Nature
In this book, the author extends the conceptual framework of his scientific world view, by introducing concepts of 'morphic resonance' and 'morphic fields' that are evolving self-organising wholes.
1988, Collins, London, ISBN 0-00-217785-4 (hbk) xx + 391 pp.

833 Sheldrake, Rupert
The Rebirth of Nature: The Greening of Science and God
This book traces the mythological and historical roots of two different attitudes to nature that we inherit. Our ancestors assumed that the world was 'alive', but the official mechanistic scientific view of nature has dominated the last few centuries and brought unprecedented abuse of the living world. The author explores how the implications of new developments in science question the validity of the mechanistic viewpoint. He indicates a new understanding of nature, where traditional wisdom, personal experience, and scientific insight, can enrich each other. He offers us the opportunity to re-establish conscious connections with the living world around us.
1990, Rider, London, ISBNs 0-7126-3775-3 (hbk) & 0-7126-4650-7 (pbk) xv + 215 pp.

834 Sherlock, Harley
Cities Are Good for Us: The Case for Close-Knit Communities, Local Shops and Public Transport
This book presents a vision of high-density cities, where people can have a high quality of life, keep their vitality, lose their traffic congestion, and again become places where everyone has the necessities and pleasures of life nearby.
2nd. Ed. 1991, Paladin, Harper-Collins, London, ISBNs 0-586-09092-4 (hbk) & 0-586-09054-1 (pbk) 255 pp.; 1st. Ed. 1990, with the subtitle "The Case For High Densities, Friendly Streets,...".

835 Shiva, Vandana
Staying Alive: Women, Ecology and Development
This book explores the role of women in relation to nature, and advocates the rediscovery of the 'feminine principle' in human interaction with the natural world, and as a way of viewing the world.
1988, Zed Books, London & Atlantic Highlands, NJ, USA, ISBNs 0-86232-822-5 (hbk) & 0-86232-823-3 (pbk) xx + 224 pp.

836 Shiva, Vandana
The Violence of the Green Revolution: Ecological Degradation and Political Conflict
In this book, the author considers the impact of the first Green Revolution in agriculture on India, and shows its catastrophic effects.
1991, Zed Books, London & Atlantic Highlands, NJ, USA, ISBNs 0-86232-964-7 (hbk) & 0-86232-965-5 (pbk) 192 pp.

837 Shiva, Vandana et al. (Eds.)
Biodiversity: Social & Ecological Perspectives
The contributors to this collection of essays argue that the roots of the crisis of genetic erosion and species extinction lie in the international economic system.
1991, Zed Books, London & Atlantic Highlands, NJ, USA, ISBNs 1-85649-053-X (hbk) & 1-85649-054-8 (pbk) 123 pp.

838 Shoard, Marian
The Theft of the Countryside
This influential book alerted the British public to how much of their countryside was being destroyed. It studies this destruction in terms of the agricultural revolution, its impact on the countryside, the failure of control methods, and the economic impact of farming and tourism. It presents several suggestions for halting this destruction.
1980, Temple Smith, London, ISBNs 0-85117-200-8 (hbk) & 0-85117-201-6 (pbk) 272 pp.

839 Silver, Cheryl Simon Silver & Ruth S. DeFries
One Earth, One Future: Our Changing Global Environment
This book provides a brief, non-technical presentation of the current state of scientific knowledge about the changes occurring in the global environment. It is based on the Forum on Global Change and Our Common Future, held in Washington in May 1989.
1990, National Academy Press, Washington, ISBN 0-309-04141-4 (hbk) xiii + 196 pp.

840 Silver, Debbie & Bernadette Vallely
The Young Person's Guide to Saving the Planet
This book for young people has entries arranged in alphabetical order, each indicating what action

to take. The book is especially useful, because the next generation may save our planet.
1990, Virago, London, ISBN 1-85381-148-3 (pbk) xxii + 120 pp.

841 Silvertown, Jonathon & Philip Sarre (Eds.)
Environment and Society
This book explains some of the fundamental environmental processes, and shows how humanity has transformed the surface of the Earth. Its papers cover: general problems, the biosphere, the impact of human societies, natural hazards, and how attitudes are changing.
1990, Hodder and Stoughton, London, ISBN 0-340-53359-5 (hbk) 280 pp.

842 Simon, Julian L.
The Ultimate Resoure
The author of this book argues that there is no scarcity of raw materials, nor will there be one in the foreseeable future; this is evidenced by long-term decreasing trends in the prices of material resources. He perceives *none* of the strict 'limits to growth' taken so seriously by most environmentalists and global modellers.
1981, Princeton University Press, Princeton, NJ, USA & Martin Robertson, Oxford, ISBN 0-85220-440-0 (hbk) x + 415 pp.

843 Simon, Julian L. & Herman Kahn
The Resourceful Earth: A Response to Global 2000
The authors of this book come to almost diametrically opposite conclusions to those of *Global 2000* (No. 50). They think that, if present trends continue, the world in 2000 will be less crowded though more populated, less polluted, more stable ecologically, and less vulnerable to disruption of resource supplies than the world in 1984.
1984, Blackwell, Oxford & New York, ISBN 0-631-13467-0 (hbk) viii + 585 pp.

844 Simpson, Struan
The Times Guide to the Environment: A Comprehensive Handbook to Green Issues
This book surveys a wide range of environmental problems, including: resource depletion, global warming, the threat to the ozone layer, acid rain, other forms of pollution, radiation threats, population, and economic and industrial growth. It proposes a wide range of possible solutions, including: sustainable economic growth on a global scale, international agreement, industrial cooperation, resource conservation, waste management, sanitation, alternative energy, and scientific research. It provides informative and objective viewpoints about various aspects of the state and

future prospects of our planet, and contains many tables with useful statistical information. It includes a description of United Nations and other international organisations, together with useful lists of intergovernmental agencies, conservation and environmental organisations, and contact addresses for further information.
1990, Times Books, London, ISBN 0-7230-0347-5 (pbk) 224 pp.

845 Skolimowski, Henryk
Living Philosophy: Eco-Philosophy as a Tree of Life
The author of this book presents his eco-cosmology as a foundation for a new world view and civilisation, and as a framework for our action. Ecological individuals understand the connection between all things, feel responsible for future generations, attempt to cultivate frugality, and have reverence for life. Eco-ethics extends and articulates this reverence, valuing the pursuit of wisdom as oppsed to the pursuit of information. Many topics are discussed.
1992, Arkana Penguin, London & Viking Penguin, New York, ISBN 0-14-019308-1 (pbk) 255 pp.

846 Slaughter, Richard A.
Future Concepts and Powerful Ideas
This resource pack is designed

"for anyone wishing to understand or utilize some of the conceptual riches of the futures field". About half of the pack presents summations of key points in large type, to be used with overhead projectors. It has brief sections on various basic ideas of futures studies, and discusses a wide range of powerful ideas on many subjects. The author's agenda for the 21st century includes: (1) repairing the damage: (2) creating sustainable economies; (3) releasing human potential; (4) creating institutions and processes of foresight; (5) finding new purposes and meanings; (6) reinventing culture.
1991, Australian Future Studies Centre, Kew, Victoria, Australia, (hbk) 186 pp.

847 Smith, Cyril W. & Simon Best
Electromagnetic Man: Health & Hazard in the Electrical Environment
This book presents new evidence to show that human beings can be sensitive to minute electromagnetic fields, generated by a wide variety of electrical systems. It is becoming increasingly clear that there are biological effects, some of them potentially harmful, from types of radiation which nearly everyone has hitherto considered insignificant.
1989, Dent, London, ISBN 0-460-04698-5 (hbk) vii + 344 pp.

848 Smith, Hedrick
The New Russians
This book provides a detailed and comprehensive account of the changes in Russia during the Gorbachev era, leading eventually to the attempted Communist coup and the break-up of the USSR.
2nd. Ed., 1991, Vintage, London, ISBN 0-09-986210-7 (pbk) xxxiii + 734 pp.; 1st. Ed. 1990, Hutchinson, London.

849 Smith, James Alan
The Idea Brokers: Think Tanks and the Rise of the New Policy Elite
This book presents a detailed history of 'think tanks', and discusses the tremendous growth in their number, size, and influence since 1945. There are over 1,200 in the USA alone, of which about 30 are described, and thousands of university-based research institutes. The author proposes a classification of the people working in them, and draws conclusions about their effects on US public policy making.
1991, Free Press, New York, ISBN 0-02-929551-3 (hbk) 313 pp.

850 Smith, Keith
Environmental Hazards: Assessing Risk and Reducing Disaster
This book covers all types of rapidly occurring major events which directly threaten human life on a community scale; they include seismic, mass movement, atmospheric, flooding, and technological hazards. It integrates insights from both earth sciences and social sciences, and uses a global range of case studies, to provide a comprehensive introduction to these environmental hazards.
1991, Routledge, London & New York, ISBNs 0-415-01216-3 (hbk) & 0-415-01217-1 (pbk) 352 pp.

851 Smith, Paul M. & Kiki Warr (Eds.)
Global Environmental Issues
This book analyses global environmental issues, and advocates building an integrated set of policies, based on ideas of sustainable development, for all major environmental issues. It explains the scientific problems in predicting the content of change, and the political problems involved in negotiating international responses.
1991, Hodder and Stoughton, London, ISBN 0-340-53362-5 (pbk) vi + 294 pp.

852 Smith, Zachary A.
The Environmental Policy Paradox
This book studies government policy in the USA, and shows how environmental policy is formed. Examples of policies are discussed for: air, water, energy, land use, agriculture, waste disposal, and general environmental problems. The author concludes

that we often understand the best short-term and long-term solutions to these problems, but that their implementation is often not done or too late to be effective.

1992, Prentice-Hall, Englewood Cliffs, NJ, USA & London, ISBN 0-13-282344-6 (pbk) xix + 266 pp.

853 Smuts, Jan Christiaan
Holism and Evolution
This book formulates the original view of 'holism', that emerged from new thinking and convergence in science and philosophy.

Latest (undated) reprint of 1st. Ed., Greenwood Press, Westport, CT, USA, (hbk) 362 pp.; 1st. Ed. 1926, Macmillan, London.

854 Smyke, Patricia (Prepared by)
Women & Health
This book asks some key questions about different aspects of the inadequate health of many women worldwide. It presents many facts and case studies, and shows what is achieved as women mobilise for change.

1991, Zed Books, London & Atlantic Highlands, NJ, USA, ISBNs 0-86232-982-5 (hbk) & 0-86232-983-3 (pbk) ix + 182 pp.

855 Smyth, Angela & Caroline Wheater
Here's Health: The Green Guide
This book discusses a wide range of environmental problems, and

indicates a variety of actions that we can take to alleviate them.

1990, Argus Books, Hemel Hempstead, Herts., 1-85486-005-4 (pbk) 155 pp.

856 Snacre, Francis
The Nature of Moral Thinking
This book surveys the foundations of ethics, and considers some philosophical questions about our thinking. It describes and discusses the main types of meta-ethical and normative ethical theories.

1992, Routledge, London & New York, ISBNs 0-415-04708-0 (hbk) & 0-415-04709-9 (pbk) vi + 187 pp.

857 Snow, Chet
Dreams of the Future: A Preview of the Futures that Lie Before Us
This book reveals the visions of the future obtained by hundreds of people, when they were regressed by hypnosis into their 'future lives', at specific periods. Although a few of these regressions were into the late 1990s or the 21st century, most of them were to specific periods between 2100 and 2500. The resulting visions varied between those of the 'Piecemeal' and 'Optimistic' scenarios of *Into the 21st Century* (No. 144), and indicated four basic types of community for people living in the future; it was not clear how far those four types would coexist, or how far they would be alternative futures. The

most obvious interpretation is that they are projections of human hopes and fears for the future, but one can speculate that they might also include some psychic 'pre-memories' of those lives. *2nd. Ed. 1991, The Aquarian Press, Wellingborough, Northants., ISBN 1-85538-027-7 (pbk) xix + 346 pp.; 1st. Ed. 1989, McGraw-Hill, New York.*

858 Solo, Robert
The Philosophy of Science, and Economics
This book is a collection of miscellaneous thoughts on philosophy of science, falsification, paradigms, mathematics, economics, social sciences, and value judgements, in relation to each other.
1991, Macmillan, London & New York, ISBN 0-333-54709-8 (hbk) vi + 138 pp.

859 The South Commission
The Challenge to the South: The Report of the South Commission
This report is by the South Commission, a high-level, independent group of people with different backgrounds and persuasions, from the developing countries (the 'South'), whose Chairman was Julius K. Nyerere of Tanzania. It considers the struggle to overcome poverty and improve the quality of life in the South. It examines the South's development record, assesses its

achievements, analyses its findings, and suggests directions for reform. Despite the difficult conditions faced by the South, the report is hopeful, and argues strongly for reliant, people-centred development strategies.
1990, Oxford University Press, Oxford & New York, ISBN 0-19-877311-0 (pbk) xv + 325 pp.

860 Spellerburg, Ian F.
Monitoring Ecological Change
This book provides a practical, thought provoking introduction to effective biological and ecological monitoring programmes, to detect changes in living communities and understand what influences them.
1991, Cambridge University Press, Cambridge & New York, ISBN 0-521-36662-3 (hbk) & 0-521-42407-0 (pbk) vi + 334 pp.

861 Spiecker, Ben & Roger Straughan (Eds.)
Philosophical Issues in Moral Education and Development
This book addresses basic questions in moral education from an international perspective. The authors find that these issues indicate the complexity of moral education and development. They question whether easy answers and straightforward solutions are available for teachers and educationalists working in this area.

1988, Open University Press, Buckingham, ISBN 0-335-15851-X (hbk) & 0-335-15850-1 (pbk) vii + 126 pp.

862 Sprent, Peter
Taking Risks: The Science of Uncertainty
This book provides a stimulating survey of the whole field of uncertainty. It explores subjective attitudes to risk, and describes practical techniques applicable to a wide range of real-life problems.
1988, Penguin Books, London & Viking Penguin, New York, ISBN 0-14-022777-6 (pbk) viii + 264 pp.

863 Sproull, Lee & Sara Kiesler
Connections: New Ways of Working in the Networked Organization
This book discusses some possible future organisational and social impacts and implications of computer-based communication, which may prove to be very significant.
1991, MIT Press, Cambridge, MA & London, ISBN 0-262-19306-X (hbk) 296 pp.

864 Starke, Linda
Signs of Hope: Working toward Our Common Future
This book records progress in the implementation of the recommendations of *Our Common Future* (No. 988). It examines positive initiatives being taken throughout the world by governments, industry, scientists, non-governmental organisations, the media, and young people. It points to areas where progress has been made, and examines the priorities for further change during the 1990s. It is not meant to be a comprehensive review of developments since the publication of *Our Common Future* in 1987, but it provides a 'snapshot' of some of the progress made in the directions that it advocated. By emphasising positive aspects, it complements the valuable publications of the United Nations Development and Environment Programmes, UNICEF, the International Institute for Environment and Development, the World Resources Institute, Worldwatch, and other organisations, which document the increasing pressure on the world resources base and the worsening standards of living of most people in the world. It describes the work of the Centre for Our Common Future and its Working Partners throughout the world.
1990, Oxford University Press, Oxford & New York, ISBNs 0-19-212993-7 (hbk) & 0-19-285225-6 (pbk) xv + 192 pp.

865 Stenson, Kevin & David Cowell (Eds.)
The Politics of Crime Control
This book contains contributions to the current debate on the prevention and control of crime, by criminologists of different schools

of thought and political outlook.
*1991, Sage Publications, London &
Newbury Park, CA, USA, ISBNs
0-8039-8341-7 (hbk) &
0-8039-8342-5 (pbk) xiv + pp.*

**866 Stern, Paul C., Oran R.
Young & Daniel Druckman
(Eds.)**
**Global Environmental Change:
Understanding the Human
Dimensions**
This book examines what is
known about the human sources,
consequences, and responses of
the present hydrological, climatic, and biological global changes.
It discusses their driving forces,
and recommends a comprehensive US national research
programme.
*1992, National Academy Press,
Washington, DC, USA, ISBN
0-309-04494-4 (hbk)308 pp.*

**867 Sternberg, Robert J. & John
Kalligian, Jr.**
Competence Considered
This book presents recent
research on real and perceived
competence from childhood
through adulthood. Its contributors apply their insights to questions of intelligence, and show
how our perceptions of our own
abilities affect our performance
and mental health.
*1990, Yale University Press, New
Haven, CT, USA & LOndon, ISBNs
0-300-04567-0 (hbk) &
0-300-05228-6 (pbk) xvii + 420 pp.*

**868 Stockholm International
Peace Research Institute**
**SIPRI International Yearbook
1993: World Armaments and
Disarmament**
This book contains up-to-date
information about weapons, military technology, military expenditure, the arms trade, armed
conflicts, developments in arms
control, and treaty negotiations.
*1993, Oxford University Press,
Oxford & New York, ISBN
0-19-829166-0 (ISSN 0953-0282)
(hbk) about 600 pp.*

869 Stokke, Olav (Ed.)
**Sustainable Development: A
Special issue of the** *European Journal of Development Research*
This selection of papers from a
1990 conference in Oslo, covers
the whole range of issues arising
in debates on sustainable
development.
*1991, Frank Cass, London, ISBN
0-7146-3449-2 (pbk).*

870 Storr, Anthony
**Human Destructiveness: The
Roots of Genocide and Human
Cruelty**
This book explores why the
human species can be capable of
the most appalling cruelty, and
the origins of its capacity for
destructiveness. Its author
attempts to throw light on genocide, racial conflict, and other
large-scale manifestations of violence, and cautions against easy

extrapolations from individual behaviour to the behaviour of groups and nations. His views are profoundly anti-Utopian, and he sees little hope for changes in human nature, beyond some slight modifications of its nastier personality traits in the light of increased understanding.

2nd. Ed. 1991, Routledge, London & New York, ISBNs 0-415-07178-X (hbk) & 0-415-07170-4 (pbk) & Grove Weidenfeld, New York; 1st. Ed. Heinemann & Chatto & Windus London & Basic Books, New York.

871   Streuffert,   Siegfried   & Robert W. Swezet
Complexity,   Managers   and Organizations
This book attempts to contribute to resolving the mystery of the highly successful organisation, and to develop a more universally applicable approach, that can point towards optimal functioning of organisation and management across a wide range of situations.

1988, Academic Press, New York & London, ISBN 0-12-673370-8 (hbk) xii + 260 pp.

872 Sumner, David
Radiation Risks: An Evaluation
This book is a balanced guide to the possible health hazards of low-level radiations from various radioactive substances.

1987, The Tarragon Press, Glasgow, ISBN 1-870781-00-7 (pbk) vi + 198 pp.

873 Sutela, Pekka
Economic Thought and Economic Reform in the Soviet Union
This book presents a detailed analysis of recent and contemporary Soviet economic thought and theory.

1991, Cambridge University Press, Cambridge & New York, ISBNs 0-521-38020-0 (hbk) & 0-521-38102-X (pbk) x + 197 pp.

874 Suter, Keith
Antarctica: Private Property or Public Heritage?
The author of this book explains why Antarctica is under threat, and why its resources are so important for humankind. He describes political manoeuvrings that might affect its future, and provides the facts for those exerting pressure before decisions are made in secret.

1991, Zed Books, London, ISBNs 0-86232-846-2   (hbk)   & 0-86232-847-0 (pbk) xi + 209 pp.

875 Suter, Keith
Global Change: Armageddon and the New World Order
The author of this book sees the collapse of Communism in the USSR and Eastern Europe, and other similar dramatic events, as part of a wider global pattern in which power is devolving from the centre to the edges. Political changes of this magnitude can bring great harm, but could also

bring benefits. The author describes four key aspects of the period in which we now live: (1) the new global *reality* ('the new world order'); (2) the new global *governance* ('peop;e power' organisations, the United Nations, and transnational corporations); the new global *warfare* (and its threat to world peace); (4) the new global *agenda* (the ideas and values needed to implement any concerted national strategy).

1992 Albatross Books, Sutherland, NSW, Australia & Claremont, CA, USA, & Lion Publishing, Oxford, ISBNs 0-86760-135-3 (AB) & 0-6459-2191-4 (LP) (pbk) 360 pp.

876 Suzuki, David
Inventing the Future: Reflections on Science, Technology, and Nature
In this collection of critical writings, the author presents several challenges to other scientists, big business and government alike. Can science solve today's global problems, or will it be misused? The author argues that scientific insights into the world around us present a fractured mosaic rather than a seamless whole. We must recognise the profound limits to science, if we are to minimise the destructive consequences of the impacts of scientific discoveries. He tackles several such issues directly, and argues that we must re-invent a future that is free of our present obsession with consumption and profit. We must restructure our priorities and our world view, through a profound shift in our attitudes to that natural world and the spiritual values that we place on other organisms.

1992, Adamantine Press, London, ISBN 0-7449-0047-6 (pbk) (ISSN 0954-6103) viii + 248 pp.

877 Suzuki, David & Peter Knudson
Genethics: The Ethics of Engineering Life
This book describes genetic structures and how man can now intervene to alter them. The authors examine some of the most important areas and applications of genetic engineering, and present a set of nine ethical principles to guide the course of genetic engineering.

2nd. Ed. 1989, Unwin Hyman, London, ISBN 0-04-440623-1 (pbk) xii + 372 pp.; 1st. Ed. 1988, Stoddart Publishing Co., Canada.

878 Swann, Dennis (Ed.)
The Single European Market and Beyond: A Study of the Wider Implications of the Single European Act
This book examines what the Single European Market actually involves, and considers the economic, monetary, social, external, and political dimensions of European unification.

1992, Routledge, London & New

*York, ISBNs 0-415-06160-1 (hbk) &
0-415-06161-X (pbk) xiv + 299 pp.*

**879 Swanson, Timothy & Edward
B. Barbier (Eds.)**
**Economics for the Wilds: Wild-
life, Wildlands, Diversity and
Development**
The authors of this book argue
strongly and in careful detail that
an economics, that properly
values the resources of the wild,
offers the best long-term solution
for the future of these resources.
Most of the world's wild lands
have always been used by local
societies that have managed their
resources sustainably; we need
the continuing participation of
these communities, and an adequate reward to them for their
management. Chapters 1-4 look
at the complexity and global
nature of the issues, and appropriate policies for their conservation   and   sustainable
management. Chapters 5-9 examine specific forms of use of wild
species and habitats. Chapter 10
argues the need for a comprehensive utilisation strategy for wildlife resources, to ensure their
continued existence and continuing benefits from them.
*1992, Earthscan, London, ISBN
1-85383-124-7 (pbk) xii + 226 pp.*

**880 Swimme, Brian & Thomas
Berry**
The Universe Story: From the

**Primordial Flaming Forth to the
Ecozoic era: A Celebration of the
Unfolding of the Cosmos**
This book "celebrates the total
community of existence as it
unites science and the humanities
through a profound and prophetic modern myth." Its authors
combine the findings of contemporary sciences with the human
search for meaning in their reflection on this interpretation of the
universe, aiming for a balanced
account of our common story.
*1992,   HarperSanFrancisco,   San
Francisco, ISBN 0-06-250826-1 (hbk)
x + 305 pp.*

**881 Tait, Joyce, Andrew Lane &
Susan Carr**
**Practical   Conservation: Site
Assessment and Management**
This   book   advocates   using
resources to be economic and at
the   same   time   enhance   the
environment.   It   emphasises
maintaining areas that are of high
value, enhancing areas of low
value, creating new conservation
areas, and integrating all these
areas into the commercial and
recreational uses of land.
*1988,   Hodder   and   Stoughton,
London, ISBN 0-340-49003-9 (pbk)
184 pp.*

882 Tawney, R. H.
The Acquisitive Society
This book is a critical review of the
principles on which 'Western'

capitalist society rests. It continues to be influential, as it is concerned with problems that we still face.
2nd. Ed. 1961, Fontana Books, London, (pbk) 191 pp.; 1st. Ed. 1921.

883 Tawney, R. H.
Religion and the Rise of Capitalism: A Historical Study
This book describes medieval theories of social ethics, and traces their influence on capitalist ideas. They prohibited usury and insisted on a just price; capitalism later largely abandoned these principles.
2nd. Ed. 1938, with later reprints, Penguin Books, London & New York, ISBN 0-14-018424-4 (pbk) 309 pp.; 1st. Ed. 1926.

884 Tawney, R. H.
Equality
This classic book addresses fundamental questions about modern Western society, discussing various aspects of equality and inequality, and the conditions of economic freedom, democracy, and socialism.
3nd. Ed. 1964, Allen & Unwin, London, (hbk) 255 pp.; 1st. Ed. 19??.

885 Taylor, Ann
Choosing Our Future: A Practical Politics for the Environment
The author of this book presents a personal but rigorous analysis of the environmental problems facing us, and of how they can be

tackled. She is concerned that they should be seen as immediate problems, not just as a topic of political debate.
1992, Routledge, London & New York, ISBNs 0-415-07945-4 (hbk) & 0-415-07846-2 (pbk) 208 pp.

886 Teilhard de Chardin, Pierre
The Phenomenon of Man
This well-known and influential book is the most important of its author's published works. Its author views all knowable reality as a process, not a static mechanism, and emphasises the absolute necessity of adopting an evilutionary viewpoint. He introduces the important concepts of 'noosphere', 'convergence', 'complexification', and the convergent integration of humanity to a final state, the 'Omega Point'.
4th. Ed. 1970, Fontana Books, Collins, London, ISBN 0-00621153-4 (pbk) 352 pp.

887 Temple, William
Christianity and Social Order
This book presents both primary and derivative social principles that, in the author's view, represent the main trend of Christian social teaching. He gives four reasons why the Church should intervene in economics and politics, and suggests how it could usefully intervene. The issues that the book addresses are very relevant today.
1942, Penguin Books, Harmondsworth, Midd., (pbk) 90 pp.

888 Temple, William (Baker, A. E. (Ed.))
William Temple and His Message
This selection of essays by William Temple, who was Archbishop of Canterbury from 1942 to 1944, covers Christianity, science, liberty, fellowship, service, the power of sacrifice, democracy, education, toleration, and sex. Their ideas are still relevant today.
1946, Penguin Books, Harmondsworth, Midd., (pbk) 235 pp.

889 Testor, Jefferson W. & Nancy A. Ferrari (Eds.)
Energy and Environment in the 21st Century
This book includes contributions on: transport, industrial processes, building systems, electrical power, and new energy technologies. Its editors conclude that the world needs another technological revolution, if the efficient and sustainable use of energy and other resources for social well-being are to be assured. A new perspective is needed on our energy and environmental future, viewing it as requiring a set of interconnected trade-offs that must be made between sufficient energy and environmental and other costs.
1991, MIT Press, Cambridge, MA, USA & London, (hbk) 1006 pp.

890 Therivel, Riki et al.
Strategic Environmental Assessment
This book studies Strategic Environmental Assessment (SEA) in depth. This new method of environmental assessment aims to ensure that development and other projects, involving strategic decisions, are based on a full understanding of likely environmental consequences. The authors argue that SEA is likely to become the most direct method for implementing sustainability, and examine the difficulties in policy impact appraisal and implementation, with case studies showing the wide variety of existing and proposed SEA systems worldwide.
1992, Earthscan, London, ISBN 1-85383-147-6 (pbk) 160 pp.

891 Thomas, Alan M.
Beyond Education: A New Perspective on Society's Management of Learning
The author of this book points out that the need for deliberate and intelligent management of learning is becoming more and more vital to governments and individuals, Therefore, we must recognise the increasing importance of learning outside formal education.
1991, Jossey-Bass Publishers, San Francisco, ISBN 1-55542-311-6 (hbk) 201 pp.

892 Thomas, Caroline
The Environment in International Relations
This book is aimed at students and others interested in the environment and international relations. It covers a wide and shifting set of problems and issues, explaining broad developments, and offering signposts for deepening understanding and further study. The author considers that the challenge for international relations and diplomacy is to promote global sustainability over parochial national and state interests. The international political and economic systems need to change, as they are a major part of the environmental problem. The situation requires unprecedented levels of cooperation, not only between countries, but also between industries, scientists, local communities, and non-governmental organisations. At the national level, both rich and poor countries must face up to their respective responsibilities, and adapt their local behaviour. Unless consumption is limited in the developed countries and population controlled in the Third World, efforts towards sustainability will be negated.
1992, Royal Institute of International Affairs, London, ISBN 0-905013-46-6 (hbk) & 0-905031-45-8 (pbk) xix + 291 pp.

893 Thompson, Grahame, Jennifer Frances, Rosalind Levacic & Jeremy Mitchell (Eds.)
Markets, Hierarchies & Networks: The Coordination of Social Life
This book offers an original approach to the central questions of coordinating social, economic and political life. It examines three very different models of social coordination: markets, hierarchies, and networks. It illustrates the characteristics of each of these models: how they work, their strengths, and their limitations. It places them in a comparative framework, offering a comprehensive overview of the mechanisms of coordination at work in different social contexts.
1991, Sage Publications, London & Newbury Park, CA, USA, ISBNs 0-8039-8589-4 (hbk) & 0-8039-8590 (pbk) v + 306 pp.

894 Thompson, William Irwin (Ed.)
Gaia: A Way of Knowing: Political Implications of the New Biology
This book brings together the new biology and the Gaia Hypothesis, and presents a picture of them and their cultural implications.
1987, Lindisfarne Press, Great Barrington, MA, USA, ISBN 0-940262-23-1 (pbk) 217 pp.

895 Timberlake, Lloyd
Only One Earth: Living for the Future

This book offers dramatic examples, from around the world, of people and communities who are learning to use the Earth's resources without despoiling them for future generations.
1987, BBC & Earthscan, London, ISBNs 0-563-20548-2 (hbk) & 0-563-20549-0 (pbk) 168 pp.

896 Tinbergen, Jan (Coordinator) & Ed. Anthony J. Dolman (Ed.)
Reshaping the International Order: A Report to the Club of Rome
This book maps out some of the major problems facing mankind and assesses the progress made in tackling them. It clarifies concepts such as 'development', and discusses how the developing countries could escape from their poverty. It presents an agenda for international action, with specific proposals and recommendations for change.
2nd. Ed. 1977, Hutchinson, London, ISBN 0-09-129051-1 (pbk) vii + 325 pp.; 1st. Ed. 1976, Dutton, New York.

897 Todar, Michael P.
Economics for a Developing World: An Introduction to Principles, Problems and Policies for Development
This book is oriented to teaching economics within the context of major problems of development and underdevelopment in the Third World, and fills a major gap

in teaching materials for this purpose.
3rd. Ed. 1992, Longman, London & New York, ISBN 0-582-07136-4 (pbk) xxxi + 517 pp.; 1st. Ed. 1977.

898 Toffler, Alvin
Future Shock
This book studies mass bewilderment in the face of accelerating change. Its author argues that the future is arriving too fast for most people to handle and that its arrival is not properly coordinated.
2nd. Ed. 1971, Bantam Books, New York & Pan Books, London, ISBNs 0-553-24649-6 (B) & 0-330-02861-8 (P) (pbk) 517 pp.; 1st. Ed. 1970, Random House, New York & The Bodley Head, London (hbk).

899 Toffler, Alvin
The Eco-Spasm Report
This book is an interim report on the condition and crisis of industrial society, and presents some approaches for dealing with it.
1975, Bantam Books, New York, ISBN 0-553-02221-1 (pbk) ix + 117 pp.

900 Toffler, Alvin
The Third Wave
This book details the new 'Third Wave' civilisation that is being created all around us, including its lifestyles, sexual attitudes, work roles, and new economic, political and social structures.
2nd. Ed. 1981, Pan Books, London,

ISBN 0-330-26337-4 (pbk) 544 pp.; 1st. Ed. 1980, Collins, London.

901 Toke, Dave
Green Energy: A Non-Nuclear Response to the Greenhouse Effect
This book presents a programme for encouraging energy efficiency with a wide range of regulatory devices and tax incentives. It concludes that a sustainable energy strategy is best implemented in a context of decentralisation, with emphasis on cooperative forms of social ownership, protecting the interests of the poor.
1990, Green Print, London, ISBN 1-85425-035-3 (pbk) viii + 136 pp.

902 Tolba, Mostafa K.
Saving Our Planet: Challenges and Hopes
This book analyses the positive and negative changes that have occurred in the environment for the past two decades. It focusses, not only on the state of the environment, but also on the interactions between development activities and the environment. It is based on a wide-ranging review of the scientific literature, and has an extensive bibliography. Part 1 describes the state of the world environment, and summarises the environmental problems. Part 2 highlights environmental impacts of development economics. Part 3 describes the impact of the state of the environment and development activities on human conditions and well-being. Part 4 reviews people's perceptions, attitudes and responses towards environmental problems. Part 5 outlines challenges and priorities for action to protect the environment and conserve natural resources; it quotes often from official environmental reports.
1992, Chapman & Hall, London & New York, ISBN 0-412-47370-4 (pbk) x + 287pp.

903 Tolba, Mostafa Kamal & Anit K. Biswas (Eds.)
Earth and Us: Population – Resources – Environment – Development
This book brings together the ideas and thoughts of some of the leading international statesmen, political leaders, economists, and environmentalists on the complex interconnections between earth and us. It examines aspects of the links between population, resources, environment and development. It outlines the authors' views on what can be done in the future. It is designed to complement Our Common Future (No. 988), and other important publications by various UN agencies, the World Bank, the World Resources Institute, and the Worldwatch Institute (No. 1651). It contributes to the overall debate for further clarification of various complex environmental

issues, and helps to increase the world's environmental consciousness.

1991, Butterworth-Heinemann, Oxford, ISBN 0-7506-1049-2 (hbk) viii + 107 pp.

904 Tolley, Rodney (Ed.)
The Greening of Urban Transport: Planning for Walking and Cycling in Western Europe

This book presents the case for incorporating Green transport modes in modern city transport plans. It outlines the principles of Green transport planning, and its environmental, social and economic issues.

1990, Belhaven Press, London & New York, ISBN 1-85293-092-6 (hbk) xviii + 309 pp.

905 Tompkins, Peter & Christopher Bird
Secrets of the Soil

This important book on the vast potential of organic methods of growing plants covers a wide variety of techniques, some of which could transform the world food situation.

2nd. Ed. 1992, Arkana Penguin, London, ISBN 0-14-019311-1 (pbk) 480 pp.; 1st. Ed. 1989, Viking Penguin, New York (hbk).

906 Tough, Allen
Crucial Questions About the Future

This book asks important questions about the future of humankind, presents different possible futures, and discusses what *we* can do to improve our prospects for the future. It aims to present and explore some fundamental questions about the future of human civilisation. Eight fundamental questions (or clusters of questions) are identified as particularly important, and each of them is the central focus for one chapter. The questions are: (1) What is most important of all? (2) Why do we act in ways that hurt our future? (3) What will our actual future turn out to be? (4) How can we achieve a satisfactory future? (5) What catastrophes will be the worst, and how can we prevent them? (6) What role will intelligent life and civilisations elsewhere in the universe play in our future? (7) From which aspects of reality can we gain meaning and purpose? (8) How can each person contribute to a better future?

The book concludes that: (1) striving to contribute to the long-term future of our civilisation can be demanding, risky, and costly, with eventual outcomes that will always be uncertain; (2) we are engaged in the most important struggle that has occurred during the last few centuries; its result could enhance or impoverish human life for the next 100 years; (3) each of us can play a significant positive role in this extraordinary drama. Each of us can contribute vigorously to the world-wide effort to improve humanity's prospects, which could even change the course of human history. Each of us can choose this effort as our highest purpose in life and as our supreme challenge. This book is a *very significant* contribution to the general discussion now in progress about the future of humankind and the planet. Its approach is innovative, in that it raises many questions for its readers, who will thereby be encouraged to think for themselves and to participate more actively, with a significant degree of hope.

2nd. Ed. 1993, Adamantine Press, London, ISBN 0-7449-0083-2 (pbk) (ISSN 0954-6103) 144 pp.; 1st. Ed.,1991, University Press of America, Lanham, MD, USA, ISBNs 0-8191-8274-5 (hbk) & 0-8191-8275-3 (pbk).

907 Toynbee, Arnold
A Study of History

This monumental classic presents a single continuous argument about the nature and pattern of the historical experience of humanity, and a panorama of the whole of human evolution, since civilisations first appeared. It provides a wide variety of illustrations from all periods of history up to modern times, and from most parts of the world. Some of these illustrations are presented in great detail. The history as such

is given in volumes 1-10, completed in 1954; volume 11 (1959) is a historical atlas and gazetteer, and volume 12 (1961) presents the author's additional thoughts since completing the original work.

2nd. Ed. 1935-1961, 12 volumes, Oxford University Press, (hbk); 1st. Ed. 1934, 3 volumes; Abridged Ed. (by D. C. Somervell) in two volumes (1946 & 1957, hbk), Oxford University Press, covering volumes 1-6 and 7-10, respectively.

908 Trainer, Ted
Abandon Affluence: Sustainable Development and Social Change

This book seeks to bring *Limits to Growth* (No. 642) up to date. Its author reviews evidence on major global problems, and argues that our commitment to affluence and growth is magnifying all of them, as they are mainly due to the state of our economic system. They cannot be overcome without fundamental change to simpler, more self-sufficient and cooperative lifestyles. The author outlines his vision of an alternative and sustainable society that could provide a high quality of life, despite low per capita resource usage rates, and allow a peaceful and just world order to emerge.

1985, Zed Books, London, ISBNs 0-86232-311-8 (pbk) & 0-86232-312-6 (pbk) xii + 308 pp.

909 Trainer, Ted
Developed to Death: Rethinking Third World Development
The author of this book provides a devastating yet constructive critique of what most contemporary economists and politicians say about development. He argues that: (1) development must be globally sustainable; (2) our commitment to affluence, economic growth and 'market economics' are the main cause of worldwide poverty, scarcity and injustice; (3) Third World starvation is caused by poverty and poor food distribution, rather than food shortages or natural disasters; (4) global justice must be obtained before world peace can be achieved; (5) Third World living standards can rise only if rich countries consume less; (6) conventional development has brought about violence and repression in its defence. He proposes a Green approach to development that could lead to a just, peaceful, and sustainable world.
1989, Green Print, London, ISBN 1-85425-008-6 (pbk) 230 pp.

910 Tudge, Colin
Global Ecology
This book offers an informal introduction to the principles of ecology, a rapidly advancing and urgent modern science. It has many colour plates, and is written to enhance and complement the *Ecology* exhibition at the Natural History Museum. It examines the roles of adaptation, competition and cooperation, chemistry and genetics, behaviour, evolution and population dynamics, together with the often disturbing human impacts on nature's delicate balance.
1991, Natural History Museum Publications, London, ISBN 0-565-01173-1 (pbk) x + 173 pp.

911 Turner, B. L., II et al. (Eds.)
The Earth as Transformed by Human Action: Global and Regional Changes in the Biosphere over the Past 300 Years
This book begins with an introductory chapter on the great transformation being undergone by the Earth's biosphere. The four main parts of the book cover the following topics: (1) changes in population and society; (2) transformations of the global environment; (3) regional studies of transformation; (4) understanding transformations.
1990, Cambridge University Press, Cambridge & New York, ISBN 0-521-36257-8 (hbk) xvi + 713 pp.

912 Turner, Kerry & Tom Jones (Eds.)
Wetlands: Market & Intervention Failures: Four Case Studies
This book, commissioned by the OECD, explores the damaging consequences for some of the world's most vulnerable ecosystems of both market-based

and interventionist policies, based on too narrow an economic framework. It presents four case studies of the management policies of wetland environments in the USA, UK, France, and Spain; these studies show how wrong policies have resulted in failure, severely reducing the amount of wetland and threatening the remainder. The authors propose measures to reduce damage in the future.
1991, Earthscan, London, ISBN 1-85383-102-6 (pbk) vi + 202 pp.

913 Tylecote, Andrew
The Longwave in the World Economy: The Present Crisis in Historical Perspective
This book discusses whether economic long waves exist and, if so, what economic and other social effects they have. Part 1 covers the theory of long waves. Part 2 relates the history of successive long wave periods, projects the possible nature of the next 'upswing', due to begin in the early 1990s and proceed into the 21st century.
1992, Routledge, London & New York, ISBN 0-415-03690-9 (hbk) xiv + 338 pp.

914 UNESCO
Vancouver Declaration: Final Report of the UNESCO Symposium on Science and Culture for the 21st Century: Agenda for Survival
This Declaration states that the present requirements for planetary survival include urgent measures in all sectors – scientific, cultural, economic, and political – and a greater sensitisation of all humankind. It considers why we have arrived at our present predicament, and states the need for new visions and paradigms rooted in a variety of cultures.
1990, Canadian Commission for UNESCO, (hbk) 314 pp.

915 UNESCO
Facing the Future: Young People and Unemployment Around the World
This book discusses the problem of youth unemployment in Europe and parts of the Third World, and raises some urgent questions about work.
1991, UNESCO, Paris & Orient Longman, Hyderabad, India, distributed by HMSO, London, ISBNs 92-3-102492-2 (U) & 0-86311-251-X (OL) (pbk) viii + 250 pp.

916 UNESCO
World Education Report
This report considers several aspects of basic education, and includes tables of world and regional educational indicators.
1991, UNESCO, Paris, distributed by HMSO, London, ISBN 92-3-102751-4 (pbk) 149 pp.

917 Union of International Associations (Ed.) (No. 1646)
Encyclopedia of World Problems and Human Potential

This book presents a monumental compendium of information about world problems, human problems, human potential, and the whole human being. It spans the whole spectrum from inner to outer limits to growth and from outer to inner potentialities. It has 20,958 entries, linked by 114,395 relationships, together with detailed divisions of problems and potentials. Its web of ideas provides a unique and powerful tool for transcending the human and world problematique, and it would become even more powerful if a new multimedia version were developed.

3rd. Ed. 1991, K. G. Saur, Munich, Germany, London & New York, ISBN 3-598-10842-7 (hbk) 2 volumes, 2133 pp.; 1st. Ed. 1976, Union of International Associations & Mankind 2000, Brussels, Belgium, with the title "Yearbook of World Problems and Human Potential".

918 United Nations
Global Outlook 2000: An Economic, Social and Environmental Perspective

This book reports a study undertaken for the United Nations General Assembly. It represents the UN Secretariat's efforts to revive and assess ongoing changes in global economic and social conditions and the global outlook for the 1990s, draws on researches and projections prepared by the UN specialised agencies and other parts of the UN system, and contains extensive statistical tables. It explores some possible policy changes and initiatives, which could be taken by individual countries and international bodies, to reverse the process of stagnation or decline in the Third World and promote long-term sustained and sustainable development there. It assesses the long-term global trends from 1960 to 1990, and considers: environment, energy, agriculture, new technologies, production and trade, population, human settlements, education, health, and social policies.

1990, United Nations Publications, New York & Geneva, distributed by HMSO, London, ISBN 92-1-109118-7 (pbk) xiii + 339 pp.

919 United Nations
Review and Appraisal of the World Population Plan of Action
(UN Population Studies, No. 115)

This report summarises the evolution of global population policies and programmes, discusses some key social and economic sectors, summarises demographic variables and trends from 1974 to 1985, and recommends enhancements to the World Population Plan of Action.

1991, United Nations, New York, distributed by HMSO, London, ISBN 92-1-151185-0 (pbk) vii + 45 pp.

920 United Nations
World Population Monitoring 1991 with Special Emphasis on Age Structure (*UN Population Studies*, No. 126)
Part 1 is a special report on population age structure, Part 2 presents the results of monitoring population trends and policies, and Part 3 summarises the larger context of current social and economic conditions relevant to global population trends and policies.
1992, United Nations, New York, distributed by HMSO, London, ISBN 92-1-151240-9 (pbk) xii + 241 pp.

921 United Nations
Nations of the Earth Report: Vol. 1: UNCED National Reports Summaries
This volume summarises the first 47 of the national government reports submitted in preparation for the UNCED Earth Summit Conferene in June 1992; a companion volume will contain further national reports. The themes are: demographic pressures, agrochemicals, deforestation, soil erosion, water shortages, waste management, and pollution.
1992, UN Publications, New York, distributed by HMSO, London, (pbk) 322 pp.

922 United Nations Development Programme (UNDP)
Human Development Report
This report is the most important

annual survey of how countries worldwide are treating their people. It provides invaluable advice on how the affluent countries can best assist the developing nations.
1992, Oxford University Press, Oxford & New York, ISBNs 0-19-507772-5 (hbk) & 0-19-507773 (pbk).

923 United Nations Environmental Programme (UNEP)
Environmental Data Report
This report brings together the best available data on the global environment, with many tables and charts. Explanations accompany the data, to aid interpretation, highlight trends, and identify issues.
3rd. Ed. 1991, Blackwell, Oxford & Cambridge, MA, USA, ISBN 0-631-18083-4 (pbk) (ISSN 0956-9324) viii + 408 pp.; 1st. Ed. 1987.
.ff

924 Vale, Brenda & Robert Vale
Green Architecture: Design for a Sustainable Future
This book provides a balanced international overview of the development of the Greening of architecture.
1991, Thames & Hudson, London, ISBN 0-500-34117-6 (hbk) 192 pp.

925 Vale, Brenda & Robert Vale
Towards a Green Architecture: Six Practical Case Studies

This book presents case studies of six sets of buildings, all deliberately designed to minimise environmental impact.
1991, RIBA Publications, London, ISBN 0-947877-47-9 (pbk) 78 pp.

926 Vallely, Bernadette
1001 Ways to Save the Planet: How You Can Change Your Lifestyle Today to Create a Greener World
This book practical guide to Green action and living describes many simple changes that we can all make in our lives today to bring about a Greener environment. It suggests actions that *we* can take now quite easily, to make a difference to the future of our planet.
1990, Penguin Books, London & New York, ISBN 0-14-013301-1 (pbk) xiii + 336 pp.

927 Vallely, Bernadette, Felicity Aldridge & Lorna Davies
Green Living: Practical Ways to Make Your Home Environment Friendly
This book explains how we can manage our homes holistically, so that all the small practical decisions and daily actions of housekeeping combine to create a whole Green system.
1991, Thorsons, London, ISBN 0-7225-2470-6 (pbk) 208 pp.

928 Van de Weyer, Robert
The Health of Nations: Politics and Morality for the 21st Century
This book presents a political, moral, and economic philosophy for a sustainable, humane, and prosperous society in the 21st century.
1991, Green Books, Hartford, Bideford, Devon, ISBN 1-870098-39-0 (pbk) 160 pp.

929 Vayrynen, Raimo (Ed.)
New Directions in Conflict Theory: Conflict Resolution and Conflict Transformation
The papers in this book focus on issues of conflict resolution from historical, structural, legal, psychological, international, economic, and environmental perspectives. They also examine social conflict resolution and urban-rural divisions between state and society.
1991, Sage Publications, London & Newbury Park, CA, USA, ISBNs 0-8039-8435-9 (hbk) & 0-8039-8437-5 (pbk) viii + 232 pp.

930 Vellve, Renee
Saving the Seed: Genetic Diversity and European Agriculture
Genetic diversity supports the security of agriculture, because crops are vulnerable to change if they have no varieties adaptable to different conditions. This book traces the decline of crop varieties in European farming, and describes what is being done to safeguard genetic resources for the future. It contains up-to-date information on genetic resources

in Europe and on those working to save them.

1993, Earthscan, London, ISBN 1-85383-150-6 (pbk) 160 pp.

931 Vickers, Jeanne (Prepared and compiled by)
Women & the World Economic Crisis
This book looks at the impact of the Third World debt crisis and the effects of economic adjustment policies. It examines the serious consequences for millions of women, including adverse effects on their health, nutrition, and children, as well as problems of unemployment, homelessness, and illiteracy. It explains the financial and trade policies that have led to these problems, then describes international, governmental, and nongovernmental organisation policies and programmes designed to alleviate these effects. It presents detailed case studies of how women themselves are confronting these problems.

1991, Zed Books, London & Atlantic Highlands, NJ, USA, ISBNs 0-86232-974-4 (hbk) & 0-86232-975-2 (pbk) xi + 46 pp.

932 Vittachi, Anuradha
Earth Conference One: Sharing a Vision for Our Planet
This book is an inspiring eyewitness account of an important event offering hope for the future of our planet. It describes the public presentations and some of the informal discussions at the Global Survival Conference at Oxford in April 1988, attended by spiritual and parliamentary leaders from all parts of the world, together with leading scientists, academics, and business executives. They listened to and learned from each other, and began to transform their outlooks and lives. Many quotations from the participants are included.

1989, Shambhala, Boston, USA & Shaftesbury, Dorset, ISBN 0-87773-503-4 (pbk) 168 pp.

933 von Bertalanffy, Ludwig
General System Theory
This book presents the fundamental ideas of 'general system theory', which emerged as a unifying discipline in the late 1940s. It later became known as 'general systems theory', which identifies certain concepts, principles and methods that have been shown not to depend on the specific nature of the phenomena involved, so that they apply broadly to every branch of knowledge. Its new, holistic approach can be used as a basis for the better organisation of society.

2nd. Ed. 1971, Allen Lane, London, ISBN 0-7139-0192-6 (hbk) xxii + 311 pp.; 1st. Ed. 1968, Braziller, New York & Allen Lane.

934 von Neumann, John & Oscar Morgenstern
Theory of Games and Economic Behaviour
This book is the classic original presentation of game theory, a remarkable mathematical theory of economics and social organisation, based on a theory of games and strategy. It contains many applications to games of strategy, and to economic and social theory, which have been extended greatly during the decades since the book was written.
3rd. Ed. 1953, Wiley, London & New York, ISBN 0-471-91185-2 (pbk) xx + 641 pp.; 1st. Ed. 1944, Princeton University Press, Princeton, NJ, USA.

935 Waddington, C. H.
Tools for Thought
This book presents a wide array of models and techniques of thinking about the world, which is viewed as a series of interacting systems, not only governed by simple cause and effect relationships. It draws ideas from general systems theory, operational research, cybernetics, futures research, and catastrophe theory, to provide revolutionary tools of thought and equipment for perceiving and improving the world as it really is.
2nd. Ed. 1977, Paladin/Granada Publishing, St. Albans, Herts., ISBN 0-586-08254-9 (pbk) xiii + 250 pp.; 1st. Ed. 1977, Cape, London.

936 Wagar, W. Warren
The Next Three Futures: Paradigms of Things to Come
For those who wish to know if they can help make the future better than it might otherwise be, this book briefly sums up forward-looking thinking and writing over the past 25 years. It also champions the integrative approach of 'alternative futures' thinking. It draws its methods and data from environmental studies, peace and war studies, macroeconomics, technology assessment, and several other fields, but seeks to synthesise these parallel approaches into one holistic vision of the human future. As the author says, "The future, like the past, must be continually re-examined and rediscovered."
The book is not a guided tour of the 21st century. It surveys the possibilities apparent here and now, and is designed to counterbalance the all-too-common preoccupation with the present, which seems to the author to defy the fact that a vast part of all our lives is lodged in the future. As he says, "Living for the present can mean failing to leave behind a safe and habitable world for our children."
U.K. Ed. 1992, Adamantine Press, London, ISBN 0-7449-0043-3 (pbk) (ISSN 0954-6103) 192 pp.; U.S. Ed. 1991, Greenwood Press, Westport, CT, USA.

937 Wagar, W. Warren
A Short History of the Future
This book charts trends and possibilities for humankind and our planet from the 1990s to the opening of the 23rd century in the tradition of H. G. Wells. It is cast as a history book, a memoir of postmodern times by an elderly scholar. It is an attempt by a professional historian to apply the methods and perspectives of his craft to study world futures. It is not a forecast, but a survey of alternatives, and an inventory of the hopes and anxieties of our time.
2nd. Ed. 1992, Adamantine Press, London, ISBN 0-7449-0074-3 (pbk) (ISSN 0954-6103) xiv + 325 pp.; 1st. Ed. 1989, University of Chicago Press, Chicago.

938 Walgate, Robert
Miracle or Menace? Biotechnology and the Third World
This book explains some of the promises offered by biotechnology, especially for the Third World, and outlines some of its associated risks and changes.
1990, Panos Institute, London, ISBN 1-870670-18-3 (pbk) viii + 199 pp.

939 Walker, Peter
Famine Early Warning Systems: Victims and Their Destitution
This book discusses requirements for local, national and global famine early warning systems,

describes the major existing systems, and gives some recommendations for their improvement.
1989, Earthscan, London, ISBN 1-85383-047-X (pbk) x + 196 pp.

940 Wall, Derek et al.
Getting There: Steps to a Green Society 1990
This book discusses how to achieve a cooperative, decentralised society, not obsessed with growth, where the elimination of the defects of consumer capitalism would allow a redistribution of the world's resources without any reduction in quality of life.
1990, Green Print, London, ISBN 1-85425-034-5 (pbk) 149 pp.

941 Ward, Barbara
The Home of Man
This book addresses the needs common to all – access, food, water, sanitation, energy, shelter, and a sense of community – and detailed ways of meeting them, with pointers to the solution of these problems.
1976, Penguin Books, Harmondsworth, Midd. & New York, London, ISBN 0-14-021942-0 (pbk) xii + 297 pp., & Andre Deutsch, London.

942 Ward, Barbara
Progress for a Small Planet
This book shows how new attitudes, supported by new and

more conserving technologies, can take us beyond our environmental and poverty crises. It outlines the planetary bargain between the world's nations that remove poverty and keep our biosphere in working order.
1979, Penguin Books, Harmondsworth, Midd. & W. W. Norton, New York, ISBN 0-14-022255-3 (pbk) xiii + 305 pp.

943 Ward, Barbara & Rene Dubos
Only One Earth
This unofficial report, commissioned for the 1972 United Nations Conference on the Human Environment at Stockholm, was prepared with the help of many experts. It aimed to obtain the best available advice to provide a conceptual framework for the conference's participants and the public, and to provide background information relevant to official policy decisions. Its five parts are concerned with the planet's unity, the unities of science, problems of high technology, the developing countries, and how to achieve an appropriate planetary order.
1972, Penguin Books, Harmondsworth, Midd., ISBN 0-14-021601-4 (pbk) 304 pp.

944 Waterstone, Marvin (Ed.)
Risk and Society: The Interaction of Science, Technology and Public Policy
This book provides a theoretical and applied exploration of the ways in which risks arise out of the 'normal' day-to-day functioning of society. Each author offers important insights on the ways in which risk issues arise, mature, and have profound effects on our society.
1991, Kluwer Academic Publishers, Dordrecht, Netherlands, London & Boston, MA, USA, ISBN 0-7923-1370-4 (hbk) xix + 178 pp.

945 Wathern, Peter (Ed.)
Environmental Impact Assessment: Theory and Practice
This book provides a comprehensive presentation of environmental impact assessment reviews its theory and practice during the 1980s.
1990, Routledge, London & New York, ISBNs 0-0-445042-7 (hbk) & 0-4445816-9 (pbk) 352 pp.

946 Webb, Adrian
The Future for UK Environment Policy: Key Players, Issues and Implications
This report discusses the role of non-governmental organisations, government bodies, and political parties in considering environmental issues and formulating environmental policies for the UK. It also looks at the European and global aspects. The appendix details some important British organisations concerned with the environment.
1991, The Economist Intelligence

Unit, London, Special Report No. 2182, ISBN 0-85858-562-7 (pbk) v + 88 pp.

947 Webber, David J. (Ed.)
Biotechnology: Assessing Social Impact and Policy Implications
This book covers different aspects of the impact of biotechnology: (1) social and political dimensions; (2) institutional responses; (3) assessing potential impacts; (4) public policy responses.
1990, Greenwood Press, Westport, CT, USA, ISBN 0-313-27454-1 (ISSN 0149-1066) (hbk) xiv + 239 pp.

948 Webster, Andrew
Science, Technology, and Society: New Directions
This book is an independent overview of trends towards big science, government involvement in funding and control of science and technology, and corporate research and development.
1991, Rutgers University Press, NJ, USA, ISBNs 0-8135-1722-2 (hbk) & 0-8135-1723-0 (pbk) 181 pp.

949 Wedgwood Benn, David
From Glasnost to Freedom of Speech: Russian Openness and International Relations
This book explores the opening up of Soviet society after 1985, the rise of glasnost, the loss of ideology, and the break-up of the USSR.
1992, RIIA/Pinter, London, ISBNs

0-85567-006-2 (hbk) & 0-85567-007-0 (pbk) vi + 106 pp.

950 Weinberg, Bill
War on the Land: Ecology and Politics in Central America
This book, based on local investigations, shows dramaticaly how war between people has become war against the land in Central America.
1991, Zed Books, London, ISBNs 0-86232-946-9 (hbk) & ISBN 0-86232-947-7 (pbk) xv + 203 pp.

951 Weiner, Jonathan
The Next One Hundred Years: Shaping the Fate of Our Living Earth
This book analyses the current state of our planet, its possible futures, and the power that we have to influence the future. It gives details of the warning signs of planetary trouble, and presents recent scientific facts and proposals for appropriate action there is time.
2nd. Ed. 1991, Bantam Books, NY, ISBN 0-553-35228-8 (pbk) 312 pp; 1st. Ed. 1990, Rider, London.

952 Weiss, Carol H.
Organizations for Policy Analysis: Helping Government Think
This book presents profiles of five external 'think tanks', four executive government units, and five legislative units that contribute to policy formulation in the USA.
1992, Sage Publications, Newbury

Park, CA, USA & London, (hbk) 289 pp.

953 Wells, H. G.
The Outline of History: From Primordial Life to Nineteen-Seventy-One
This is the first book of H. G. Wells' *encyclopedic educational trilogy* (see also Nos. 957 & 961), and it has since been revised several times. It is the first book ever written on the history of the world as a whole, and a broad review of humankind's place in space and time.
10th. Ed. 1972 (Revised and updated by Raymond Postgate & G. P. Wells), Cassell, London, ISBN 0-304-93856-4 (hbk) xviii + 1103 pp.; 1st. Ed. 1919, Newnes, London.

954 Wells, H. G.
A Short History of the World
This book is not a shortened version of *The Outline of History* (No. 953), but was written afresh for 'busy' general readers.
Latest Ed. 1976 (Revised and adapted by Raymond Postgate & G. P. Wells), Collins, London (hbk) xii + 370 pp.; 1st. Ed. 1922, Cassell, London.

955 Wells, H. G.
The Open Conspiracy: Blue Prints for a World Revolution
This book presents H. G. Wells' concept of the 'Open Conspiracy' as a possible worldwide network of thinking people, who would provide the essential framework for a politically, socially, and economically unified world order. The eventual 'world state' that Wells envisages there would be run by a process of social networking, with impartial worldwide controls for certain specific functions, and not by a 'parliament of mankind' and its accompanying bureaucracy.
2nd. Ed. 1930, Hogarth Press, London, (hbk) 243 pp.; 1st. Ed. 1928, Gollancz, London.

956 Wells, H. G.
What Are We to Do with Our Lives
This is the final version of *The Open Conspiracy* (No. 955).
2nd. Ed. 1935, revised 1938, The Thinker's Library, Watts, London, (hbk) vi + 150 pp.; 1st. Ed. 1931, Heinemann, London.

957 Wells, H. G.
The Work, Wealth and Happiness of Mankind
This is the third and last book of H. G. Wells' *encyclopedic educational trilogy* (see also Nos. 953 & 961). Its comprehensive account of economic life also introduces concepts of functional economic planning at the global level.
3rd. Ed. 1954, Heinemann, London (hbk) xiii + 867 pp.; 1st. Ed. 1931, Doubleday, Doran & Co., New York, 2 vols.

958 Wells, H. G.
World Brain
This book outlines H. G. Wells'

ideas of 'World Brain' and 'World Encyclopaedia', for organising libraries and information services into one integrated international system covering all fields of knowledge. It aims to provide a worldwide source of up-to-date information in a form accessible to general readers as well as specialists.

1938, Methuen, London, (hbk) xvi + 130 pp.

959 Wells, H. G.
The Rights of Man, or What Are We Fighting For

This book discusses, and includes a text of, the Sankey Declaration of the Rights of Man, whose formulation was largely H. G. Wells' work.

1940, Penguin Books, Harmondsworth, Middlesex, (pbk) 128 pp.

960 Wells, H. G.
The Outlook for Homo Sapiens: An Amalgamation and Modernization of Two Books: 'The Fate of Homo Sapiens' and 'The New World Order'

This book contains many of H. G. Wells' ideas about how to realise a world civilisation. It outlines what was happening to humankind during World War 2, and what the future prospects seemed to be.

1942, Secker & Warburg, London, (hbk) 288 pp.

961 Wells, H. G. in collaboration with Julian Huxley & G. P. Wells
The Science of Life

This is the second book of H. G. Wells' *encyclopedic educational trilogy* (see also Nos. 953 & 957). It is a comprehensive survey for general readers of biology as it was at the time.

2nd. Ed. 1934-37, Cassell, London, (hbk) 9 vols., 1575 pp.; 1st. Ed. 1929-30, Amalgamated Press, London, 3 vols.

962 Wells, Phil & Mandy Jetter
The Global Consumer: Best Buys to Help the Third World

This book presents various ways in which consumer power can help the Third World. Its authors consider that few people yet realise their power to contribute to permanent change in the global situation.

1991, Gollancz, London, ISBN 0-575-05000-4 (pbk) xii + 340 pp.

963 Weston, Victoria (Suzzanne Bentz (Ed.))
Into the Future and the Next Millennium, Part 1

This book provides many provocative and 'wild' or 'not-so-wild' 'psychic' predictions, perhaps in some case extrapolations, about future trends in the USA and worldwide, well into the 21st century.

1991, Oscar Dey Publishing, USA, ISBN 0-962318-1-3 (pbk) 250 pp.

964 Wheale, Peter R. & Ruth McNally
Genetic Engineering: Catastrophe or Utopia?
This book clearly explains the dangers inherent in the diffusion of new technologies of genetic engineering, and critically evaluates the ways in which society should control it.
1988,   Harvester   Wheatsheaf, London, ISBN 0-7450-0010-X (hbk) xvi + 332 pp.

965 Wheale, Peter & Ruth McNally
The Bio Revolution: Cornucopia or Pandora's Box?
This book considers genetic engineering in relation to the environment, animal welfare, ethics, acceptability, and control.
2nd. Ed. 1990, Pluto Press, London, ISBN 0-7453-0338-2 (pbk) 304 pp.; 1st. Ed. 1989, ISBN 0-7453-0337-4 (hbk) 192 pp.

966 *Which?*
Gardening without Chemicals
This report gives guidance on many easy and practical ways of effective   gardening   without resort to chemicals.
1990,   Consumers   Association, Hertford.

967 Whiston, Thomas G. & Roger L. Geiger
Research and Higher Education: The United Kingdom and the United States
This collection of original papers by major researchers in higher education, helps to fill the gap in published information about that research, and focusses on the UK and US experience.
1992, Open University Press, Milton Keynes, ISBN 0-335-15641-X (pbk) xiv + 205 pp.

968 White, Frank
The SETI Factor: How the Search for Extraterrestrial Intelligence Is Changing Our View of the Universe and Ourselves
This book presents 12 plausible scenarios for the search for extraterrestrial intelligence for the next 25 years.
1990, Walker and Company, New York, ISBN 0-8027-1105-7 (hbk) 250 pp.

969 Whitehead, A. N.
Science and the Modern World
In this book, the author first presented a version of his holistic philosophy, including his theory of Organic Mechanism
6th. Ed. 1975, Fontana Books, Collins, London, ISBN 0-00-643641-2 (pbk) 252 pp.; 1st. Ed. 1925, Cambridge   University   Press, Cambridge & Macmillan, New York.

970 Whitehead, A. N.
Adventures of Ideas
This book presents the most developed form of its author's integrated and holistic philosophical approach, which was one of

the forerunners of 'systems thinking'.

2nd. Ed. 1942, Penguin Books, Harmondsworth, Midd., (pbk) 349 pp.; 1st. Ed. 1933, Cambridge University Press, Cambridge.

971 Whyte, William Foote
Social Theory for Action: How Individuals and Organizations Learn to Change

This book describes the evolution of participatory structures and processes in agricultural research and development and in industrial relations, with special emphasis on the diffusion of social inventions through organisational learning.

1991, Sage Publications, Newbury Park, CA, USA & London, ISBN 0-8039-4166-8 (hbk) vii + 301 pp.

972 Wiener, Norbert
Cybernetics: The Science of Communication and Control in the Animal and the Machine

This book is the original formulation of 'cybernetics' by its originator; the definition in the subtitle still conveys a good idea of much of its subject matter. Cybernetics is concerned especially with identifying negative and positive feedback loops in the processes that systems undergo during their evolution. Though it evolved independently of general systems theory, it has many ideas in common with it.

2nd. Ed. 1961, MIT Press, Cambridge, MA, USA & Wiley, New York, (hbk) xvi + 212 pp.; 1st. Ed. 1948, Wiley, New York.

973 Wiener, Norbert
The Human Use of Human Beings

This book discusses some of the human and social implications of cybernetics (see No. 972).

3rd. Ed. 1968, Sphere Books, London, SBN 7221-9134-0 (pbk) 186 pp.; 1st. Ed. 1950, Eyre & Spottiswoode, London.

974 Wijkman, Anders & Lloyd Timberlake
Natural Disasters: Acts of God or Acts of Man?

This book shows how people, especially in the Third World, make their environment more liable to natural disasters, and themselves more vulnerable to them. It calls for longer-term appropriate development to reduce the risks and effects of disasters.

1984, IIED/Earthscan, London, ISBN 0-905347-54-4 (pbk) 146 pp.

975 Wike, Soeren & Tom Jones (Eds.)
Forests: Market & Intervention Failures

This book presents five case studies of forest management failures in Europe. It shows how market and intervention policies can increase forest destruction, and future policies can avoid past mistakes.

1992, Earthscan, London, ISBN 1-85383-101-8 (pbk) xiv + 204 pp.

976 Wilber, Ken
Eye to Eye: The Quest for the New Paradigm

This book takes a first step towards a comprehensive model of consciousness and reality, comprising science, psychology, philosophy, and religion. It examines the empirical realm of the senses, the rational realm of the mind, and the contemplative realm of the spirit.

2nd. Ed. 1990, Shambhala Publications, Boston, USA & Shaftesbury, Dorset, ISBN 0-87773-549-2 (pbk) ix + 340 pp.; 1st. Ed. 1983.

977 Wilkinson, James
Green or Bust

This book is a popular but comprehensive review of the major environmental problems and issues facing us today.

1990, BBC Books, London, ISBN 0-563-36032 (pbk) vi + 202 pp.

978 Williams, Allan M.
The European Community: The Contradictions of Integration

This book surveys and analyses the aims, operations and policies of the European Community (EC), and their consequences for individual countries, Europe, and the global political economy.

1991, Blackwell, Oxford & Cambridge, MA, USA, ISBNs 0-631-17476-1 (hbk) & 0-631-18088-0 (pbk) xiii + 185 pp.

979 Williams, Frederick
The New Telecommunications: Infrastructure for the Communications Age

This book shows how telecommunications can be applied to create new competitive advantages in business, increased productivity in services, and economic development in cities, states, and nations.

1991, Free Press, New York, ISBN 0-02-935281-9 (hbk) viii + 247 pp.

980 Williamson, J. J.
An Outline of the Principles and Concepts of the New Metaphysics

This book outlines its author's broad philosophical approach, and presents a set of principles, underlying all human experience, intended to clarify transcendental awareness and indeed all empirical knowledge. It presents an important example of a philosophy which attempts to provide a universal framework covering all aspects of life, including human affairs and the physical and other sciences.

1967, Society of Metaphysicians, Hastings, (pbk) vi + 40 pp.

981 Wilson, Graham K.
Business and Politics: A Comparative Introduction

This book explains the relationships between business, government and politicians in the USA, Europe and Japan, and assesses

how far business dominates government in various parts of the world.
1985, Chatham House Publications, Chatham, NJ, USA, ISBNs 0-934540-51-9 (hbk) & 0-934540-50-0 (pbk) xii + 155 pp.

982 Winpenny, J. T.
Values for the Environment: A Guide to Economic Appraisal
This practical guide to the economic treatment of the environment in project appraisal uses cost-benefit analysis as its decision framework, and is based on extensive case material.
1991, HMSO, London, ISBN 0-11-580257-6 (pbk) x + 277 pp.

983 Wistrich, Ernest
After 1992: The United States of Europe
This book presents a picture of a united and unified Europe after 1992, including a vision of Europe in 2014. Its author advocates federalism and the principle of 'subsidiarity', where any issue is addressed at the lowest level at which it can be handled effectively.
2nd. Ed. 1991, Routledge, London & New York, ISBNs 0-415-04451-0 (hbk) & 0-415-06457-0 (pbk) x + 160 pp.; 1st. Ed. 1989.

984 Wolfe, Alan (Ed.)
America at Century's End
In this book, leading sociologists consider recent changes in US lifestyles, in families, cities, competitiveness, shortage of time, new immigration patterns, and the enduring race dilemma.
1991, University of California Press, Berkeley, CA, USA, ISBN 0-520-07476-9 (hbk) 579 pp.

985 Wolfe, Joan
Making Things Happen: How to Be an Effective Volunteer
This guide for volunteers, with many valuable points, includes a checklist of how to be effective, and advice on being persuasive.
2nd. Ed. 1991, Island Press, Washington, DC, USA, ISBNs 1-55963-127-9 (hbk) & 1-55963-126-0 (pbk) 208 pp.; 1st. Ed. 1981.

986 World Bank, The
World Development Report 1990
This report examines what has gone right and wrong in previous policies for developing the Third World, and looks towards the future. It recommends that both developed and developing countries pursue a twofold strategy to reduce world poverty: (1) promoting broad-based growth that will generate income-earning opportunities for the poor; (2) ensuring that the poor can benefit from such opportunities, by improving access to education, health care, and other social services. This strategy would enable the poor to use effectively their principal asset, their ability to work. The

report advocates allocating larger shares of aid budgets to countries which have shown in their policies a genuine commitment to reduce poverty.

It concludes that: (1) there were gradual improvements in living standards during the 1980s for most, though by no means all, developing countries; (2) there is a reasonable chance of laying a good economic foundation to meet the challenge of poverty and improving the prospects for many if not most people in the Third World during the 1990s. It includes very extensive statistical tables, covering all countries.

1990, Oxford University Press, Oxford & New York, ISBNs 0-19-520850-1 (hbk) & 0-19-520851-X (pbk) xiii + 260 pp.

987 World Bank, The
World Tables 1991

For each of 139 countries, this book gives four pages of important economic and financial quantities and indicators, from 1969 to 1989.

1991, The Johns Hopkins University Press, Baltimore, MD, USA & London, ISBN 0-8018-4252-2 (pbk) (ISSN 1043-5573) xvi + 655 pp.

988 World Commission on Environment and Development, The
Our Common Future

This book, also known as *The Brundtland Report*, presents the results of an examination of the critical environment and development problems of our planet. It attempts to formulate realistic proposals to solve these problems, and to ensure that human progress can be achieved through sustainable development without sacrificing the prospects of future generations. It advocates a marriage of economics and ecology, so that governments and individuals can take responsibility, not only for environmental damage, but also for policies causing such damage.

It urges us to act now to start changing these harmful policies. The common challenges that it presents are in the areas of: population and human resources, food and agriculture, species and ecosystems, energy, industry, and the urban environment. The common endeavours that it calls for include: managing our common natural resources in the oceans, space, and Antarctica; the achievement of peace, security, and sustainable development; and institutional and legal changes for common action. This very important report has had considerable influence on later studies and initiatives for improving the human and planetary situation.

1987, Oxford University Press, Oxford & New York, ISBN 0-19-282080-X (pbk) xv + 400 pp.

989 World Commission on Environment and Development, The, Advisory Panel on Food, Security and the Environment
Food 2000: Global Policies for Sustainable Agriculture
This report argues that the present harmful trends of world food production and distribution, leading to widespread hunger and environmental degradation, are not inevitable. It proposes new agricultural techniques and land reform strategies in a Seven Point Action Plan, whose innovative recommendations for global action aim simultaneously to meet the growing demand for food by raising yields, and to improve the environment in areas being lost to agriculture.
1987, Zed Books, London & Atlantic Highlands, NJ, USA, ISBNs 0-86232-708-3 (hbk) & 0-86232-709-1 (pbk) xi + 131 pp.

990 World Development Movement
Costing the Earth: Striking the Global Bargain at the 1992 Earth Summit
This book presents the World Development's Costing the Earth Programme, that was an important part of the UK's preparation for the UNCED Earth Summit, held in Rio de Janeiro, Brazil, in June 1992. It looks at the hard questions that need to be resolved in striking the 'global bargain' of environmental security: (1) how

the necessary resources will be generated; (2) who should pay the price. It brings together the views of experts from all sectors of political, economic and social life, from both 'North' and 'South' countries, as well as members of the general public, to address the crucial issues of resources and responsibilities. It presents reports of two preparatory London conferences in April 1991, a public conference and an experts' conference. It also includes submissions of views on: aid, debt and investment, trade, and investing in technologies and human resources.
1991, World Development Movement, London, ISBN 0-903272-29-6 (pbk) 173 pp.

991 World Health Organisation
Health Promotion: A Discussion Document on the Concepts and Principles
This report presents WHO's 1984 'Health Promotion' programme, which agreed five principles of health promotion, formulated five priorities for the development of policies in health promotion, and identified four political and moral dilemmas facing health promotion policies.
1984, WHO Regional Office for Europe, Copenhagen.

992 World Resources Institute, The
World Resources 1992-93
The *World Resources* series is

intended to meet the critical need for accessible, accurate information on urgent global issues of our time. It provides an objective and up-to-date report of conditions and trends in the world's natural isues and in the global environment. This book has a special focus on sustainable development, in support of the June 1992 UNCED Conference at Rio de Janeiro, Brazil. Specific areas covered include: population and human development, food and agriculture, basic economic indicators, forests and rangelands, wildlife and habitat, fresh water, oceans and coasts, atmosphere and climate, energy, land cover and human settlements, and policies and institutions, including non-governmental organisations (NGOs).

1992, Oxford University Press, Oxford & New York, ISBNs 0-19-506230-2 (hbk) & 0-19-506231-0 (pbk) (ISSN 0887-0403) xiv + 385 pp.

993 World Resources Institute, The, The World Conservation Union (IUCN), and the UN Environment Programme (UNEP)
Global Biodiversity Strategy: Guidelines for Action to Save, Study, and Use Earth's Biotic Wealth Sustainably and Equitably
This report shows that conserving biodiversity is essential for sustainable development, and proposes 85 specific actions.

1992, World Resources Institute Publications, Washington, ISBN 0-915282-74-0 (pbk) 260 pp.

994 Yearley, Steven
The Green Case: A Sociology of Environmental Issues, Arguments and Politics
This book describes the basis of Green arguments, and discusses their social and political implications. It considers why the wave of Green opinion has started moving, and examines the successes of leading campaign groups and the forces shaping the future of Green politics and policies worldwide. It shows the systematic connections between aid, debt, development, and urgent environmental problems.

1991, HarperCollins, London & New York, ISBN 0-04-445752-9 (pbk) ix + 197 pp.

995 Yudkin, John
The Penguin Encyclopaedia of Nutrition
This encyclopedia not only discusses proteins, vitamins, and other nutrients, and which foods contain them. It covers all aspects of people in relation to their food, including: food supplies and population, food distribution and preservation, the functions of the nutrients and the effects of too little and too much of them.

1985, *Penguin, London & Viking Penguin, New York, ISBNs 0-670-80111-9 & 0-7139-1662-1 (hbk) 431 pp.*

996 Zander, Michael
A Bill of Rights?
This book contributes to the debate on whether the UK should have a Bill of Rights. It provides a detailed history of the debate, and a comprehensive survey of the arguments for and against a Bill of Rights.
3rd. Ed. 1985, Sweet & Maxwell, London, ISBN 0-421-34510-1 (pbk) viii + 105 pp.

997 Zerzan, John & Alice Carnes (Eds.)
Questioning Technology: Tool, Toy or Tyrant?
This book encourages its readers to think critically about technology and their everyday lives. All its contributors face the consequences of dependence on technology, and some suggest solutions.
1991, New Society Publishers, Philadelphia, PA, USA, ISBNs 0-86571-204-2 (hbk) & 0-86571-205-0 (pbk) 222 pp.

998 Zimbardo, Philip G. & Michael R. Leippe
The Psychology of Attitude Change and Social Influence
This book covers many aspects, principles, and applications of major social influences, including those affecting attitudes.
1991, McGraw-Hill, New York, ISBN 0-07-072877-1 (pbk) xxi + 438 pp.

999 Zohar, Danah
The Quantum Self: A Revolutionary View of Human Nature and Consciousness Rooted in the New Physics
The author of this book argues that the insights of quantum physics can improve our understanding of everyday life, and develops a whole new quantum psychology.
1990, Bloomsbury Publishing, London, ISBN 1-7475-0271-4 (pbk) xiv + 245 pp.

1000 Zukav, Gary
The Dancing Wu Li Masters: An Overview of the New Physics
The author of this book conveys the ideas that have revolutionised modern physics very directly, for the benefit of general readers, and compares quantum mechanics, modern psychology, and Eastern mysticism.
1979, Rider/Hutchinson, London, ISBNs 0-09-139400-7 (hbk) & 0-09-139401-5 (pbk) 355 pp.

2

PERIODICALS

Introductory Note

178 periodicals relevant to 21st century studies, including a few book series, are listed here in alphabetical order of title. Each entry has as many as possible of the following lines:

(1) title in bold type, but any subtitle is omitted;

(2) any former title(s) or other relevant miscellaneous information (not present in all entries);

(3) any sponsoring organisation(s), if not the same as the publisher (not present in all entries);

(4) brief description of content, subject matter, aims and nature;

(5) publisher(s);

(6) International Standard Serial Number (ISSN), if assigned and

known, followed by frequency code;

(7) information about book reviews if any.

The frequency code is either a letter (A for annual, Q for quarterly, M for monthly, W for weekly, V for variable) or it has the form "n pa", where n is the number of times it currently appears per year.

The last line indicates the approximate number and lengths of book reviews in a typical issue, or is "No reviews." if the periodical's policy is not to review books. It is omitted if the periodical's reviewing policy was not known at the time of going to press. This line is especially useful for readers who wish to locate additional books on the subjects covered by the periodical. Approximate number indications range from "a few", "several", "quite a lot", "many", to "very many". Approximate length indications range from "short", "medium", "medium-long", to "long", and beyond that to "essay review" or "review essay"; a range of lengths is indicated if different reviews may have very different lengths. The assignment of these terms can be only approximately correct.

Up-to-date information about many periodicals not listed here is given in the current version of *Ulrich's International Periodicals Directory* (No. 1162), which has listings by subject as well as in title order. See also *Gale Directory of Publications* (No. 1052). It should be noted that some periodicals are *not* included in these directories.

List of Periodicals

1001 Adamantine Studies on the 21st Century

Series of books on the emerging discipline of '21st century studies', which is uniquely devoted to viewing the world in an integrated way; the strands in this series include business, environment, futures and global policy.
Adamantine Press, London
ISSN 0954-6103 15-20 books pa

1002 Alternatives

Centre for the Study of Developing Societies, International Peace Research Institute, and World Order Models Project
Covers the implications of global change, exploring emerging worldwide processes and creative energies from local communities.
Lynne Rienner Publishers, Boulder, CO, USA
ISSN 0304-3754 Q
No reviews.

1003 American Book Publishing Record (ABPR)

This lists most of the books published in the USA in a given month or year. It has indexes and a subject guide.
Bowker, New Providence, NJ, USA
ISSN 0002-7707 M & A (Cumulative)
No reviews.

1004 Annals of the American Academy of Political and Social Science

American Academy of Political and Social Science
Forum for the interdisciplinary discussion of single problems and policy issues, affecting the USA and the world.
Sage Publications, Newbury Park, CA, USA & London
ISSN 0002-7162 6 pa

1005 Applied Economics

Papers on applications of economics to specific problems in public and private sectors,

especially practical quantitative studies.
Chapman & Hall, London
ISSN 0003-6846 M
No reviews.

1006 Appropriate Technology
Papers on a wide range of alternative technology topics.
IT Publications, London
ISSN 0305-0920 Q
A few medium reviews.

1007 BBC Wildlife
Articles, news, and views on wildlife and conservation issues.
World Publications, BBC Enterprises, London
ISSN 0265-3656 M
A few short and medium reviews.

1008 Beshara Magazine
Feature articles on new paradigms and other new developments in science, spirituality, current affairs, culture and the arts.
Silverdove, Frilford, Abingdon, Oxford
ISSN 0954-4584 V
Many short to long reviews.

1009 BioScience
Papers on the advancement of the biological sciences and their medical, agricultural, environmental, and other applications.
American Institute of Biological Sciences, Washington, DC, USA
ISSN 0006-3568 11 pa
Several medium-long to long reviews.

1010 Books in Print
The current issue of this directory lists all books currently published or distributed in the USA, with information about price, ISBN, publisher, and other bibliographic information.
Bowker, New Providence, NJ, USA
ISSN 0068-0214 A
No reviews.

1011 Books in Print Supplement
Mid-year supplement to *Books in Print* (No. 1010), listing new and forthcoming books, price changes, out-of-print titles, etc.
Bowker, New Providence, NJ, USA
ISSN 0000-0310 A
No reviews.

1012 Books in Series
Lists in-print, reprinted and out-of-print popular, schlolarly and professional series of books, published or distributed in the USA.
Bowker, New Providence, NJ, USA
ISSN 0000-0906
No reviews.

1013 Books Out of Print
Lists publisher-verified entries for out-of-print titles, with full bibliographic details, together with other relevant information.
Bowker, New Providence, NJ, USA
V (latest Ed. 1984-88)
No reviews.

1014   British   Journal   of Management
Research papers on management-

oriented theories and topics from the whole range of business and management disciplines.
Wiley, Chichester & New York
ISSN 1045-3172 Q
No reviews.

1015 Bulletin of Science, Technology & Society
National Association for Science, Technology & Society
Designed to provide communication between as wide a range as possible of those interested in science, technology and society.
STS Press, University Park, PA, USA
ISSN 0270-4676 6 pa
Many short reviews.

1016 Bulletin of the Atomic Scientists, The
Articles on military, nuclear, environmental and other impacts of science and technology; also news items.
Educational Foundation for Nuclear Science, Chicago, IL, USA
ISSN 0096-3902 M
No reviews.

1017 Business Week
Articles on business, economics, current affairs, government, information processing, science, technology, etc.
McGraw-Hill, New York
ISSN 0007-7136 W
No reviews.

1018 Caduceus
Articles on health, healing, complementary medicine, and holistic medicine and science, with some news items.
Caduceus, Leamington Spa, Warks.
ISSN 0952-4584 Q
A few short to medium reviews.

1019 Choices
Articles and commentaries on human development, Third World, environmental, and other issues; includes news of creative initiatives.
United Nations Development Programme, New York
Q
No reviews.

1020 Community Development Journal
Articles on community development, seen as political, economic, and social programmes linking people with institutions and governments.
Oxford University Press, Oxford & New York
ISSN 0010-3802 Q
Medium to long reviews.

1021 Conflict Studies
Studies covering aspects of the causes, manifestations and trends of political instability and conflict in all parts of the world.
Research Institute for the Study of Conflict and Terrorism, London
ISSN 0069-8792 M
No reviews.

1022 Courier
(Formerly *UNESCO Courier*)
Semi-popular articles on education, science, and culture in most countries of the world.
UNESCO, Paris
ISSN 0041-5278 M
No reviews.

1023 Design Spirit
Articles and news items on various aspects of environmentally responsible design, including ecotechnology.
Design Spirit, New York
ISSN 1050-8252 Q
A few medium-long reviews.

1024 Development
Monitors current interdisciplinary debates on development thinking, and explores issues on the cutting edge of development.
Society for International Development (SID), Rome
ISSN 1011-6370 Q
Several short reviews.

1025 Development and Change
Interdisciplinary journal, with critical analyses and discussion papers on current issues of economic, social and cultural development.
Sage Publications, London & Newbury Park, CA, USA
ISSN 0012-155X
No reviews.

1026 Earthwatch
Aims to improve understanding of the Earth, the diversity of its inhabitants, and the processes affecting the quality of life on Earth.
Earthwatch Expeditions, Watertown, MA, USA
ISSN 8750-0183 6 pa
No reviews.

1027 Ecologist, The
Feature articles on a wide range of environmental, and also developmental, problems and issues; some articles on Green philosophy.
Ecosystems, Sturminster Newton, Dorset
ISSN 0261-3131 6 pa
Several short to medium-long reviews.

1028 Economic Policy
Authoritative analyses of the choices on current issues confronting economic policy makers.
Cambridge University Press, Cambridge & New York.
ISSN 0266-4658 2 pa
No reviews.

1029 Economist, The
Articles on world politics, current affairs, business, finance, science and technology; also has economic and financial statistics.
The Economist, London
ISSN 0113-0613 W
A few medium to medium-long reviews.

1030 Economy and Society
Papers discussing economics in

relation to society and politics.
Routledge, London
ISSN 0308-5147 Q
A few review articles only.

1031 Energy Economics
Articles on economic theory and its application to energy issues.
Butterworth-Heinemann, Oxford
ISSN 0140-9883 Q
No reviews.

1032 Energy Management
Energy Efficiency Office
Articles on energy conservation and energy efficiency, and global energy and environmental issues.
Bofoers Publishing, West Drayton, Midd.
6 pa
No reviews.

1033 Energy Policy
Balanced articles rigorously analysing economic, environmental, political, planning, and social aspects of energy supply and use.
Butterworth-Heinemann, Oxford
ISSN 0301-4215; M
A few long reviews.

1034 Environment
Articles, surveys, commentaries and special reports on a wide range of environmental and allied issues, written in an accessible style.
Heldref Publications, Washington, DC, USA
ISSN 0013-9157 10 pa
A few short reviews.

1035 Environment Business
Articles for senior management in industry, covering pollution control, environmental protection, and sustainable resource management.
Information for Industry, London
ISSN 0959-7042 24 pa

1036 Environment Today
Articles for executives, scientists, and engineers, operating facilities for controlling pollution and hazardous waste.
Enterprise Communications, McLean, VA, USA
9 pa

1037 Environmental Action
Action- and consumer-oriented articles, news, and information on key US environmental issues.
Environmental Action, Washington, DC, USA
ISSN 0013-922X 6 pa
A few medium to long reviews.

1038 Environmental Health
Articles on aspects of environmental health and public health.
The Institution of Environmental Health Officers, London
ISSN 0013-9297 M
No reviews.

1039 Environmental Pollution
International journal concerned with biological, chemical and physical aspects of environmental pollution and pollution control.
Elsevier Science Publishers, Barking,

Essex
ISSN 0264-7491 M
A few medium-long reviews.

1040 Environmental Science and Technology
Feature articles, precis articles, regulatory focus articles, and specialist research reports on environmental science and technology.
American Chemical Society, Washington, DC, USA
ISSN 0013-936X M
Several short reviews.

1041 Environmental Values
The White Horse Press, Isle of Harris, Scotland
Aims to justify environmental policy and clarify the relationship between practical policy issues, principles, and assumptions.
ISSN 0963-2719 Q

1042 Food Policy
Long articles and short reports on all aspects of the economics, planning and politics of agriculture, food and nutrition.
Butterworth-Heinemann, Oxford
ISSN 0306-9192 Q
Occasional medium reviews.

1043 Forbes
Covers US and international business and current affairs, finance, investment, law, computing, communications, science and technology.
Forbes Inc., New York.

ISSN 0015-6914 27 pa
No reviews.

1044 Fortune International
Articles on the economy, business, management, environment, innovation, news, and current affairs trends, in the USA and worldwide.
Time Inc., Zofingen, Switzerland
ISSN 0015-9018 M
Several short to medium reviews.

1045 Fourth World Review
Articles and letters on 'Fourth World', alternative, and human issues, for small nations, small communities, and the human spirit.
Fourth World Educational Research Association Trust, London
5 pa
Several short to long reviews.

1046 Future Survey
Summaries of books, articles, and reports concerning forecasts, trends, and ideas about the future, covering most aspects of life.
World Future Society, Bethesda, MD (No. 1647)
ISSN 0190-3241 M
Very many short to medium-long reviews.

1047 Future Survey Annual
Annual compilation of one year's reviews, with more than 600 entries originally published in *Future Survey* (No. 1046).

World Future Society, Bethesda, MD (No. 1647)
ISSN 0190-3241 A
Very many short to medium-long reviews.

1048 Futures
Multidisciplinary journal on long-term forecasting for decision making and policy making on all aspects of the future of humanity.
Butterworth-Heinemann, Oxford
ISSN 0016-3287 M
Several medium-long to long reviews.

1049 Futures Research Quarterly
Long articles reporting the findings of researchers and practitioners in futures research and allied fields.
World Future Society, Bethesda, MD, USA (No. 1647)
ISSN 8755-3317 Q
Occasional reviews only.

1050 Futurist, The
Articles on issues and possible futures in current affairs, the environment, science, technology, and other aspects of human life.
World Future Society, Bethesda, MD, USA (No. 1647)
ISSN 0016-3317 6 pa
Many short and a few medium-long reviews.

1051 Gaia Magazine
Popular articles covering all issues of interest to the Gaian movement, including harmonious and sustainable living.
Gaia Books, London & Stroud, Glos.
ISSN 0956-6449 V
A few short reviews.

1052 Gale Directory of Publications
Annual guide to newspapers, magazines, journals, and related publications. Coverage of US publications is especially comprehensive.
Gale Research, Detroit
ISSN 0892-1636 A
No reviews.

1053 Geoforum
International multidisciplinary journal, mainly concerned with geosciences and management of all aspects of the human environment.
Pergamon Press, Oxford & New York.
ISSN 0016-7185 Q
No reviews.

1054 Geographical Magazine
The Royal Geographical Society
Articles on physical and human geography, the environment, energy, economics, and politics; also briefings for A-level geography students.
World Publications, BBC Enterprises, London
ISSN 0016-741X M
No reviews.

1055 Geography
Papers on various aspects of

human geography, including population, agriculture, transport, and global inequalities.
Geographical Association, Sheffield (No. 1479)
ISSN 0016-7487 Q
Many short to medium reviews.

1056 Global Environmental Change
International journal on aspects of the environmental process, which are threatening the sustainability of life on Earth.
Butterworth-Heinemann, Oxford
ISSN 0959-3780 Q

1057 Green Magazine for Our Environment
Articles and news on important environmental issues and various aspects of the environment in relation to everyday life.
The Green Magazine Company, London
M
Several short reviews.

1058 Green Teacher
Cross-curricular international journal about good practice in Green school education, sharing experience on environmental issues.
Centre for Alternative Technology, Machynlleth, Powys, Wales
6 pa
Several medium to long reviews.

1059 Greenwave
Magazine presenting a 'genuine

Green alternative' viepwoint, and giving full support to many Green campaigning groups.
Greenwave, Wirral, Cheshire
Q
A few short to long reviews.

1060 Harvard Business Review
Articles on major US and international business themes; also case studies and debates.
Harvard Business Review
ISSN 0017-8012 6 pa
No reviews.

1061 Health Policy and Planning
London School of Hygiene and Tropical Medicine, London
Papers on various aspects of health policy and planning, concentrating on health policy issues in the developing world.
Oxford University Press, Oxford & New York
ISSN 0268-1080 Q
Several medium to medium-long reviews.

1062 Health Promotion International
World Health Organisation
Promotes the move to a new approach to public health worldwide, and action outlined in the Ottawa Charter for Health Promotion.
Oxford University Press, Oxford & New York.
ISSN 0957-4824 Q
A few short to medium-long reviews.

1063 Human Ecology
Interdisciplinary forum for papers on the complex and varied systems of interactions between people and their environment.
Plenum Press, New York & London
ISSN 0300-7839 Q
Several medium to long reviews.

1064 Human Relations
International interdisciplinary forum across the social sciences, covering work, home, community and other aspects of organised life.
Plenum Press, New York & London
ISSN 0018-7267 M
No reviews.

1065 IEEE Technology and Society Magazine
IEEE Society on Social Implications of Technology
Forum for free, informal aspects of social implications of technology, including environment and sustainable development.
IEEE, New York.
ISSN 0278-0097 Q
No reviews.

1066 Impact of Science on Society
UNESCO
Articles on important aspects of the impact of science and technology on humankind and the planet; many issues are on themes.
Taylor & Francis, London
ISSN 0019-2872 Q
No reviews.

1067 Intelligent Enterprise, The
Thought-provoking articles on policy issues relating to information management, policy making, business and information strategies, etc.
Aslib, London
ISSN 0964-4113 M (now discontinued)
A few long reviews.

1068 International Affairs
Royal Institute of International Affairs (No. 1519)
An impartial forum for discussion and debate on current international problems and issues.
Cambridge University Press, Cambridge
ISSN 0020-5850 Q
Many short to medium-long reviews.

1069 International Journal of Ambient Energy, The
Disseminates information on renewable energy sources and their potential use to serve human needs.
Ambient Press, Lutterworth, Leics.
ISSN 0143-0750 Q
Several short to long reviews.

1070 International Journal of Environment and Pollution
The European Centre for Pollution Research
Papers on all aspects of pollution, hazards and risks, and their management and policies; also covers 'clean' technologies.

Interscience Enterprises, Geneva, Switzerland
ISSN 0957-4352 Q
No reviews.

1071 International Journal of Environmental Studies (Sections A & B)
International forum for most aspects of environmental studies.
Gordon and Breach Science Publishers, London & New York
ISSN 0020-7233 8 pa

1072 International Journal of Urban and Regional Research
Articles and debates on urban and regional issues worldwide.
Blackwell, Oxford & Cambridge, MA, USA
ISSN 0309-1317 Q
Several medium to medium-long reviews, and a few review articles.

1073 International Management
Articles on business, finance, trade, and industry in Europe.
Reed Business Publishing, London
ISSN 0020-7888 M
No reviews.

1074 International Review of Applied Economics
Papers on applications of economic ideas to the real world, both empirical studies and economic policies, and their mutual interaction.
Arnold, London
ISSN 0269-2171 3 pa
A few review articles.

1075 International Security
Long articles on specific international security issues.
Center for Science and International Affairs, Harvard University, Cambridge, MA, USA
ISSN 0162-2889 Q
No reviews.

1076 International Social Science Journal
UNESCO
Theme-oriented issues on important world issues, including conflict research, democracy, Europe, and global environmental change.
Blackwell, Oxford & Cambridge, MA, USA
ISSN 0020-8701 Q
No reviews.

1077 Issues in Science and Technology
Aims to inform public opinion about scientific and technical issues, and raise the quality of public and private decision making.
National Academy of Sciences, Washington, DC, USA
ISSN 0748-5492 Q
Several long reviews.

1078 Journal of Applied Philosophy
Society for Applied Philosophy
Focus for philosophical research relevant to areas of practical concern, including politics, economics, science policy, and

education.
Carfax Publishing Co., Abingdon, Oxon.
ISSN 0264-3758 2 pa
Several long reviews.

1079 Journal of Applied Systems Analysis
(Formerly *Journal of Systems Engineering*)
Papers on various aspects of systems analysis, especially 'soft systems methodology' as applied to decision making, practical policies, etc.
Department of Systems and Information Management, University of Lancaster, Lancaster.
ISSN 0308-9591 A
Several long reviews.

1080 Journal of Business Ethics
Papers providing a wide variety of methodological and disciplinary perspectives on ethical issues related to business in the widest sense.
Kluwer Academic Publishers, Dordrecht, Netherlands, London & Boston, MA, USA
ISSN 0167-4544 M
A few medium reviews.

1081 Journal of Conflict Resolution
Peace Science Society
International interdisciplinary journal of social scientific theory and research on human conflict, mainly international conflicts.
Sage Publications, Newbury Park,
CA, USA & London
ISSN 0022-0027 Q
No reviews.

1082 Journal of Consumer Policy
Papers on a wide range of consumer issues in law, economics and behavioural sciences, and in other aspects of consumer affairs.
Kluwer Academic Publishers, Dordrecht, Netherlands, London & Boston, MA, USA
ISSN 0342-5843 Q
Quite a lot of short reviews and a few long reviews.

1083 Journal of Creative Behavior, The
Papers on human creativity, and its applications to education, the arts, invention, planning, etc., also on creativity in small groups.
Creative Education Foundation, Buffalo, New York.
ISSN 0022-0175 Q
No reviews.

1084 Journal of Development Studies
Articles and discussions on various aspects of Third World development.
Frank Cass, London
ISSN 0022-0388 Q
Several medium to long reviews.

1085 Journal of Environmental Management
Specialist papers on environmental impact assessment, environmental management,

environmental regulation, land use, etc.
Grange-over-Sands, Cumbria
ISSN 0301-4717 M
No reviews.

1086 Journal of Environmental Quality
Articles on environmental quality in natural and agricultural eco-systems, also on environmental issues and technological impacts.
Madison, WI, USA
ISSN 0047-2715 Q
Quite a lot of short to medium-long reviews.

1087 Journal of Environmental Systems
Papers on the analysis and solution of problems relating to the system-complexes, which constitute our total societal environment.
Baywood Publishing Company, Amityville, NY, USA
ISSN 0047-2473 Q

1088 Journal of Forecasting
International multidisciplinary journal, covering theoretical, practical, computational, and methodological aspects of forecasting.
Wiley, Chichester & New York
ISSN 0277-6693 6 pa
No reviews.

1089 Journal of International Affairs
(Formerly *Columbia Journal of International Affairs*)

Papers on the wide range of issues confronting contemporary international policy makers.
Columbia University, New York
ISSN 0022-197X 2 pa
Several long to very long reviews.

1090 Journal of Law and Society
(Formerly *The British Journal of Law and Society*)
Papers on the legal aspects of social problems, especially problems relevant to law and order.
Blackwell, Oxford & Cambridge, MA, USA.
ISSN 0263-323X Q
A few long reviews.

1091 Journal of Management Studies
Papers on research and practice in organisation theory, strategic management and human resource management.
Blackwell, Oxford & Cambridge, MA, USA
ISSN 0022-2380 6 pa
Several short to long reviews.

1092 Journal of Moral Education, The
Interdisciplinary forum for considering all aspects of moral education and development across the human lifespan.
Carfax Publishing Company, Abingdon, Oxon.
ISSN 0305-7240 3 pa
Several medium-long reviews.

1093 Journal of Peace Research
International   Peace   Research

Association
Articles with theoretical rigour, methodological sophistication and policy orientation, that make original contributions to peace research.
Sage Publications, Newbury Park, CA. USA & London
ISSN 0022-3433 Q
Many short reviews and occasional review essays.

1094 Journal of Public Policy
Articles, using concepts from any of the social sciences to analyse significant problems facing contemporary governments.
Cambridge University Press, Cambridge & New York
ISSN 0143-814X Q
Several medium-long to long reviews.

1095 Journal of Social Issues
Society for the Psychological Study of Social Issues
Each issue covers a specific social problem area or issue, often froma psychological viewpoint.
Plenum Press, New York & London
ISSN 0022-4537 Q
No reviews.

1096 Journal of Social Policy
Social Policy Association
Papers analysing social policy and administration, especially integrating theoretical ideas and concepts with empirical evidence.
Cambridge University Press, Cambridge & New York

ISSN 0047-2794 Q
Several medium-long to long reviews.

1097 Journal of the Institution of Water and Environmental Management
Specialist papers and comments on water and environmental management, and public planning issues.
Institution of Water and Environmental Management, London
ISSN 1951-7359 6 pa
Has reviews.

1098 Journal of Urban Economics
Research papers in the rapidly expanding field of urban economics.
Acadamic Press, San Diego, New York & London
ISSN 0094-1190 6 pa
No reviews.

1099 Knowledge
International interdisciplinary journal, covering the nature of expertise and the translation of knowledge into practice and policy.
Sage Publications, Newbury Park, CA, USA & London
ISSN 0164-0259 Q
No reviews.

1100 Long Range Planning
Strategic Planning Society
A leading journal on strategic planning, and on developing and implementing strategies and

plans, especially over the longer term.
Pergamon Press, Oxford
ISSN 0024-6301 6 pa
Many short and a few long reviews.

1101 Management Today
Articles on various aspects of management and business, also on related current affairs and technological issues.
Management Publications, London
ISSN 0225-1925 M
A few medium-long reviews.

1102 Natural Hazards
International Society for the Mitigation and Prevention of Natural Hazards
Papers on physical aspects of natural hazards, statistics of forecasting catastrophes, risk assessment, and precursion of hazards.
Kluwer Academic Publishing, Dordrecht, Netherlands, London & Boston, MA, USA
ISSN 0921-030X 3 pa
A few medium-long reviews.

1103 Nature
General articles, commentary, news, views and correspondence on major scientific topics, both research and wider impacts.
Macmillan, London
ISSN 0028-0836 W
Several short to medium reviews.

1104 Neometaphysical Digest
Articles seeking fundamental

solutions to human problems, and the understanding of consciousness and its manifestations.
Society of Metaphysicians, Hastings, E. Sussex
V
A few short reviews.

1105 New Dimensions
Articles on a variety of 'New Age', esoteric, and psychological themes, including 'mind versus matter' and symbolism.
New Dimensions, Irchester, Northants.
ISSN 0960-5959 M
Has several medium to long reviews.

1106 New Dimensions
Articles focussing on the real issues underlying major news events.
New Dimensions Publishing Company, Grants Pass, OR, USA
M

1107 New Frontier
Informative articles by, and interviews with, internationally respected leaders in the evolution of human consciousness.
New Frontier Education Society, Philadelphia, PA, USA
ISSN 0086-4616 M

1108 New Humanity, The
The world's first politico-spiritual journal, with no party political or religious affiliation, with a wide range of themes.

The New Humanity, London
ISSN 0307-0980 6 pa

1109 New Internationalist, The
Articles and news, concerned with how to achieve radical changes to meet the basic material, mental and spiritual needs all people.
New Internationalist Publications, London
ISSN 0305-9529 M
A few short reviews.

1110 New Paradigms
(Sequel to *New Paradigms Newsletter* (No. 1204))
Journal of fundamental conceptual change, with articles about new paradigms – fundamental new ways of thinking – and other new ideas and possibilities in science, philosophy, religion, current affairs, the humanities, the arts, indeed all aspects of life. Each issue is devoted mainly to a specific theme or group of related themes.
Adamantine Press, London
ISSN 0951-6026 Q
Very many short to long reviews.

1111 New Renaissance
Renaissance Universal
Articles on new economics, human rights, health, science, philosophy, and spirituality, with a holistic approach.
New Renaissance, Mainz, Germany
ISSN 0939-1657 Q
A few medium to long reviews.

1112 New Scientist
Many semi-popular features, articles and news items on advances in science and technology, and their applications and impact.
New Science Publications, London
ISSN 0262-4079 W
Several short to medium reviews.

1113 Noetic Sciences Review
Articles, interviews, research updates, and review essays on the people and ideas in the forefront of consciousness research.
Institute of Noetic Sciences, Sausalito, CA, USA
ISSN 0897-1005 Q

1114 Nonprofit and Voluntary Sector Quarterly
(Formerly *Journal of Voluntary Action Research*)
Association for Research on Nonprofit Organizations & Voluntary Action
Articles on various aspects and issues of voluntary organisations and other nonprofit organisations.
Jossey-Bass, San Francisco, CA, USA
ISSN 0899-7640 Q
No reviews.

1115 One Earth
Findhorn Foundation and Findhorn Community (No. 1578)
Articles on various 'New Age' themes and events; aims to share inspirational energy, leading to spiritual awareness.
One Earth, Findhorn, Forres,

Scotland
ISSN 0143-8247 Q
A few medium-long reviews.

1116 Orbis
Papers on international affairs and issues worldwide.
Foreign Policy Research Institute, Philadelphia, PA, USA
ISSN 0030-4387 Q
Many short reviews and a few essay reviews.

1117 Our Planet
(Formerly *UNEP News* and *Uniterra*)
United Nations Environment Programme
Articles and news promoting international awareness of all aspects of the present environmental and conservation crisis.
UNEP Information Service, Nairobi, Kenya
ISSN 1013-7394 Q

1118 Oxford Review of Economic Policy
Has special issues on important economic themes
Oxford University Press, Oxford & New York
ISSN 0266-903X Q
No reviews.

1119 Peace & Change
Scholarly articles on peace research and conflict resolution, related to achieving a peaceful, just, humane society.
Sage Publications, Newbury Park,

CA, USA & London
ISSN 0149-0508 Q

1120 Peace Courier
Articles on human rights violations, worldwide ecological issues, and nuclear weapons testing; news of activities fostering peace.
World Peace Council, Helsinki, Finland
ISSN 0031-594X M

1121 Peace Research
Canadian Peace Research & Education Association
Scientific and scholarly papers on world peace, focusing on the problems of war, armaments, peace movements, and human rights.
Brandon University, Brandon, Manitoba, Canada
ISSN 0008-4697 3 pa

1122 Philosophy & Public Affairs
Papers discussing the philosophical aspects of specific legal, social, and political problems and related philosophical issues.
Princeton University Press, Princeton, NJ, USA
ISSN 0048-3915 Q
No reviews.

1123 Planning Perspectives
School of Geography, University of Birmingham
Historical and perspective articles on many aspects of plan making, implications of planning, and

environmental issues.
Spon, London
ISSN 0266-5433 Q
No reviews.

1124 Policy Studies
Factual articles on studies of economic, industrial and social policy, and the workings of political institutions.
Policy Studies Institute, London
ISSN 0144-2782 Q
No reviews.

1125 Political Quarterly
Non-technical articles on issues of public policy, on the basis of knowledge of the most authoritative sources and expert opinions.
Political Quarterly Publishing Company, London
ISSN 0032-3179 Q
Several medium to long reviews.

1126 Population Studies
Specialist articles on demography and population studies, *not* discussion of population issues.
Population Investigation Committee, London School of Economics, London
ISSN 0032-3292 3 pa
Quite a lot of short to long reviews.

1127 Public Choice
Papers on the intersection between economics and political science.
Kluwer Academic Publishing, Dordrecht, Netherlands, London & Boston, MA, USA

ISSN 0048-5829 Q
A few long reviews.

1128 Public Interest, The
Articles about a wide range of issues of public interest, especially in the USA and Canada.
National Affairs, Washington, DC, USA
ISSN 0033-3357 Q

1129 Reconciliation Quarterly
Articles on pacifism and international peace issues.
Fellowship of Reconciliation, London
ISSN 0034-1479 Q
A few short to medium reviews.

1130 Regional Studies
Regional Studies Association
Original research papers and reviews on urban and regional development, with a broad, cross-disciplinary approach.
Carfax Publishing, Abingdon, Oxford
ISSN 0034-3404 7 pa
No reviews.

1131 Religion
Articles on all kinds of religions, and on religion in relation to contemporary global and ethical issues.
Academic Press, San Diego, New York & London
ISSN 0048-721X Q
Several medium to medium-long reviews.

1132 RERIC International Energy Journal

(Formerly *Renewable Energy Review Journal)*
Articles on solar energy and other forms of renewable energy, at an intermediate academic level, aimed at a broad range of readers.
Regional Energy Resources Information Centre (RERIC), Institute of Technology, Bangkok, Thailand (No. 1517)
ISSN 0857-6173 2 pa
No reviews.

1133 Research Policy
Papers on research policy, research management, research planning, research and development, innovation, and technology assessment.
North-Holland, Elsevier Science Publishers, Amsterdam
ISSN 0048-7333 6 pa
A few medium to long reviews.

1134 Resources, Conservation and Recycling
Papers and news on inter-disciplinary aspects of the management, and especially the conservation, of renewable and non-renewable resources.
Elsevier Science Publishers, New York & Amsterdam
ISSN 0921-3449 Q
No reviews.

1135 Resurgence
Articles on environment, sustainable economics and business, Green living, Green philosophy, spirituality, and other alternative themes.
Resurgence, Hartland, Bideford, Devon
ISSN 0034-5970 6 pa
Many short to medium-long reviews.

1136 Revision
Helen Dwight Read Educational Foundation
Articles on consciousness and change, including holistic, 'New Age', and global and personal transformation perspectives.
Heldref Publishers, Washington, DC, USA
ISSN 0275-6935 Q

1137 Risk Analysis
Society for Risk Analysis
Papers on new developments in risk analysis, including their mathematical, engineering, social, and psychological aspects.
Plenum Press, New York & London
ISSN 0272-4332 Q

1138 Science
Research news and reports, also policy discussions and news on the social impact of science and technology and on global issues.
American Association for the Advancement of Science, New York
ISSN 0036-8075 W
No reviews.

1139 Science & Public Affairs
The Royal Society (No. 1618) & the British Association for the Advancement of Science (No. 1451)

Popular articles on science and technology and their applications and implications.
The Royal Society, London
ISSN 0268-490X Q
A few medium to long reviews.

1140 Science and Public Policy
International Science Policy Foundation, London (No. 1590)
Papers on science and technology policy, and public issues about the impact of science and technology, in all parts of the world.
Beech Tree Publishing, Guildford, Surrey
ISSN 0302-3427 6 pa
Several medium to long reviews.

1141 Science as Culture
Articles on the social consequences of the natural sciences, especially in relation to global issues.
Free Association Books, London
ISSN 0950-5431 Q
No reviews.

1142 Science of the Total Environment, The
Papers on research about a variety of changes in the environment caused by human activities.
Elsevier Science Publishers, Amsterdam
ISSN 0042-9697 42 pa

1143 Scientific American
Semi-popular articles on recent advances in science and technology and their applications and impact; special issues each September, whose themes have included planet management (No. 756), energy (No. 757), computer communications and networks (No. 758) and mind and brain (No. 759).
Scientific American, New York
ISSN 0036-8733 M
Quite a lot of short to medium reviews.

1144 Scientific World, The
Articles on the social impact of science and technology, including global issues.
World Federation of Scientific Workers, Croydon, Surrey
ISSN 0036-8857 Q
No reviews.

1145 Social & Legal Studies
Critical papers and debates on law, jurisprudence and the criminal justice agenda, with a wide interdisciplinary perspective.
Sage Publications, Newbury Park, CA, USA & London
ISSN 0964-6639 Q
No reviews.

1146 Social Inventions
Presents social inventions on many aspects of life, including human relationships, education, economics, politics, and the environment.
Institute for Social Inventions (No. 1642)
ISSN 0954-206X Q
Several short to long reviews.

1147 Sociology
Debates and articles on sociological themes and sociology, in relation to social problems and issues, especially in the UK.
British Sociological Association (BSA) Publications, Solihull, West Midlands
ISSN 0038-0385 Q
Many medium-long reviews.

1148 Sociology and Social Research
Articles on sociology in relation to social problems, especially in the USA.
University of Southern California, Los Angeles, CA, USA
ISSN 0038-0393 Q
No reviews.

1149 Solar Energy
International Solar Energy Society
International journal for scientists, engineers and technologists, working on solar energy and its practical applications.
Pergamon Press, Oxford & New York
ISSN 0038-092X M
No reviews.

1150 South
Articles and news on economic development, especially in the Third World but also in Eastern Europe, and business and development issues.
South Media and Communications, london
ISSN 0260-6976 M
A few short reviews.

1151 Survival
International Institute for Strategic Studies
Articles on strategic studies and international studies.
Brassey's, London
ISSN 0039-6338 Q
Several medium-long reviews and a few essay reviews.

1152 Systems Practice
Articles on the development of systems thinking, and the use of systems methods to improve decision making and practical management.
Plenum Press, New York & London
ISSN 0894-9859 Q
Several medium to long reviews.

1153 Technological Forecasting and Social Change
Papers on technological forecasting as a social planning tool, and on the interaction of technology with integrative planning.
Elsevier Science Publishing, New York
ISSN 0040-1625 M
A few medium to medium-long reviews.

1154 Technology Analysis & Strategic Management
International research journal, linking the analysis of science and technology with the strategic needs of policy makers and management.
Carfax Publishing Company, Abingdon, Oxon.

ISSN 0953-7325 Q
A few medium reviews.

1155 Technology in Society
Papers on technology assessment, technology transfer, management of technology, technology and policy, economics and social aspects.
Pergamon Press, New York & Oxford
ISSN 0160-791X Q
No reviews.

1156 Technovation
Papers on technological innovation and entrepreneurship, with international coverage.
Elesevier Advanced Technology, Oxford & New York
ISSN 0166-4972 8 pa
No reviews.

1157 Teilhard Review, The
Journal bridging science and religion, also covering human issues, devoted to the thought of Teilhard de Chardin and its implications.
The Teilhard Centre for the Future of Man, London
ISSN 0050-2184 Q
Several short to medium reviews.

1158 Theory and Society
Papers, reviews, and critiques on current affairs and social issues.
Kluwer Academic Publishing, Dordrecht, Netherlands, London & Boston, MA, USA
ISSN 0304-2421 6 pa
Several long reviews.

1159 Time International
News articles and commentaries on current affairs, business, science, technology and culture.
Time International, Amsterdam
ISSN 0040-781X W
No reviews.

1160 Transnational Associations (Formerly *International Associations*)
Articles on miscellaneous aspects of international affairs, international associations, and new paradigms.
Union of International Associations, Brussels (No. 1557)M
ISSN 0250-4928 6 pa

1161 Transnational Perspectives (Formerly *World Federalist*)
International journal of world concerns, analysing the major trends in world society, and making policy suggestions for common interests.
Transnational Perspectives, Geneva, Switzerland
ISSN 0252-9505 3 pa
Many short to long reviews and a few essay reviews.

1162 Ulrich's International Periodicals Directory
The most comprehensive annual compilation of information about international periodicals, in three volumes; Vols. 1 and 2 contain a classified list of periodicals in alphabetical order of subject name, with alphabetical listing of

periodical name for each subject, and Vol. 3 has a list of periodicals in title order and various other lists.
Bowker, New Providence, NJ, USA
ISSN 0000-0175 A
No reviews.

1163 UNESCO Yearbook on Peace and Conflict Studies
Annual compilation of information about peace research and human coflict research.
Greenwood Press, Westport, CT, USA
ISSN 0250-779X A
No reviews.

1164 Urban Affairs
Interdisciplinary research papers and scholarly analyses on various urban themes, including urban life, economics, and policies.
Sage Publications, Newbury Park, CA, USA & London
ISSN 0042-0816 Q
A few medium-long reviews.

1165 Urban Studies
International forum of social and economic aspects of and contributions to urban and regional planning.
Urban Studies, Glasgow
ISSN 0042-0980 7 pa
Quite a lot of short to long reviews.

1166 Whitaker's Books in Print
(Formerly *British Books in Print*) Comprehensive list of UK books in print, with full bibliographic details of each book listed.
Whitaker, London & Bowker, New Providence, NJ, USA
ISSN 0068-1360 A
No reviews.

1167 Whole Earth Review
Articles, features and news on a wide range of holistic, alternative, Green, and spiritual themes.
POINT, Sausalito, CA, USA
ISSN 0749-5056 Q
Quite a lot of short to medium-long reviews.

1168 Whole Life
Articles offering new ideas and concepts on how to improve the quality of life, focussing on the body-mind-spirit connection.
Whole Life Enterprises, New York
ISSN 0888-2061 6 pa

1169 Work, Employment & Society
Articles on various social problems and issues, especially those relating to work and employment.
British Sociological Association, London
ISSN 0950-0170 Q
Several medium to medium-long reviews.

1170 World Development
Multidisciplinary international journal, studying and promoting world development, and improving living standards and quality of life.

Pergamon Press, Oxford & New York
ISSN 0305-750X M

1171 World Economy, The
Papers on international economic relations, especially national, continental and global trade policy issues.
6 pa
ISSN 0378-5920 6 pa
Several medium to medium-long reviews.

1172 World Futures
Vienna Academy for Global and Evolutionary Studies & General Evolution Research Group
Clarifying humanity's options and improving its prospects, by studying patterns of change and development in nature and society.
Gordon and Breach Science Publishers, New York & Reading, Berks.
ISSN 0260-4027 M

1173 World Futures General Evolution Studies Book Series
Vienna Academy for Global and Evolutionary Studies & General Evolution Research Group
Books on the study of general patterns of change and development in nature and society, focussing on evolutionary theory, and perspectives.
Gordon and Breach Science Publishers, New York & Reading, Berks.
ISSN 1043-9331 V
No reviews.

1174 World Today, The
Articles on current affairs worldwide.
The Royal Institute of International Affairs, London
ISSN 0043-9134 M
A few long reviews.

1175 World Wastes
(Formerly *Solid Wastes Management*)
Articles, news and product information on practical waste management.
Communication Channels, Atlanta, GA, USA
ISSN 0161-035X M
No reviews.

1176 World Watch
Promotes a vision of sustainable society, from an interdisciplinary viewpoint, and covers the whole range of global environmental issues.
Worldwatch Institute, Washington, DC, USA (No. 1651)
ISSN 0896-0615 6 pa

1177 Worldwatch Papers
Each issue is a paper, addressing some important environmental or other world issue, from a global, interdisciplinary viewpoint.
Worldwatch Institute, Washington, DC, USA (No. 1651)
V
No reviews.

1178 Zygon
Chicago Center for Religion and Science and 3 other organisations
Explores ways to unite what has

been disconnected in modern times: values and knowledge, goodness and truth, religion and science.

Blackwell, Oxford & Cambridge, MA, USA
ISSN 0591-2385 Q
A few long reviews.

3

NEWSLETTERS

Introductory Note

46 newsletters relevant to 21st century studies are listed here in alphabetical order of title. This is probably only a very small sample of all the relevant newsletters. Newsletters range from short newssheets to quite substantial journals which also have extensive news items; thus there is no hard and fast line between newsletters and periodicals. Each entry has as many as possible of the following lines: (1) title in bold type, but any subtitle is omitted; (2) any former title(s); (3) sponsoring organisation(s); (4) brief description of content, subject matter, aims and nature; (5) International Standard Serial Number (ISSN), if any, followed by frequency code; (6) number and frequency of book reviews, coded as for periodicals.

The frequency code is either a letter (A for annual, Q for quarterly, M for monthly, W for weekly, V for variable) or it has the form "n pa", where n is the number of times it currently appears per year.

List of Newsletters

1179 Brain/Mind Bulletin and Common Sense
Interface Press, Los Angeles, CA, USA
Ground-breaking news from the clinical, social, educational, creative and spiritual sciences, especially the study of the human brain, human potentialities, and the mind-body interface. It also includes articles on new paradigms and on politics, defined as "the realm of strategy that draws vision into the realm of matter".
ISSN 1064-671X M
Several short to medium reviews.

1180 Business Network Newsletter, The

The Business Network, London (No. 1376)
Articles, mainly about new approaches to business but sometimes about global and environmental issues; also various news items and rpeorts of the Business Network's meetings.
2 pa (not currently being published)
Several medium to long reviews.

1181 Collaborative Inquiry
Centre for the Study of Organizational Change and Development, University of Bath, Bath, Avon
Intended as a vehicle to build a community of people, who are practising research with people rather than research on people; articles, information about work in progress, and news items.
3 pa
Several reviews.

1182 Development Alternatives
Development Alternatives, New Delhi, India
Articles and news on issues of sustainable development in India and South East Asia, including environment, energy, and technology.
10 pa
No reviews.

1183 ENDS Report, The
Environmental Data Services, London (No. 1468)
Feature articles, information about environmental regulations, news of current environmental issues, market news, and other news items.
0966-4076 M
No reviews.

1184 Energy Report
Springfield Information Services, Peterborough
Many short to medium news items and reports on energy policy and technology, including new developments in energy sources, fuel and power, energy conservation and efficiency, and uses of energy.
ISSN 0093-7657 M
No reviews.

1185 Friends of the Earth Magazine
(Formerly *Not Man Apart*)
Friends of the Earth, Washington, DC, USA (No. 1478)
Informative articles and action news for environmental activists and other citizens concerned about the environment.
6 pa

1186 Gadfly
Ipswich Resource Centre, Ipswich, Suffolk (No. 1505)
Usually brief commentaries and news items about current issues, including global issues, and events.
M
No reviews.

1187 GA News
Geographical Association, Sheffield

(No. 1479)
News and views of education in geography in the UK.
6 pa
Quite a lot of short reviews.

1188 Global Family Update, The
Global Family, San Anselmo, CA, USA (No. 1394)
News items, features and contact list for an international network of individuals and groups, who are dedicated to experiencing unconditional love and peace in our lifetimes.
6 pa
A few short reviews.

1189 Globe, The
Global Partnership '92, London (No. 1395)
A networking and news bulleting for global partnerhip, and for agencies wotking for equity between peoples and for life on Earth.
V (broader launch being planned)
No reviews.

1190 Green Link
Milton Keynes Environmental Network, Milton Keynes (No. 1411)
News items about local environmental events and opportunities for enjoying and improving the natural environment in the Milton Keynes area; typical of many local environmental newsletters.
M
Very occasional short reviews.

1191 Green Paths Newsletter
Green Paths Centre, Dartington, Totnes, Devon (No. 1398)
A link for people who have attended Green Paths events, and for others concerned about Green and environmental issues; announcements of forthcoming Green Paths events, reports on past events, and news of Green Paths people.
V (started recently)
No reviews.

1192 Greenpeace
(Formerly *Greenpeace Examiner*) (No. 1485)
Greenpeace US National Office, Washington, DC, USA
Current news and articles on Greenpeace activities worldwide.
ISSN 0899-0190 6 pa

1193 Habitat
The Environment Council, London (No. 1466)
Miscellaneous news on the environment and environmental issues, especially nature conservation, and on future environmental events.
ISSN 0028-9043 M
Quite a lot of short reviews.

1194 Holistic Health Research Network Newsletter (No. 1401)
Holistic Health Research Network, Glasgow, Scotland
Articles and news items designed to develop communication between people interested in

research into holisitc health;
includes both conventional and
complementary   medical
approaches and techniques, and
emphasises a whole person
approach to health and health
care.
ISSN 0956-0483 2 pa
A few short reviews.

**1195 Ideas for Tomorrow Today
Bulletin (No. 1403)**
Ideas for Tomorrow Today, London.
Presents news of global and other
trends and events in a wider perspective, networking for a better
future, and linking with those on
the creative edge of change; by
and for people who are participating in transformation in their
work and in their lives.
6 pa
Several short reviews.

1196 IMRA News
*International Marketing Research
Association, London*
News items, announcements and
articles on qualitative and quantitative industrial marketing
research techniques and skills,
aimed at professional market
researchers.
M

1197 INES Newsletter
*International Network of Engineers
and Scientists for Global respon-
sibility (INES), Reinbek, Germany
(No. 1407)*
Short articles on global issues,

especially on war and peace and
impacts of science and technology, and news of INES and its
member organisations.
V (started in 1992)
No reviews.

**1198 International Social Science
Council Bulletin**
*International Social Science Council,
Paris (No. 1503)*
News of International Social Sci-
ence Council projects and
members.
Q
Quite a lot of short reviews.

1199 Living Green
*The LIFE STYLE Movement, Little
Gidding, Cambs. (No. 1506)*
Articles and news on mis-
cellaneous alternative themes,
and on the benefits, complexities
and difficulties of a less wasteful
and more sustainable way of
living.
3 pa
Several short to medium-long
reviews.

**1200 National Home Energy Rat-
ing Newsletter, The**
*National Energy Foundation, Milton
Keynes (No. 1602)*
News items about energy conser-
vation and energy efficiency,
especially in the home.
V
No reviews.

1201 New Consumer
New Consumer, Newcastle-upon-

Tyne (No. 1510)
Articles, commentary, news and views about special consumer issues; encourages informal debate mobilising consumer power for positive economic, social and environmental change.
ISSN 0958-7349 Q
A few medium reviews.

1202 New Economics
New Economics Foundation, London (No. 1604)
Articles and news on new economic themes, mapping out the emergence of a new economics based on democracy and sustainability.
ISSN 0951-6476 Q
A few short to medium-long reviews.

1203 New Options
New Options, Washington, DC, USA
Articles investigating new political issues beyond those of the traditional left and right, and reinterpreting national and international news; also a reader's debating forum.
ISSN 0890-1619 11 pa

1204 New Paradigms Newsletter
(Predecessor to *New Paradigms* (No. 1110)
New Paradigms Publications, Milton Keynes
Articles presenting new paradigms – fundamental new ways of thinking – and new ideas and new possibilities on many subjects and aspects of life; also a correspondence column and a few news items.
ISSN 0951-6026 V (final issue, February 1992)
Several short to long reviews.

1205 New World
United Nations Association, London (No. 1539)
News and commentary on the United Nations and on UK United Nations Association events and activities.
Q
A few short to medium reviews.

1206 Outlook on Science Policy
Beech Tree Publishing, Guildford, Surrey
News about policies for science and technology and about the International Science Policy Foundation. (No. 1590)
ISSN 0165-2818 M

1207 Planetary Connection
(Formerly *Planetary Link Up, Global Link Up & Link Up)*
Articles and news on 'New Age' and planetary themes and networking.
Planetary Networking Group, Broadway, Worcs.
ISSN 0261-5029 Q
A few short to medium-long reviews.

1208 Rainbow Ark
Rainbow Publications, London (see

No. 1424)
News and short articles, providing people with information which will enable them to consciously choose their beliefs, thoughts and customs to their effectiveness in creating a better world; has long lists of forthcoming meetings and events.
ISSN 0961-7779 Q
A few short to medium reviews.

1209 RERIC News
Regional Energy Resources Information Centre (RERIC), Asian Institute of Technology, Bangkok, Thailand (No. 1517)
Reports and news on alternative energy systems, especially using renewable energy, and their applications worldwide.
ISSN 0125-1775 Q
Several short to medium reviews.

1210 Research Council for Complementary Medicine Newsletter, The
The Research Council for Complementary Medicine, London (No. 1518)
News items about complementary medicine and especially about complementary medical research.
Q
No reviews.

1211 Reuse/Recycle
Technomatic Publishing Company, Lancaster, PA, USA
News of development in resource recycling and products made from recycling, especially in the USA but also in Europe.
ISSN 0048-7457 M
No reviews.

1212 Scientific and Medical Network Newsletter, The
The Scientific and Medical Network, Denham, Uxbridge, Midd. (No. 1419)
Articles and reports on a wide variety of scientific and medical themes, with special emphasis on complementary medicine and the relations between science and spirituality; also many news items and announcements of events of interest to Network members.
3 pa
Very many short to long reviews.

1213 Scientists for Global Responsibility
(Formerly *Electronics and Computing for Peace Newssheet*)
Scientists for Global Responsibility, London (comprising several formerly separate organisations) (No. 1524)
News, commentary, and announcements of events about science and technology in relation to issues of war and peace.
Q
Several short reviews.

1214 Social Policy
Social Policy Corporation, New York
Articles and news items on various social issues, including the environment, the workplace,

health, peace, civil rights, and policies.
ISSN 0037-7783 Q
No reviews.

1215 SPUR
World Development Movement, London (No. 1545)
News and feature articles on a wide range of Third World and environmental issues.
ISSN 0306-5367 6 pa
A few short reviews.

1216 TRANET Newsletter
TRANET, Rangeley, ME, USA (No. 1422)
Lists appropriate and alternative technology initiatives worldwide; each issue also contains a directory on a specific topic or region.
Q

1217 Turning Point 2000
Turning Point 2000, Cholsey, Oxon. (No. 1423)
Aims to encourage personal, local, national and international systematic programmes of change towards a one-world human community for the 21st century, that will enable people and conserve the Earth; reports activities, events and ideas contributing to these changes.
2 pa
Many short reviews.

1218 WARMER Bulletin
The WARMER (World Action for Recycling Materials & Energy from Rubbish) Campaign, Tunbridge Wells, Kent (No. 1543)
Technical features and news, acting as a worldwide information service to encourage the recycling of materials and energy from post-consumer waste; devoted to a safer and healthier environment.
Q
No reviews.

1219 Waste & Environment Today News Journal
AEA Technology, Harwell, Berks.
Articles, reports and news from the UK and abroad, on waste disposal, waste recovery, waste management, environmental hazards, and business aspects of waste.
M
A few medium to long reviews.

1220 What Doctors Don't Tell You
What Doctors Don't Tell You, London
Articles on the risks of drugs and modern medical procedures, and on latest medical research using safe alternatives to drugs and surgery; helps members of the public to take greater control over their health and that of their families, and publishes information normally confined to medical journals.
M

1221 World Federalist Bulletin

Association of World Federalists, Littlehampton, West Sussex (No. 1449)
News and commentary on international affairs and associated global issues such as war and peace.
Q
Several short to medium reviews.

1222 World Goodwill Newsletter
World Goodwill, London, Geneva & New York (No. 1426)
Combines commentary and information on world affairs and wider human and global issues with details of the work and programme of World Goodwill.
ISSN 0818-4984 Q
No reviews.

1223 World Peace News
World Peace News, New York
Worldwide report of governmental and other activities, emphasising what is and what is not happening at the United Nations, with non-governmental organisations (NGOs), and in the news media.
ISSN 0049-8130 6 pa

1224 World's Children
The Save the Children Fund, London (No. 1620)
Short articles and news of projects on behalf of children worldwide and on children's and mothers' problems and issues.
Q
A few short reviews.

4

NON-PRINT MEDIA

Introductory Note

This chapter is intended to be an introductory guide to sources of information, relevant to the 21st century, that are available on non-print media.

Section 4.1 provides some useful information on 27 items of literature *about* non-print media, including books on how to use them, a few periodicals about non-print media, and catalogues and listing of available titles in various non-print media. Its entries for books and periodicals are formatted as in Chapters 1 and 2 respectively.

The remainder of this chapter provides some useful specific examples of titles, listed in title order, in various commercially available non-print media. For each medium, the titles listed here are only a small sample of those that are relevant, but cross-references are made to the appropriate catalogues in Section 4.1. Each entry has the following lines:

(1) title and any subtitle in bold type;

(2) description of content and of any special features of the medium;

(3) date or frequency of publication and publisher.

Section 4.2 lists four microfiches, but only those that contain very extensive bibliographic material, because more modern media seem more appropriate for most other purposes. Section 4.3 gives 16 examples of relevant conventional CD-ROMs, which display textual, graphic, and pictorial information. In general, CD-ROMs have a slower but more easily understood user interface for information retrieval. Section

4.4 gives six examples of multi-media CD-ROMs, which present sound, moving pictures, and video, as well as ordinary visual information. Section 4.5 gives 84 examples of relevant online databases and database services. Section 4.6 gives one example of relevant personal computer databases, available on floppy disks. Section 4.7 gives four examples of relevant personal computer software, most of which can be run on IBM-compatible personal computers, together with five relevant books.

4.1 Literature about Non-Print Media

1225 Ambron, Sueann & Kristina Hooper
Interactive Multimedia: Visions of Multimedia for Developers, Educators & Information Providers
Computer-controlled multimedia are new and exciting, and already changing our view of traditional media and education. This book is based on one of a series of publications by Apple Computer, Inc., to provide information on how technology can be used in education. Its collection of 21 articles by leading multimedia researchers and developers is a useful source of ideas for software and hardware developers, educators, publishers and information providers. It is full of examples and pilot projects that are defining the new meaning of multimedia, and provocative philosophical and conceptual essays.
1988, Microsoft Press, Redmond, WA, USA, distributed in UK by Penguin Books, Harmondsworth, Midd., ISBN 1-55615-124-1 (pbk) xi + 340 pp.

1226 Armstrong, C. J. & J. A. Large
Manual of Online Search Strategies
The first two chapters of this book give general guidance on the use of online and CD-ROM databases, their evaluation and selection, and citation indexing. Most of the chapters provide guides to the available onlne and CD-ROM databases in specific subject areas, including: patents, natural sciences, agriculture, energy, the environment, engineering, mathematics, statistics, computer and information science and technology, social and behavioural sciences, law, business, economics, politics, the humanities, education, philosophy, and religion. The last chapter provides a list of databases for quick reference. The appendix provides a selective list of directories, bibliographies and reference works. There are indexes for: (1) databases, computer hosts and information providers; (2) source

material and search aids; (3) subjects. The book is primarily aimed at online searchers, but also CD-ROM searchers, who have some experience of information retrieval and databases in their own areas, but who need guidance on less familiar areas. The book also offers advice on databases and information retrieval to subject experts *without* knowledge of information retrieval.

2nd. Ed. 1992, Ashgate Publishing, Aldershot, Hants., ISBN 1-85742-007-1 (hbk) xv + 694 pp.

1227 Armstrong, C. J. & Large, J. A.
CD-ROM Information Products: The Evaluative Guide
This guide gives detailed evaluations of 23 CD-ROM products, Each entry gives information about a CD-ROM database and its search software, installation procedures, and documentation. For many of the products, the reviews are detailed enough to be a useful supplement to the product's user manual.

1991, Gower, Aldershot, Hants., ISBN 0-566-03645-2 (hbk) xiv + 408 pp.

1228 Barlow, Monica, John Button & Pat Fleming (Eds.)
ECO Directory of Environmental Databases 1992
This directory gives details of almost 300 environmental databases in the UK. It includes a comprehensive listing of voluntary and non-governmental organisations holding computerised environmental information, together with databases available commercially on CD-ROM and online networks. It helps to identify who holds what information on computer databases, and how access can best be obtained to these resources. It provides background information on the use and abuse of databases, going online, search services, and new projects. It has a bibliography and several indexes.

1992, ECO Environmental Education Trust, Bristol, ISBN 1-874666-00-8 (pbk) 291 pp.

1229 Cabeceiras, James
The Multimedia Library: Materials Selection and Use
This book aims to alert and inform its readers that a library's services can extend far beyond traditional print media into a wide variety of multimedia forms. It investigates the various media forms that library collections should include, and gives information about their nature, characteristics, technologies and contributors. Issues analysed include the selection and use of various media forms, and the unique capabilities and advantages of each of them. Detailed information is given about how each media form may

be used most effectively and efficiently, and on the selection of appropriate equipment for handling the media.

Academic Press, San Diego, New York & London, ISBN 0-12-153953-4 (hbk) xvii + 317 pp.

1230 CD-ROM International
This newsletter gives details of international developments in CD-ROM, CD-I, CDV and DVI media and technologies.
Transtex International, Paris
ISSN 0987-8238 18 pa
No reviews.

1231 CD-ROM Librarian
Critical, unbiased, informative analyses of CD-ROM producs, services and market developments; news, reports, articles, bibliographies and many other features to keep users up-to-date. It includes the monthly supplement to *CD-ROMs in Print* (No. 1235).
Meckler, Westport, CT, USA & London
ISSN 0893-9934 11 pa
Many reviews.

1232 Cox, John
Keyguide to Information Sources in Online and CD-ROM Database Searching
This book provides an integrated guide to the literature, reference aids, and key organisational sources of information about online and CD-ROM searching

worldwide. Part 1 gives an evaluative survey of the major information sources. Part 2 provides an annotated bibliography. Part 3 is an international directory of relevant organisations. There is an extensive index.
1991, Mansell, London & New York, ISBN 0-7201-2093-4 (hbk) xxiv + 247 pp.

1233 Desmarais, Norman (Ed.)
CD-ROM Local Area Networks
This book discusses the set-up, management, hardware and software issues, and copyright and licensing implications of installing a CD-ROM local area network in a library or information centre.
1991, Meckler, Westport, CT, USA & London, ISBN 0-88736-700-3 (hbk) 175 pp.

1234 Desmarais, Norman (Ed.)
CD-ROM Reviews 1987-1990: Optical Product Reviews from *CD-ROM Librarian*
This book is a compilation of the analytical reviews of CD-ROM products originally appearing in FICD-ROM Librarian (No. 1231).
1991, Meckler, Westport, CT, USA & London, ISBN 0-88736-733-X (hbk) 100 pp.

1235 Desmarais, Norman (Ed.)
CD-ROMs in Print 1992
This catalogue is an internationally compiled listing, providing an accurate and comprehensive guide to over 2700

commercially available CD-ROM titles. It has up to 26 items of information on each CD-ROM, including: title, provider, publisher, US and non-US distributors, format, computer and software details, price, date coverage, subject(s), equivalent online services, and description. It provides access to its data via eight indexes. The CD-ROM version of this book is No. 1258.
5th. Annual Ed. 1992, Meckler, Westport, CT, USA & London, ISBN 0-88736-780-1 (pbk) (ISSN 0891-8191) 450 pp.; 1st. Ed. 1988.

1236 ECO Environmental Education Trust
Surviving the Computer Jungle: Uses and Abuses of Computer in Environmental Information Provision
This is the report of the one-day seminar organised by the ECO Trust in Bristol in November 1991. It incldues: presentations by key speakers, workshop discussions of the issues involved in applying computer technology to environmental information, and listings of participants and relevant organisations.
1992, ECO Environmental Education Trust, Bristol, ISBN 1-874666- – (pbk) 32 pp.

1237 Ensor, Pat (Ed.)
CD-ROM Research Collections: An Evaluative Guide to Bibliographic and Full Text CD-ROM

Databases
This catalogue lists 114 CD-ROM databases chosen for their usefulness to bibliographic research. Each entry includes: an evaluative review, publisher, equipment requirements, price, review citations, arrangement and control, scope and content, and additional information.
1991, Meckler, Westport, CT, USA & London, ISBN 0-88736-779-8 (hbk) 302 pp.

1238 Finlay, Matthew (Ed.)
The CD-ROM Directory 1993
This comprehensive international guide to CD-ROM is rapidly expanding with successive editions. Its detailed information on products and services includes: (1) 3,597 CD-ROM and multimedia titles in every field of interest; (2) many multimedia titles, in CD-I, CDTV, CD-ROM XA, VIs and Data Discman; (3) CD-ROM hardware; (4) CD-ROM software and applications; (5) over 2,800 worldwide company profiles; (6) books and journals on CD-ROM; (7) conferences and exhibitions on CD-ROM. There are six extensive indexes. The directory is invaluable for all CD-ROM industry professionals, current users of CD-ROM products, and those anticipating using CD-ROM. Its CD-ROM version is No. 1257.
9th. Ed. Dec. 1992, TFPL Publishing, London, ISBN 1-870889-30-4 (pbk) over 1,000 pp; 1st. Ed. 1986. A

new printed edition is published every year.

1239 CD-ROM Librarian
Covers all aspects of CD-ROM and CD products, including the latest developments in technology and applications, product reviews, and a monthly update to *CD-ROMs in Print.*
Meckler, Westport, CT, USA & London
ISSN 0893-6634 11 pa
*No reviews.

1240 CD-ROM Professional
Provides information on what really works in CD-ROM equipment and software, announces new CD-ROM products; its coverage includes multimedia.
TFPL Publishing, London
ISSN 1049-0833 6 pa
Long reviews of many commercial CD-ROM titles.

1241 Information World Review
A leading European information industry newsletter for the information community, with news for both users and producers of electronic information and allied services, including new product announcements, database developments, electronic and optical publishing, multimedia, software companies, and computer networking.
Learned Information, Oxford
ISSN 0950-9879 M
A few short reviews.

1241 Jones, Claire & William Dowland
Online Sources of European Information: Their Development and Use
This book contains information about the European information industry, the business use of online information, and European Community information, telecommunications and policies. It contains eight appendices and a bibliography.
1990, Avebury, Aldershot, Hants., ISBN 1-85628-010-1 (hbk) viii + 125 pp.

1243 Kessler, Jack
Directory to Fulltext Online Resources 1992
This book is an introduction to and a directory of current fulltext online services and CD-ROMs. It also describes specific fulltext projects in detail, together with online public access catalogues (OPACs), electronic conferences, and electronic bulletin boards, that also contain fulltext material. Each entry describes the origination and content of a given service, and provides access instructions and other background information.
1992, Meckler, Westport, CT, USA & London, ISBN 0-88736-833-6 (hbk) 165 pp.

1244 Marcaccio, Kathleen Young et al.
Directory of Online Databases, Vol. 13, No. 1

This directory includes descriptions of 4,447 online databases, 2,034 online database producers, and 772 online database services. Each entry for a database includes available information on: alternate name and former name (if any), database type (e.g., bibliographic, fulltext, numeric), subject(s) covered, producer, online service, content and description, language(s), coverage (e.g., international, US), time span, updating, and alternate formats. Each entry for a database producer includes a list of the databases that it publishes. Each entry for a database service includes a list of its available databases. There are geographic, subject, and master (alphabetical name) indexes.

Jan 1992, Cuadra/Gale, Gale Research, Detroit, MI, USA, ISBN 0-8103-8428-0 (pbk) (ISSN 0913-6840) ix + 914 pp.

1245 Meadow, Charles T.
Text Information Retrieval Systems
The processes of seeking and finding information in a datbase, using a computer, all too often seems excessively complex. This book demystifies the process, and makes it accessible to all users. It concentrates on *interactive* information retrieval. It presents the workings of actual information retrieval systems, discusses the ambiguity of language and systems, explains techniques for overcoming problems, and considers the roles of users, the representation of information, database content, and computer aspects. Although some of its sections need some knowledge of programming languages and elementary algebra, its general meaning can be appreciated sufficiently well with only a little knowledge of information technology.

1992, Academic Press, San Diego, New York & London, ISBN 0-12-487410-X (hbk) xix + 303 pp.

1246 Meckler
CD ROM Market Place 1991: An International Guide
This book is an invaluable guide for librarians and industry professionals in CD-ROM publishing. It enables the user to locate any product, organisation or individual in the CD-ROM industry, and provides details of over 800 CD-ROM publishers and distributors worldwide. Each entry includes: name of firm, address, telephone number, fax number, key personnel, product line(s), date of first CD-ROM publication, and date of firm's foundation.

1991, Meckler, Westport, CT, USA & London, ISBN 0-88736-680-5 (pbk) (ISSN 1047-966X) 176 pp. A new edition is published each June.

1247 Multimedia
Articles, commentary, and news on all forms of multimedia; topics

covered include: events, new products, production, applications and issues.
EMAP Maclaren, London
Q
Reviews of products, but not books.

1248 Oppenheim, Charles (Ed.)
CD-ROM: Fundamentals to Applications
This is one of the first books on CD-ROM technology, and is written by international experts who cover many aspects of CD-ROMs and include several case studies. Topics covered include: optical storage media, current CD-ROM products, CD-ROM publishing, user-interface design, applications of CD-ROM, and the marketing of CD-ROM. Sources of information are given, together with a guide to keeping up-to-date.
1988, Butterworths, London & Boston, ISBN 0-408-00746-X (hbk) xii + 308 pp.

1249 Optech
Optech CD-ROM Product Descriptions & Price List
This dealer's catalogue contains a useful list of representative conventional CD-ROMs and multimedia CD-ROMs on a wide variety of topics, and even games, that can be obtained through Optech; some of the more relevant items listed here are reviewed in Sections 4.3 and 4.4. Each entry

gives a comprehensive outline, name of publisher, price, and type(s) of personal computer which can play the CD-ROM. Details are also given of CD-ROM computers and workstations, CD-ROM drives, and CD-ROM networking equipment.
1992, Optech, Farnham, Surrey; updated several times a year.

1250 Science Reference Information Service (SRIS)
Online Databases Available at SRIS
This booklet lists online database services, accessible from SRIS' Online Search Centre by people who visit it at 25 Southampton Buildings, London WC2A 1AW. It includes information about costs of use and gives some preliminary guidance to users.
British Library, London, updated from time to time.

1251 Sibley, J. F. (Ed.)
UK Online Search Services
This directory lists British organisations offering an online database brokerage service. It includes information on availability, charges, speed of service, expected enquirer participation, CD-ROM availability, etc. It also lists for each entry the host database computers accessed, the subject specialisations offered, and the availability of services for selective dissemination of information (SDI). It has geographical,

online database, and CD-ROM indexes.
1991, Aslib, London, ISBN 0-85142-288-8 (pbk) viii + 154 pp.

See also Handbook of Special Librarianship and Information Work (No. 276).

4.2 Microfiches

1252 Books in Print on Microfiche
Microfiche version of *Books in Print* (No. 1010 & 1011).
Bowker, New Providence, NJ, USA, annual.

1253 Books in Print on Microfiche: Subjects
Microfiche version of *Books in Print: Subjects* (No. 1010).
Bowker, New Providence, NJ, USA, annual.

1254 British Books Out of Print on Microfiche
Microfiche version of *Books Out of Print* (No. 1013).
Whitaker, London, annual.

1255 Whitaker's Books in Print on Microfiche
Microfiche version of *Whitaker's Books in Print* (No. 1166).
Whitaker, London, annual.

4.3 CD-ROMs

For guides to the use of CD-ROMs, see Nos. 1226, 1227, 1229-1233, *1238 & 1248. For recent information about products, see Nos. 1234, 1238, 1240 and 1241. For recent information about publishers, see 1246. For catalogues and lists of CD-ROMs, see Nos. 1226-1228, 1231, 1232, 1235, 1237 & 1239-1241. Vol. 3 of Ulrich's* International Periodicals Directory *(No. 1162) lists about 500 titles of periodicals available on CD-ROM.*

1256 Books in Print
CD-ROM version of *Books in Print* (Nos. 1010 & 1011).
Bowker, New Providence, NJ, USA, annual.

1257 CD-ROM Directory on Disc, The (Finlay, Matthew (Ed.))
This comprehensive international guide to CD-ROM is rapidly expanding with successive editions. Its detailed information on products and services includes: (1) 2,200 CD-ROM titles in every field of interest; (2) many multimedia titles, in CD-I, CDTV, CD-ROM XA, and DVI; (3) CD-ROM hardware; (4) CD-ROM software and applications; (5) 2,600 worldwide company profiles; (6) books and journals on CD-ROM; (7) conferences and exhibitions on CD-ROM. The directory is invaluable for all CD-ROM industry professionals, current users of CD-ROM products, and those anticipating using CD-ROM. Its printed version is No. 1238.
9th. Ed. Dec. 1992, TFPL Publishing, London, ISBN 1-870889-32-0;

1st. Ed. 1986. A new CD-ROM edition is published every six months.

1258 CD-ROMs in Print 1992 (Desmarais, Norman (Ed.))
This catalogue is an internationally compiled listing, providing an accurate and comprehensive guide to over 2700 commercially available CD-ROM titles. It has up to 26 items of information on each CD-ROM, including: title, provider, publisher, US and non-US distributors, format, computer and software details, price, date coverage, subject(s), equivalent online services, and description. It provides access to its data by searching the whole database. Users can search for any combination of words in any field or combination of fields. The printed version of this book is No. 1235.
5th. Annual Ed. 1992, Meckler, Westport, CT, USA & London, ISBN 0-88736-812-3; 1st. Ed. 1988.

1259 CIA World Factbook, The
A yearly world almanac, compiled by the Central Intelligence Agency, that provides extensive information about the geographic, economic and political characteristics of all the countries of the world, together with maps, national statistics and much additional information.
Quanta Press, USA, distributed in the UK by Optech, Farnham, Surrey, updated annually. For the multimedia version, see No. 1275.

1260 Digital Chart of the World
Detailed atlas of the world, on four disks, using digitised maps based on air charts at a scale of 1:1,000,000. The software allows a personal computer screen to become a window onto the atlas, which can be scrolled all round the world and zoomed in and out, adding or removing detail as required. Maps can also be printed out in special formats.
Chadwyck Healey, Cambridge.

1261 Economist on CD-ROM, The
Up to four years of *The Economist* (No. 1029) are contained on one CD-ROM, one covering 1987 to 1990, and a second covering 1990 onwards.
FT Profile & Chadwyck Healey, Cambridge, updated quarterly.

1262 Encyclopedia of Associations
CD-ROM version of the *Encyclopedia of Associations* (No. 1428), providing detailed descriptions of over 87,000 organisations, including over 10,000 international associations.
Gale Research, Detroit & London, updated every six months.

1263 Facts on File,
Draws from 75 leading US and non-US sources to survey significant events and trends worldwide; contains full-test news for ten years.

Facts on File, USA, distributed in the UK by Optech, Farnham, Surrey.

1264 Financial Times on CD ROM
Contains all the *Financial Times* news, features, editorial comment, and surveys for one year, in fully searchable format, but *excludes* tables of financial data and other statistics, graphics, pictures, and news agency and other copyright materials.
Chadwyck Healey, Cambridge, updated annually.

1265 Guardian, The
Contains all the *Guardian* news, features and editorials for one year. The whole text can be searched easily, quickly and comprehensively, or browsed. The distinctive design and appearance of the newspaper is preserved.
Chadwyck Healey, Cambridge, updated annually.

1266 Hutchinson's Encyclopedia
This is the CD—ROM version of the oldest established one-volume encyclopedia in the UK. It is very easy to use. For the multimedia version, see No. 1274.
1991, Attica Cybernetics, Oxford.

1267 Independent, The
Contains all the *Independent* news, features and editorials for one year. Information can be retrieved from it very easily.

Chadwyck Healey, Cambridge, annual and also three-year CD-ROMs.

1268 New Grolier Electronic Encyclopedia, The: All 21 Volumes of the *American Academic Encyclopedia* on a Single CD-ROM
This is an example of a conventional CD-ROM encyclopedia. It comes in a pack with CD-ROM, user's guide booklet and floppy disk. Over 30,000 articles on a wide variety of subjects are included, which can be searched by title, word or phoneme, using pull-down menus. Up to ten articles can be displayed at once on the screen window. For an up-to-date multimedia version, see No. 1276. For the online version, see No. 1280.
1988, Grolier Electronic Publishing, Danbury, CT, USA, ISBN 0-7122-3818-0.

1269 Times, The & Sunday Times
Contains all the *Times* and *Sunday Times* news, features and editorials for one year. Information can be retrieved from it very easily.
Chadwyck Healey, Cambridge, annual, various prices.

1270 Time Magazine Compact Almanac, The
The second edition has over 10,000 articles from *Time* from 1989 to well into 1991, and major stories and selected articles from

1923 to 1988. See also No. 1277 for multimedia version.
Compact Publishing, USA, distributed in the UK by Optech, Farnham, Surrey.

1271 Whitaker's Books in Print on CD-ROM

CD-ROM version of *Whitaker's Books in Print* (No. 1166).
Whitaker, London, annual.

4.4 Multimedia CD-ROMs

For guides to the use of multimedia CD-ROMs, see Nos. 1225, 1229, 1230, 1238 & 1247. For recent information about products, see Nos. 1230, 1234, 1238, 1240, 1241, 1247 & 1249. For recent information about publishers, see 1246. For catalogues and lists of CD-ROMs, see Nos. 1235, 1238, 1239 & 1249.

1272 Compton's Multimedia Encyclopedia

This is a multimedia CD-ROM with text, graphics, animation and sounds. It includes the 8-million word text of *Compton's Encyclopedia*, published by Encyclopedia Britannica, and is very enjoyable and easy to use. For example, it can be used to 'travel back' to any point in American history to about 500 years ago, or explore in depth any of about 1,000 locations in the world.
1991, Britannica Software, USA, distributed in the UK by Optech, Farnham, Surrey.

1273 Ecodisc

Simulation of a real nature reserve, allowing users to explore, experiment, observe, collect information, consult experts, view species data projections up to 50 years' ahead, and manage the reserve. Slide shows can be made of pictures and text collected or created during a session. The simulation provides good opportunites for problem solving, and is especially suitable for older schoolchildren. There is a multilingual presentation in English and eight other European languages.
1991, Multimedia Corporation, USA, distributed in UK by Optech, Farnham, Surrey.

1274 Hutchinson's Multimedia Encyclopedia

This is the first British fully multimedia encyclopedia. Besides texts of articles, there are illustrations, colour photographs, moving pictures, and also high-quality sounds, including famous speeches and music. For a conventional CD-ROM version, see No. 1266.
1991, Attica Cybernetics, Oxford.

1275 Multimedia World Factbook & CIA World Tour, The

A yearly world almanac, compiled by the Central Intelligence

Agency, that provides extensive information about the geographic, economic and political characteristics of all the countries of the world, together with maps, national statistics and much additional information. It includes actual performances of national anthem segments.

Bureau Development, USA, distributed in the UK by Optech, Farnham, Surrey, updated annually. For the conventional CD-ROM version, see No. 1259.

1276 New Grolier Multimedia Encyclopedia 1992, The
This multimedia CD-ROM encyclopedia features all 21 volumes of the *Academic American Encyclopedia*, with over 30,000 articles, some of them very up-to-date, on a wide variety of subjects. Besides texts of articles, there are illustrations, colour photographs, moving pictures, and also high-quality sounds, including famous speeches, music, animal voices and bird calls. There are over 250 high resolution colour maps, and over 3,000 pictures, many of which can be printed. For an earlier conventional CD-ROM version, see No. 1268. For the online version, see No. 1280.

1992, Grolier Electronic Publishing, Danbury, CT, USA, distributed in UK by Optech, Farnham, Surrey.

1277 Time Magazine Compact Almanac, The

The second edition has over 10,000 articles from *Time* (No. 1159) from 1989 to well into 1991, and major stories and selected articles from 1923 to 1988. On a multimedia personal computer, motion and sound videos can be played, showing important events from 1989 to 1991. See also No. 1270 for conventional CD-ROM version.

Compact Publishing, USA, distributed in the UK by Optech, Farnham, Surrey.

4.5 Online Databases

For guides to the use of online databases, see Nos. 1226, 1229, 1232, 1242, 1243 & 1245. For catalogues and lists of online databases, see Nos. 1226, 1228 (environmental data), 1232, 1242, 1243, 1244 (the most comprehensive catalogue), 1250 & 1251. Vol. 3 of Ulrich's International Periodicals Directory (No. 1162) lists about 3,000 titles of periodicals available on online databases.

1278 ABI/INFORM
Mainly American coverage of methods, techniques and strategies of value in business decision making, management, marketing, finance, economics, business data processing, etc.

UMI/Data Courier, Louisville, KY, USA

1279 Abstracts on Tropical Agriculture

Online version of *Abstracts on Tropical Agriculture*, giving a worldwide coverage of tropical and subtropical agricultural literature.
Royal Tropical Institute, Information and Documentation, Amsterdam

1280   Academic   American Encyclopedia
Online version of the printed *Academic American Encyclopedia*. For conventional and multimedia CD ROM versions, see Nos. 1268 and 1276, respectively.
Grolier Electronic Publishing, Danbury, CT, USA

1281 Acid Rain
Covers causes, effects and countermeasures to this major environmental problem.
Bowker A & I Publishing, New Providence, NJ, USA

1282 Acompline/Urbaline
Covers urban affairs, especially aspects of interest to social scientists, natural scientists and engineers.
London Research Centre, London

1283 AGRICOLA
Covers worldwide literature on agriculture, food, nutrition, forestry, natural resources, agricultural economics, rural sociology, ecology, and biological sciences.
US National Agricultural Library, Beltsville, MD, USA

1284 Air Pollution Technical Information Center File (APTIC)
Covers literature on the source, effects, prevention, and control of air pollution, and on its economic, social, political, legal, and administrative aspects.
Library Services Office, US Environmental Protection Agency, Research Triangle Park, NC, USA

1285 Applied Science and Technology Index
Online version of *Applied Science and Technology Index*; covers English-language publications in the applied sciences and technology.
H. W. Wilson Company, New York

1286 Aqualine Abstracts
Covers worldwide literature on every aspect of water, wastewater, and the aquatic environment.
Library and Information Services, WRC, Marlow, Bucks.

1287 Aquatic Sciences and Fisheries Abstracts
Covers worldide literature on the science, technology, and management of marine, brackish, and fresh water environments, including their pollution, environment, and fisheries.
Aquatic Sciences and Fisheries Information System, FAO, Rome,

1288 BBC Summary of World Broadcasts
Fulltext database, containing the

complete texts of English translations of radio and television broadcasts and news agency reports from about 130 countries; covers political and economic news, and information on trade, industry, comunications, transport, energy, agriculture, fisheries, etc.
FT PROFILE, London

1289 Bibliographic Index
Online version of *Bibliographic Index*, covering many bibliographies on academic and general interest topics.
H. W. Wilson Company, New York

1290 Biology Digest
Online version of *Biology Digest*, covering worldwide life science literature, oriented to life science courses.
Plexus Publishing, Medford, NJ, USA

1291 BIOSIS Previews
Online version of *Biological Abstracts*, covering all aspects of biology, and including US biological patents.
BIOSIS, Philadelphia, PA, USA

1292 Biotechnology Abstracts
Covers journal articles and patents on all scientific and technical aspects of biotechnology.
Derwent Publications, London

1293 Book Data
Covers citations and abstracts of currently available and forthcoming English-language books and non-serial publications from about 900 British, European, and North American publishers.
Book Data, Twickenham, Midd.

1294 Book Review Digest (BRD)
Online version of *Book Review Digest*, giving extracts from and references to current reviews of English-language books.
H. W. Wilson Company, New York

1295 Book Review Index (BRI)
Online version of *Book Review Index*, giving citations of reviews of books and periodicals in about 500 journals, mainly in the humanities, social sciences, education, and library science.
Gale Research, Detroit, MI, USA

1296 Books in Print
Online version of *Books in Print* (Nos. 1010 & 1011), issued annually.
Bowker, New Providence, NJ, USA

1297 Bowker's International Serials Database
Online version of *Ulrich's International Periodicals Directory* (No. 1162) and associated publications.
R. R. Bowker, New Providence, NJ, USA

1298 CAB Abstracts
Provides online access to worldwide literature in agriculture, forestry, veterinary medicine,

nutrition, agricultural biotechnology, agricultural economics, rural development, rural sociology, etc.
CAB International, Wallingford, Oxon.

1299 CD-ROM Databases
Covers currently marketed databases on CD-ROM.
Worlwide Videotex, Boston, MA, USA

1300 Chemical Exposure
Covers literature on toxicology and biomedical effects of pesticides, drugs, and other chemicals.
Science Applications International Corporation, Oak Rideg, TN, USA

1301 COMPENDEX
Covers world literature in engineering and technology, including managerial aspects.
Engineering Information, New York

1302 COMPUSCIENCE
Covers European and North American literature on computer science.
FIZ Karlsruhe, Eggenstein-Leopoldshafen, Germany

1303 Computer Database
Covers literature on computers, computer applications, the computer industry, telecommunications, electronics, and related fields.
Information Access Company, Foster City, CA, USA

1304 Conference Papers Index (CPI)
Covers papers presented at scientific and technical conferences worldwide.
Cambridge Scientific Abstracts, Bethesda, MD, USA

1305 CSA Life Sciences Collection (LSC)
Provides online access to 15 abstracts journals in genetics, microbiology, ecology, etc.
Cambridge Scientific Abstracts, Bethesda, MD, USA

1306 Cumulative Book Index
Online version of *Cumulative Book Index*, covering English language books published worldwide.
H. W. Wilson Company, New York

1307 Current Biotechnology Abstracts (CBA)
Covers the scientific, technical and technocommercial literature of biotechnology, including journal articles, patents, and news items.
Royal Society of Chemistry, Information Services, Cambridge

1308 Current Research in Britain
Covers research carried out in British universities and government departments.
British Library, Document Supply Centre, Boston Spa, Wetherby, W. Yorks.

1309 Current Research Information System (CRIS)

Covers ongoing and recently completed research projects in agriculture, forestry, and biology. *Cooperative State Research Service, US Department of Agriculture, Beltsville, MD, USA*

1310 Current Technology Index (CTI)
Indexes leading British journals in all fields of modern technology and applied science.
Bowker-Saur, London

1311 Dissertation Abstracts Online
Covers accepted dissertations at accredited North American institutions and over 200 institutions elsewhere.
University Microfilms International, Ann Arbor, MI, USA

1312 Educational Resources Information Centre (ERIC)
Covers literature in all education and education-related areas.
US Department of Education, Washington

1313 Electronic Publishing Abstracts (EPUBS)
Provides technical and market coverage of information technology, in cluding transmission, storage, input and retrieval of text and pictures.
Pira International, Information Services, Leatherhead, Surrey

1314 Encyclopedia of Associations

Online version of the *Encyclopedia of Associations* (No. 1428), providing detailed descriptions of over 87,000 organisations, including over 10,000 international associations.
Gale Research, Detroit & London, updated every six months.

1315 Energy Information Data Base
Covers literature on all types and all aspects of energy and energy systems.
International Research and Evaluation, Eagan, MN, USA

1316 Energy Science and Technology
Covers the world's scientific and technical information on energy.
US Department of Energy, Washington, DC, USA.

1317 Energyline
Covers energy issues and problems, including economic aspects, research and development, unconventional enrgy sources, conservation, consumption, and environmenal impact.
Bowker A & I Publications, New Providence, NJ, USA

1318 Enviroline
Covers the major international environmental and resources literature, including sections on management, law and political science.
Bowker A & I Publishing, New Providence, NJ, USA

1319   Environmental Bibliography
Online version of *Environmental Periodicals Bibliography*; covers literature on the environment, including pollution, waste management, recycling, health hazards, alternative energy, and urban planning.
International Academy at Santa Barbara, Santa Barbara, CA, USA

1320 Foreign Trade and Economics Abstracts
Covers worldwide literature on foreign trade, economics, and management, including: markets, import regulations, investment, economic situations, and country-specific data.
Netherland Foreign Trade Agency Library, The Hague

1321   FOREST   Information Retrieval System
Covers various aspects of forest and wood products, ranging from harvesting trees to marketing.
Forest Products Research Society, Madison, WI, USA

1322   Gale's   Publications Database
Online version of *Gale Directory of Publications* (No. 1052) providing descriptive and content information about newsletters, magazines, directories, publishers, etc.
Gale Research, Detroit, MI, USA

1323 GEOBASE
Covers worldwide literature on geology, mineralogy, geophysics, climatology, geography, ecology, environment, natural resources, Third World studies, and transport.
Elsevier/Geo Abstracts, Norwich, Norfolk

1324 GreenNet
Worldwide online information and electronic mail system for Green activists and others involved or interested in environmental and Green issues.
GreenNet, London

1325 Health and Safety Science Abstracts
Covers the interdisciplinary science of safety and hazard control, in relation to identifying, controlling or eliminating hazards.
Cambridge Scientific Abstracts, Bethesda, MD, USA

1326 ICONDA
Covers technical and managerial aspects of civil engineering, building, construction, architecture, and town planning, including appropriate technology, computer-aided design and project management.
Fraunhofer-Gesellschaft, Stuttgart, Germany

1327   Information   Sciences Abstracts
Online version of *Information Sciences Abstracts*, covering worldwide literature on information

science and related areas, including documentation, library science, information services, and information storage and retrieval.
IFI/Plenum Data Company, Alexandria, VA, USA

1328 INSPEC
Provides worldwide technical coverage of major aspects of physics, electrical engineering, communications, computing and information technology.
Institution of Electrical Engineers, Stevenage, Herts.

1329 International Road Research Documentation (IRRD)
Covers transport and road research, including civil engineering, vehicles, traffic, and transport economics and administration.
OECD, Road Research Transport Programme, Paris

1330 Knowledge Index
Provides especially easy and convenient user access to a wide variety of online databases, including many of those listed here.
Dialog Information Services, Palo Alto, CA, USA & Learned Information (Europe), Oxford

1331 Library and Information Science Abstracts (LISA)
Online version of *Library and Information Science Abstracts*, also covering publishing and

bookselling.
Bowker-Saur, London

1332 Magazine Index
Contains information from more than 500 popular US and Canadian magazines, with full texts of selected articles.
Information Access Company, Foster City, CA, USA

1333 Management and Marketing Abstracts
Covers international developments in theoretical and practical areas of activities in various aspects of management and marketing.
Pira International, Information Services, Leatherhead, Surrey

1334 MathSci
Covers pure and applied mathematics worldwide.
American Mathematical Society, Providence, RI, USA

1335 MEDLINE
Online version of parts of *Index Medicus*; covers journal articles in the whole medical field, including dentistry and veterinary medicine.
US National Library of Medicine, Bethesda, MD, USA

1336 Meeting Agenda
Announcements of worlwide congresses, conferences, meetings, workshops, exhibitions and trade fairs.

Commissariat a l'Energie Atomique, Centre d'Etudes Nucleaires de Saclay, Gif-sur-Yvettes, France

1337 Mental Health Abstracts

Abstracts of journals, books and reports on all aspects of mental health and illness.

IFI/Plenum Data Company, Alexandria, VA, USA

1338 Microcomputer Index

Covers the use and applications of microcomputers, personal computers, and software packages.

Learned Information, Medford, NJ, USA

1339 MOLARS

Covers worldwide literature on meteorology, climatology, surface oceanography, and planetary atmospheres.

Great Britain National Meterorological Library, Bracknell, Berks.

1340 Newspaper Periodicals Abstracts

Covers articles from over 25 major newspapers and 450 general interest, professional, and academic periodicals, on various subjects including science, economics, business, education, social problems, health, and current affairs.

1341 NTIS Bibliographic Data Base

Covers US Government reports on physical, life and social sciences, and business, management, administration, and information science.

US National Technical Information Service, Springfield, VA, USA.

1342 Oceanic Abstracts

Covers worlwide technical literature on all aspects of the oceans.

Cambridge Scientific Abstracts, Bethesda, MD, USA

1343 PAIS International

Covers social sciences, emphasising current issues and public policies.

Public Affairs Information Service, New York

1344 Pollution Abstracts

Covers environmental pollution research amd related engineering studies.

Cambridge Scientific Abstracts, Bethesda, MD, USA

1345 POPLINE

Covers worldwide literature on population demography, vital statistics, family planning, and associated policy issues.

US National Library of Medicine, Bethesda, MD, USA

1346 PsycINFO

Online version of *Psychological Abstracts*; covers worldwide literature on psychology and related social and behavioural sciences.

American Psychological Association, Arlington, VA, USA

1347 PTS New Products Announcements/Plans
Contains complete unedited texts of press releases announcing new products, especially in high technology and emerging industries.
Predicasts, Cleveland, OH, USA

1348 PTS PROMT
Covers worldwide commercially important developments in all kinds of markets and technologies; includes legislation and company news.
Predicasts, Cleveland, OH, USA

1349 Reuter Textline
Provides full abstracts of informative headlines from a wide range of UK and foreign newspapers, together with full-text specialist databases on electronics, retailing, banking, etc.; also information on companies, industries, economics, public affairs, the European Community, and Eastern Europe.
Reuters Holdings, London

1350 SciSearch
Covers general scientific information, based on citation indexing.
Institute for Scientific Information, Philadelphia, PA, USA

1351 Social Sciences Index
Covers English langugae literature on the social sciences, including economics, geography, environmental sciences, planning, public administration, political science, and psychology.
H. W. Wilson Company, New York

1352 Social SciSearch
Covers the social sciences, including all the social sciences handled by *Social Sciences Index* but also education, history, philosophy, and library and information science.
Institute for Scientific Information, Philadelphia, PA, USA

1353 Sociological Abstracts
Online version of *Sociological Abstracts*; covers sociology and related social and behavioural sciences.
Sociological Abstracts, San Diego, CA, USA

1354 SRIS Catalogue
Covers holdings of the Science Reference and Information Service (SRIS), including all periodicals and all books added since 1974.
SRIS, London

1355 TELEDOC
Covers worlwide literature in English and French on telecommunications, electronics, and allied fields.
CNET, Issy-les-Moulineaux, France

1356 Telegen
Covers journals, reports and conference proceedings in genetic engineering and other areas of biotechnology.
Bowker A & I Publishing, Providence, NJ, USA

1357 TOXLINE
Covers toxicology, pharamacol-
ogy, biochemistry, physiology,
and the environmental effects of
drugs and other chemicals.
*US National Library of Medicine,
Bethesda, MD, USA*

1358 TRANSDOC
Covers physical, economic, commercial, social, and other aspects
of transport.
CEMT-CIDET, Paris

1359 World Information Library
Covers statistical and factual
information about the economic,
socail, and political situation in
over 125 countries.
*Multinational Computer Models,
Fairfield, NJ, USA*

1360 World Patents Index (WPI)
Covers patents from 27 countries,
and European and PCT patents
and research disclosures; date
coverage varies with subject and
country.
Derwent Publications, London

1361 World Reporter
Provides back files of various
lengths for various newspapers
and periodicals.
*FT Profile, Sunbury-on-Thames,
Middlesex*

4.6 Personal Computer Databases

1362 EcoBase: The Key to the Environment

This is a personal computer
database, containing information
about all kinds of organisations,
in the UK and Ireland, which are
involved nationally, regionally
and locally in environmmental
matters. It contains over 4,600
addresses and 600 cross-references, together with over 170 different categories and locations to
choose from, to help users to find
precisely those organisations that
interest them. It is useful for
libraries, schools, voluntary
organisations, businesses, public
authorities and government
departments. It can also be used
to print address labels and lists
and to create output files of
selected addresses.
1991, Keystroke Knowledge, London.

4.7 Software

1363 Multi Country Model
This is an econometric model,
that allows trial forecasts and analyses of economic variables for 37
countries and trade shares for 50
countries. It contains historical
data for about 3,000 variables and
2,500 trade shares, back to 1957.
Macro, Southborough, MA, USA

1364 SimCity
This is an urban simulation game,
that can be used to explore and
test city design strategies.

1365 SimEarth
This is a remarkable Earth simulation game, that can be used to

explore and test planet design strategies. It is highly praised by James Lovelock, whose books are Nos. 596 to 598; it includes his Daisyworld Model. He hopes that it might develop into "a personal model that might become a guide for living right with the world, a way of testing for ourselves the long-term consequences of different ways of living".

1366 STELLA

This is the global modelling software associated with the 1992 book *Beyond the Limits* (No. 641). It is essentially a personal computer version of the World 3 software, whose results are presented in the 1972 book *Limits to Growth* (No. 642). Although few changes have been made from the World 3 model, STELLA enables its users to explore for themselves the effects a variety of possible global policies.

SimCity and SimEarth are described in the following four books:

1367 De Maria, Russell
SimEarth: The Official Strategy Guide

This comprehensive guide to SimEarth has four parts: (1) Tutorial; (2) The Reference; (3) The Simulator; (4) Appendices.
1991, Prime Publishing, Rocklin, CA, USA, ISBN 1-55958-103-4 (pbk) xxii + 314 pp.

1368 Derrick, Dan & Dennis Derrick
Master SimCity/SimEarth: City and Planet Design Strategies

This book shows how to use SimCity and SimEarth to formulate, test and explore city and planet design strategies.
1991, Sams, Carmel, IN, USA, ISBN 0-672-22787-8 (pbk) xix + 508 pp.

1369 Wilson, Johnny L.
The SimCity Planning Commission Handbook

This book shows how to use SimCity as an urban planning simulator.
1990, Osborne McGraw-Hill, New York, ISBN 0-07-881660-2 (pbk) xxvii + 193 pp.

1370 Wilson, Johnny L.
The SimEarth Bible

This book provides a guide to developing winning strategies for SimEarth, covers all the computer models that it uses, and is packed with useful, insightful tips.
1991, Osborne McGraw-Hill, New York, ISBN 0-07-881845-5 (pbk) xxii + 193 pp.

Another relevant book is:

1371 Barney, Gerald O. et al.
The Microcomputer Software Catalog

1991, Westview Press, Boulder, CO, USA, ISBN 0-8133-8297-1 (pbk) 338 pp.

5

NETWORKS

Introductory Note

54 of the many social networks contributing to a better 21st century are listed here in alphabetical order of name. This is probably only a very small sample of all the relevant networks.

Each entry has as many as possible of the following lines: (1) name in bold type; (2) brief description of aims, nature and activities; (3) title or brief indication of newsletter or other publication, with item number(s) if listed in this bibliography; (4) contact(s) if known; (5) contact address if known; (6) telephone and fax number(s) if known.

Information about networks is provided in the literature cited below, also in directories of organisations, including Nos 1428-1443.

Directories of Networks

1372 Environment Support Unit, NCVO
The NEST Directory of Environmental Networks
Directory of several dozen environmental networks in the UK, of which the Milton Keynes Environmental Network (No. 1471) is an example.
1992, National Council of Voluntary Organisations, London, (pbk).

1373 Harvey, Brian
Networking in Europe: A Guide to European Voluntary Organisations
This guide lists networks in Europe, concerned with: the position of the voluntary sector, development issues, politics, women, elderly people, consumer affairs, family issues, broad social issues, social policy, child welfare, human rights, etc.
1992, NCVO, London, ISBN 0-7199-1388-1 (pbk) xiii + 338 pp.

List of Networks

1374 Amnesty International
Network of small groups world-wide, concerned with the rights of prisoners of conscience and political prisoners in all countries. It is one of the world's most famous international organisations, well-known for its sometimes successful attempts to release and improve the conditions of prisoners.
Contact: Amnesty International rue Berkmanns 91, B.1060 Brussels, Belgium
Tel: 322-537-1302, Fax: 322-537-4750

1375 Basic Income European Network (BIEN)
Links individuals and groups interested in unconditional basic income provision for all citizens.
Contact: BIEN
Bosduifstraat 21, B. 2018 Antwerp, Belgium
Tel: 323-220-4181

1376 Business Network, The
Network aiming to link, inform, support and encourage those who wish to see business value people and the environment. It promotes a holistic approach to business, based on the interdependence of the individual, the organisation, the community and the environment. The network is now in a transitional phase, so that it has no official contact or address at the time of writing. Those interested in finding out more should contact the author.

1377 Caritas International
Aims to spread charity and justice throughout the world, especially via emergency services, helping developing countries, and challenging social exclusion.
Contact: Caritas International Palazzo San Calisto, V-00120 Citta del Vaticano, Italy
Tel: 306-698-7197

1378 Centre for Creation Spirituality
Promotes the concept of creation spirituality, as expressed in Matthew Fox's books (Nos. 343 & 344), through networking of people and holding of talks, workshops, network days, and liturgies in various parts of the UK. Builds bridges between people of different cultures and religions, and inspires compassionate action for political and environmental healing.
Publications: *InterChange* journal
Contact: Petra Griffiths, Coordinator or John Doyle, Administrator
St. James's Church, 197 Piccadilly, London W1V 9LF
Tel: 071-287 2741 & 071-734 4511

1379 Earthstewards Network, The
Network of caring people around the world, like a global family,

with a wealth of skills, resources, and interests, helping to transform negativity into positive, useful action. It joins in projects for global networking, citizen diplomacy, conflict resolution, and youth exchanges.
Publications: *Earthstewards Newsletter*
Contact: Earthstewards Network
P.O. Box 10697, Bainbridge Island, WA 98110, USA
Tel: 206-842-7986

1380 Environment and Development in the Third World (ENDA Third World)

An international associative organisation, consisting of rural and urban grassroots groups, assisted by a variety of experts. Its activities include: direct field actions, the pooling of ideas, research, training, and publications.
Contact: ENDA Third World
B.P. 3370, Dakar, Senegal, with branches in several other Third World countries
Tel: 21-60-27/22-42-29,   Fax: 22-26-95

1381 Eurolink Age

Network concerned with older people and issues of ageing, linking organisations from 17 countries around the EEC and Western Europe working for elderly people.
Contact: Eurolink Age

rue du Trone 98, B. 1050, Brussels, Belgium; 1268 London Road, London SW16 4EF
(Belgium) Tel: 322-512-9360, Fax 322-512-9360; (UK) Tel: 081-679 8000, Fax: 081-679 6727

1382 European Anti-Poverty Network (EPAN)

Pressure group to analyse the nature of poverty by researching its causes, promoting solidarity, advising on European policies and programmes, promoting contacts and exchanges, and analysing policies permitting social exclusion.
Contact: EPAN
rue Rempart des Moines 78/4, B. 1000 Brussels, Belgium
Tel: 322-512-1652

1383 European Environmental Bureau (BEE)

Network of about 130 environmental organisations, aiming to protect and conserve the environment, and ensuring that their values are reflected in European Community policies, especially in agriculture, industry, energy, and transport.
Contact: BEE
rue de la Victorie 22-26, B.1060 Brussels, Belgium

1384 European Federation for the Welfare of the Elderly (EURAG)

Broad-based organisation, representing 22 European countries, especially interested in housing

for elderly people, problems of the very old, and elderly women.
Contact: General Secretariat, EURAG
Schmiedgasse 26 (Amtshaus), A-9010 Graz, Austria
Tel: 316-872-3002, Fax: 316-872-3019

1385 European Forum for North-South Solidarity
Network of 17 European associations and voluntary organisations concerned with development issues, aiming to influence European attitudes, policies, and funding patterns towards the developing countries.
Contact: European Forum for North-South Solidarity
rue Jonquoy 25, F 75014, Paris, France (temporary address)
Tel: 331-4539-0862, Fax: 331-4539-7164

1386 European Network for the Unemployed (ENU)
Network of eight national organisations, fighting to eliminate poverty and unemployment, and building up a network for information exchange, expenditure, and solidarity.
Contact: ENU
48 Fleet Street, Dublin 2, Ireland
Tel: 1-0795-316, Fax 1-6792253

1387 European Network of Women (ENOW)
Links women's groups interested in pursuing issues at the European level; sees itself as both radical and critical.
Contact: Secretariat, ENOW
rue Blanche 29, B-1050 Brussels, Belgium
Tel: 322-537-7988, Fax: 322-537-5596

1388 Farmers' Third World Network
Raises awareness in the UK farming community about farming in developing countries, and the effect of European farming practices and policies on agricultural development and developing country farmers.
Contact: Adra Friggen, Coordinator
The Arthur Rank Centre, National Agricultural Centre, Kenilworht, Warks.
Tel: 0203-696969 ext. 338, Fax: 0203-696900

1389 Findhorn Foundation Network
Informal network of people who live or have lived in or visited or worked with the Findhorn Foundation (No. 1578), an international spiritual community dedicated to help heal the divisions within humanity and create a human society living in unity with the Universal Spirit and its whole Creation.
Publications: *One Earth Magazine* (No. 1115)
Contact: The Findhorn Foundation
The Park, Findhorn, Forres IV36 0TZ, Scotland

Tel: 0309-690311

1390 Friends of the Earth International (FoE International)
International network of organisations in various countries, working to protect the global environment and the rational use of natural resources. It campaigns for the prevention of: sea and water pollution, damage to the ozone layer, acid rain, destruction of tropical rain forests, misuse of biotechnology, etc. See also Nos. 1477 & 1478.
Contact: FoE International
PO Box 19, 100 GD Amsterdam, The Netherlands; European Coordination Office, rue Blanc 29, B. 1050 Brussels, Belgium
(Belgium) Tel: 322-537-7228, Fax: 322-537-5596

1391 Future in Our Hands (FIOH)
Network, concerned with environmental and developmental issues. It originated in Norway, and has about a dozen branches in Europe and the Third World. Two books (Nos. 735 & 736) describe its work.
Contact: FIOH Movement
120 York Road, Swindon, SN12 2TP
Tel: 0793-532353

1392 Global Business Network
International network bringing together people from many different fields, to help companies to gain insight into the future.

Details are given by one of its founders, Peter Schwartz, in his book *The Art of the Long View* (No. 822).

1393 Global Energy Network International (GENI)
Private international organisation of people, committed to solving global problems with business solutions. It proposes a cooperative, global grid of electrical transmission lines, transferring excess power during temporary low consumption periods to other locations having temporary high loads.
Contact: GENI
2141 Cardinal Drive, San Diego, CA 92123, USA
Tel: 619-565-2386

1394 Global Family, The
International network of individuals and groups, dedicated to experiencing unconditional love and peace in our lifetimes. They believe that our many environmental, social, economic and human problems have as their root cause a sense of separation.
Publications: *Global Family Update, The (No. 1188)*
Contact: Global Family
112 Jordan Avenue, San Anselmo, CA 94960, USA
Tel: 415-453-7600, Fax: 415-453-7685

1395 Global Partnership '92
Proposed network for global part-nerhip, especially for communication between development and environmental agencies working for equity between peoples and for life on Earth; a broad launch is planned for late 1992 and early 1993.
Publications: *The Globe* (No. 1189)
Contact: Benny Dembitzer, Director
PO Box 1001, London SE24 9NL
Tel: 071-924 0974, Fax: 071-738 4559

1396 Green Academic Network (GAN)
Network being established to link concerned environmentalists both within and between British higher education institutions. Its founders hope to establish a GAN group in each institution, which will provide a focus for Green curriculum development design and the formulation of institutional Greening policies. It is closely linked to GHECO (No. 1483).
Contact: GHECO
154 Buckingham Palace Road, London SW1W 0TR
Tel: 071-730 8868, Fax: 071-730 8858

1397 GreenNet
Provides a global computer communications network for environment, peace, and human affairs, together with services to countries poorly served by commercial communications services.
Contact: Viv Kenton, Coordinator
25 Downham Road, London N1 5AA
Tel: 071-923 2624, Fax: 071-254 1102

1398 Green Paths Centre
Network linking people who attend or have attended Green Paths events, and others concerned about Green and environmental issues.
Publications: *Green Paths Newsletter* (No. 1398)
Contact: Michael Kendall
Foxhole, Dartington, Totnes, Devon TQ9 6EB
Tel: 0803-867075 & 071-485 9981

1399 Greenpeace International
A leading worldwide network of non-violemt, direct action campaigning organisations in about 30 countries. It pioneers confrontational, risk-taking, attention-seeking methods, combined with hard-headed political lobbying and research. It has campaigned for: saving whales, dolphins, and seals; promoting improved water quality; and opposing toxic waste dumping. See also Nos. 1484 & 1485.
Contact: Greenpeace International
Kaizersgracht, 1016 DW, Amsterdam, The Netherlands; ave de Tervuren 36, B-1040 Brussels, Belgium.
(Belgium) Tel: 322-736-9927, Fax: 322-736-4460

1400 Holistic Education Network, The
Promotes holistic approaches to education through regular meetings, workshops and groups.
Contact: Holistic Education Network
81 Guinness Court, Mansell Street, Aldgate, London E1 8AA

1401 Holistic Health Research Network
A network of people interested in research into holistic health; including both conventional and complementary medical approaches and techniques, and emphasising a whole person approach to health and health care.
Publications: *Holistic Health Research Network Newsletter* (No. 1194)
Contact: Trevor Lakey, Coordinator
18 March Street, Strathbungo, Glasgow G41 2PX, Scotland
Tel: 041-424 0603

1402 Holistic Network, The
Publicises the working of practitioners in alternative and complementary medicine, and in personal and spiritual growth.
Publications: *The Holistic Health Network Directory* (No. 1432)
Contact: The Holistic Network
172 Archway Road, Highgate, London N6 5BB
Tel: 081-341 6789; Fax: 081-348 4579

1403 Ideas for Tomorrow Today
Network of people who are participating in transformation in their work and in their lives, networking for a better future, and linking with those on the creative edge of change; already in contact with many other networks, organisations and groups.
Publications: *Ideas for Tomorrow Today Bulletin* (No. 1195)
Contact: Will Sutherland, Director or Alan Senior, Administrator
31 Bellevue Road, Ealing, London W13 8DF
Tel: 081-997 8892

1404 International Council of Voluntary Agencies (ICVA)
Independent international association of non-profitmaking NGOs, active in the field of humanitarian assistance and development cooperation; provides a permanent international liaison structure for voluntary agency consultation and cooperation, and acts as a liaison body between the voluntary sector and the United Nations and other inter-governmental organisations.
Contact: ICVA
rue Gautier 13, CH 1201, Geneva, Switzerland
Tel: 22-732-6600/1/2, Fax: 22-738-9904

1405 International Council on Social Welfare (ICSW)
Non-governmental organisation

committed to social development, with representatives in ten countries, concerned with all aspects of social welfare and social development issues. It is a bridge between voluntary and governmental sectors, ranging from the grassroots to international levels, and an agent for their mutual cooperation and coordination.

1406 International Forum for Child Welfare (European Group)
Aims to improve conditions for and services to children, especially those whose rights are violated, and to raise the status of children in general. It also lobbies the European Community on children's issues.
Contact: Sarah Wilkins
8 Wakley Street, London EC1V 7LT
Tel: 071-833 3319, Fax: 071-833 8636

1407 International Network of Engineers and Scientists for Global Responsibility (INES)
Network of initiatives, organisations and individuals from around the world, who are working for peace and international security, justice and sustainable development, and a responsible use of science and technology.
Publications: *INES Newsletter* (No. 1147)
Contact: Naturwissenschaftler-Initiative
Lohbrueggerstr. 20, 2057 Reinbek / Hamburg, Germany

Tel: 40-7220678, Fax: 40-7220579

1408 International Union for the Conservation of Nature / World Conservation Union (IUCN)
Conservation alliance of nearly 120 nations, that initiates and supports conservation projects worldwide, bringing governments and voluntary organisations together as partners.
Publications: Various publications, e.g. No. 913.
Contact: IUCN
ave du Mont-Blanc, CH 1196 Gland, Switzerland
Tel: 22-647181, Fax: 22-642926

1409 International Workers Aid/ Entraide Ouvriere
Specialised socialist-based network, bringing together voluntary aid agencies; coordinates members' projects in developing countries, and brings the debt crisis and other development issues to the attention of European political institutions.
Contact: Bureau de Liaison, Entraide Ouvriere
rue Montagne aux Herbes Potageres 37, B-1000 Brussels, Belgium
Tel: 322-219-4682, Fax: 322-218-8415

1410 Living Economy Network, The (LEN)
International network of economists and other social and natural scientists, working for the emergence and increasing influence of

a new school of economic thought, Living Economics.
Publications: Mailings to LEN participants, normally twice a year
Contact: Paul Ekins, Coordinator
42 Warriner Gardens, London SW11 4DU
Tel: 071-498 8180, Fax: 071-498 8183

1411 Milton Keynes Environmental Network
Local network to provide publicity and support to the many groups, working in the Milton Keynes area to protect and improve the natural environment; typical of many local environmental networks in the UK.
Publications: *Green Link* (No. 1190)
Contact: MKEN
City Discovery Centre, Bradwell Abbey, Milton Keynes MK13 9AP
Tel: 0908-227229

1412 National Council for Voluntary Organisations
Provides professional advisory services to voluntary organisations, protects their interests, and promotes new social action. It works with voluntary organisations on issues of immediate concern, and creates projects to strengthen the effectiveness of voluntary action.
Contact: Sir Geoffrey Chandler, CBE, Chairman or Usha Prashar, Director

26 Bedford Square, London WC1B 3HU
Tel: 071-636 4066

1413 NGO Liaison Committee
Leading network concerned with development, bringing together up to 600 voluntary organisations from all parts of Europe, concerned with development issues; works on education of public opinion about development issues, development policy, the promotion of NGOs in developing countries, and European Communities aspects.
Contact: Comite de Liaison NGO
rue de Cortenberg 62, B.1040 Brussels, Belgium
Tel: 322-736-4087, Fax: 322-732-1934

1414 Peace Child Europe
Network aiming to draw children from Eastern and Western Europe into the international Peace Child family, by promoting local performances of the play "Peace Child", attracting European students into International Summer Programmes, and, with the United Nations, promoting the International Day of Peace each year.
Publications: Various educational materials for peace.
Contact: Peace Child Europe
Postbus 772, 1200 AT Hilversum, The Netherlands
Tel: 035-210 395, Fax: 035-231 217

1415 Probono
International network for business and human development, aiming to promote global development and corporate and individual growth through quality in the business environment. It has groups in several countries including the UK.
Publications: *Owners' Update* newsletter.
Probono
176 Sutherland Avenue, Little Venice, London W9 1HR
Tel: 071-286 1386, Fax: 071-286 2465

1416 Public Health Alliance, The
National movement of organisations and individuals to promote and defend public health in the UK, according to its Charter for Public Health. It covers the full range of public health issues.
Publications: Various publications and packs on public health.
Contact: The Administrator
Snow Hill House, 10-15 Livery Street, Birmingham B3 2PE
Tel: 021-235 3698

1417 Red Cross
Oldest of the modern networks, now working worldwide to help disaster victims and war wounded. Its national societies have expanded this work to respond to economic and social problems, and help unemployed, learning-impaired, and elderly people.
Contact: Liaison Office of the Red Cross, Communications Department
ave de la Paix 19, CH 1202 Geneva, Switzerland
Tel: 22-734-6001

1418 Salvation Army International
Multinational Chritain evangelical and welfare organisation, working in over 90 countries, especially for the homeless and for prisoners worldwide.
Contact: Salvation Army HQ
101 Queen Victoria Street, London EC4P 4EP
Tel: 071-236 5222

1419 Scientific and Medical Network, The
Informal international group, consisting mainly of qualified scientists and doctors, with some psychologists, engineers, philosophers and other professionals. It addresses a wide variety of scientific and medical themes, with special emphasis on complementary medicine and the relations between science and spirituality.
Publications: *Scientific and Medical Network Newsletter* (No. 1212)
Contact: David Lorimer, Director
Lesser Halings, Tilehouse Lane, Denham, near Uxbridge, Middx. UB9 5DG
Tel/Fax: 0895-835818

1420 Self-Esteem Network
Brings together people from a

wide range of backgrounds, who are interested in promoting self-esteem, and applying it to strengthen equal opportunities and social justice.
Contact: Self-Esteem Network
32 Carisbrooke Road, London E17 7EF
Tel: 081-521 6977, Fax: 081-521 5788

1421 Third World Information Network (TWIN)
Promotes networking between First and Third Worlds, through information exchange and active entrepreneurial intervention to facilitate trade and technology transfer.
Contact: Pauline Tiffen, Director
345 Goswell Road, London EC1V 7JT
Tel: 071-837 8222

1422 TRANET
Links people in all parts of the world, who are interested in appropriate and alternative technology initiatives, transport, communications, learning, the environment, Green politics, women's issues, and many other areas. It has an excellent library, and can refer enquirers to sources of information and advice.
Publications: *TRANET Newsletter* (No. 1216)
Contact: TRANET
Box 567, Rangeley, ME 04970, USA
Tel: 207-864-2252

1423 Turning Point 2000
Network of people aiming to encourage personal, local, national and international systematic programmes of change towards a one-world human community for the 21st century, that will enable people and conserve the Earth. It is in touch with many other networks, projects, organisations and groups.
Publications: *Turning Point 2000* (No. 1217)
Contact: James Robertson or Alison Pritchard
The Old Bakehouse, Cholsey, near Wallingford, Oxon. OX10 9NU
Tel: 0491-652346

1424 Vision Network, The
Movement for the creation of a better world, aiming to: encourage a holistic approach to creating a shared vision, promote the cooperation of all committed individuals and groups to realising this vision, and provide a platform for sharing, evaluating, and enriching all efforts to create a peaceful and wholesome world on Earth.
Publications: *Rainbow Ark* (No. 1208)
Contact: Rainbow Publications
PO Box 486, London SW1P 1AZ
Tel: 071-828 2782

1425 Women's Environmental Network
Network informing and

empowering women who care about the environment. It campaigns vigorously for higher standards of product life cycle analysis, especially in the UK, and for a holistic form of environmental auditing; it also does research and provides information.

Contact: Bernadette Vallely, Director

Aberdeen Studios, 22 Highbury Grove, London N5 2EA

Tel: 071-354 8823, Fax: 071-354 0464

1426 World Goodwill, London, Geneva & New York

International network of people endeavouring to promote goodwill for all humanity worldwide, concerned about world affairs and addressing wider human and global issues. Its founders were inspired by the holistic philosophy of Alice Bailey.

Publications: *World Goodwill Newsletter* (No. 1222)

Contact: World Goodwill

3 Whitehall Court, Suite 54, London SW1A 2EF; 1 Rue de Varembe, Case Postale 31, 1211 Geneva 20, Switzerland; 113 University Place, 11th Floor, PO Box 722 Cooper Station, New York, NY 10276, USA

1427 Worldwide Indigenous Scientific Network (WISN)

Network aiming to promote the revitalisation, exploration, growth, and exchange of traditional knowledge of indigenous peoples.

Contact: Dr Pamela Colorado, Coordinator

2500 University Drive NW, Faculty of Social Work, University of Calgary, Calgary, Alberta T2N 1N4, Canada

6

ORGANISATIONS

Introductory Note

105 of the many organisations contributing to a better 21st century are listed here in alphabetical order of name. This is probably only a very small sample of all the relevant organisations. Each entry has as many as possible of the following lines: (1) name in bold type; (2) brief description of aims, nature and activities; (3) title, and also reference number of newsletter or other publication, if listed here; (4) contact(s); (5) contact address; (6) telephone and fax number(s) if any. Although reasonable efforts have been made to check this information, **no guarantee is made that every entry is correct**, because a few organisations may have changed their contact details very recently.

Extensive information about organisations and associations in general as well as in the environmental field is published in the 16 books reviewed in the next section. The following books include lists and other details of organisations in environmental and allied fields: (1) the 1992 books *The Green Business Guide* (No. 313), *Dictionary of Environmental Science and Technology* (No. 756), and *Earth Report 3* (No. 380); (2) the 1991 books *On Alternative Ways of Studying the Future* (No. 230) and *The Future for UK Environmental Policy* (No. 946); (3) the 1990 books *The Earthscan Action Handbook for People and Planet* (No. 594), *The Times Guide to the Environment* (No. 844), *The Energy Alternative* (No. 716), and *The New Green Pages* (No. 150); (4) the 1988 book *Green Pages* (No. 310). For details of 'think-tanks', especially in the USA, see, for example, *The Idea Brokers* (No. 849) and *Helping Governments Think* (No. 952). The 1992 book *A New World Order* (No. 305) contains much useful information about grassroots movements for global change in all parts of the world. For lists of 'New Age' groups, see *The Aquarian Guide to the New Age* (No. 159).

Directories of Organisations and Associations

1428 Burek, Deborah M.
Encyclopedia of Associations 1993
This directory gives details of about 23,000 national and 53,000 regional, state, and local organisations in the USA. Many different types of organisation are covered, and the entries are arranged alphabetically in 18 chapters, each covering a type or grouping of related types. Each entry includes organisation name, address and other contact details, foundation date, aims, description, details of publications, details of annual meetings and conferences, etc. Name and keyword indexes are included, together with alphabetical listings of organisations corresponding to specific keywords. For CD-ROM and online versions, see Nos. 1262 & 1314, respectively.
27th. Ed. 1992, Gale Research, Detroit & London, ISBN 0-8103-7619-9 (hbk) (ISSN 0071-0202) 3 Vols., xxvii + 3645 pp.

1429 Deziron, Mireille & Leigh Bailey
A Directory of European Environmental Organizations

This is an extensive directory of governmental and non-governmental international and national organisations within the European Community. It is an invaluable reference source for anyone seeking information on the structure of a relevant organisation, or needing to know who to consult. It includes: (1) up-to-date names and addresses of key officials; (2) structure charts of complex organisations; (3) detailed descriptions of agencies and their functions.
1991, Blackwell, Oxford & New York, ISBN 0-631-18386-8 (hbk) xiv + 178 pp.

1430 Gause, Ken (Ed.)
Worldwide Government Directory with International Organizations 1992
Part 1, covering most of this directory, lists national government orgainsations, arranged alphabetically nation name order. Part 2 lists international organisations, and United Nations (UN) organisations, specialised agencies, affiliated agencies, regional economic commissions, and national-related agencies.
1992, Belmont Publications, Washington, ISBNs 0-9629283-2-1 (hbk) & 0-9629283-X (pbk) (ISSN 0894-1521) xii + 1190 pp.

1431 Giles, Shirley
The Third World Directory: A Guide to Organisations Working

for Third World Development
This directory consists mainly of a directory of Third World development organisations, whose entries are listed in alphabetical order; the main list is followed by a supplementary list. The directory section is preceded by articles on education and development and development values.
1990, Directory of Social Change, London, ISBN 0-907164-57-9 (pbk) vii + 128 pp.

1432 Grant, Ian (Ed.) & Philip Wylie (Compiled by)
The Holistic Network Directory: The Directory of Holistic Practitioners and Centres Covering the UK and Eire
This directory lists organisations by over 80 different classifications in holistic medicine and personal development. The section for each class is introduced by an expert and gives details of the relevant organisations and prectitioners. There are also 28 articles on different aspects of the whole area, guidance on use of the directory, a list of centres, information about products and services, an index of professional qualifications, and an alphabetical index of organisations and practitioners.
1992, The Holistic Network, London, ISBN ? (pbk) ? pp.

1433 Henderson, C. A. P. (Ed.)
Pan-European Associations: A Directory of Multi-National

Organisations in Europe
This directory lists many associations, operating in more than one European country, in alpabetical name order. Its layout is similar to that of *Directory of British Associations* (No. 1434).
2nd. Ed. 1991, CBD REserach, Beckenham, Kent, ISBN 0-900246-54-5 (hbk) x + 206 pp.

1434 Henderson, G. P. & S. P. A. Henderson (Eds.)
Directory of British Associations and Associations in Ireland
This directory lists many of the associations in the British Isles in alphabetical name order. Each entry gives organisation name, contact details, date of formation, type, brief description including aims, information about branches and groups if any, details of activities, details of publications, membership data, etc. A list of unverified and 'lost' associations is given. There are subject and abbreviations indexes.
11th Ed. 1992, CBD Research, Beckenham, Kent, ISBN 0-900246-57-X (hbk) xvi + 610 pp.; 1st. Ed. 1965.

1435 Irvin, L. & P. T. Reid (Eds.)
International Organizations
This supplement to *Encyclopedia of Associations 1993* (No. 1428) provides information about over 10,000 international nonprofit membership organisations, concerned with all subjects and areas of activity. Its entries mostly cover

organisations in the USA, but include multinational groups and national organisations based in other countries. For CD-ROM & online versions, see Nos. 1262 & 1314, respectively.

25th. Ed., Gale Research, Detroit & London, ISBN 0-8103-7406-4 (hbk) 2 vols.

1436 Keller, Hans J. & Daniel Maziarz (Eds.)
Who Is Who: In Service to the Earth
This book is a handy reference guide to the visions, people, projects, and organisations offering hope and solutions for global transformation. It includes 41 visions of a positive future, together with a directory of 5,000 people, projects, and organisations working to heal our planet. They aim to have a constructive effect on ecology, holistic health, communities, advocacy, citizen diplomacy, and many other important fields in the USA and other countries.

1991, VisionLink Education Foundation, Waynesville, NC, USA.

1437 Milner, J. Edward, Carol Filby, Marian Board & Alan Phillips (Eds.)
The Green Index: A Directory of Environmental Organisations in Britain and Ireland
This directory contains a comprehensive list of organisations in the British Isles. It contains over 5,200 entries for national, regional and local organisations, institutions, societies, groups, and companies, including: action groups, national environmental groups, natural history societies, county conservation trusts, government bodies, and 'quangos'. Each entry includes: name, address, telephone number, type of organisation, aims, publications. The directory is indexed by county or region, and grouped by activity. It is a valuable source of information and contacts for all those interested in or concerned about conservation and the environment.

1990, Cassell, London, ISBN 0-304-31882-5 (pbk) xviii + 366 pp.

1438 NCVO
The Voluntary Agencies Directory 1991
This directory gives details of UK voluntary organisations, including societies, projects and foundations, with a wide variety of purposes, in alphabetical name order. It has a fairly detailed classification, under which it indexes the organisations. It gives guidance on its use, contains a list of specially useful addresses, and lists abbreviations and acronyms of organisations.

12th. Ed. 1991, Bedford Square Press, London, ISBN 0-7199-1286-5 (pbk) xi + 238 pp.

1439 OECD
Directory of Non-Governmental Environment and Development

Organisations in OECD Member Countries

This directory is a valuable source of information on the growing number of NGOs in OECD countries, that are undertaking development activities relevant to the environment. This information is useful for governments, inter-governmental organisations, and NGOs themselves. Although it has some gaps, it aims to cover the major sectors of NGO environment and development activity. The main entries are listed alphabetically for each OECD country. Each entry has name, address, telephone number, fax number (if any), contact(s), aims, general information, development actions and education, environment actions and information, combined environment/education information, and miscellaneous comments. There are an alphabetical list and index of organisations, a list of organisations' acronyms, and indexes for environment education themes, development education themes, environment actions by subject, environment actions by country and subject, and development actions by country and subject. There are several articles.

1992,   OECD,   Paris,   ISBN 92-64-03536-2 (pbk) 409 pp.

1440 OECD
Directory of Non-Governmental Organisations in OECD Member Countries

This directory is a valuable source of information on the growing number of NGOs in OECD countries. This information is useful for governments, inter-governmental organisations, and NGOs themselves. Although it has some gaps, it aims to cover the major sectors of NGO activity. The main entries are listed alphabetically for each OECD country. Each entry has name, address, telephone number, fax number (if any), contact(s), aims, general information, details of activities, and miscellaneous comments. There are an alphabetical list and index of organisations, a list of organisations' acronyms, and subject indexes. There are several articles.

1990,   OECD,   Paris,   ISBN 92-64-03373-4 (pbk) 708 pp.

1441 Trzyna, Thaddeus C. & Roberta Childers (Eds.)
World Directory of Environmental Organizations

This directory describes over 2,600 organisations concerned with protecting the Earth's environment and resources. It includes sections on organisations working on specific problem areas, organisations covering major parts of the world, UN organisations, other inter-governmental organisations, and non-governmental organisations. It also includes an introductory guide, country and area listings, and a list of other directories and databases. The following 'generations' of organisations are covered: (1) conservation organisations, founded between the late 19th and mid-20th centuries; (2) the environmental movement, starting about 1970; (3) the sustainability movement, starting about 1980; (4) an emerging movement that may link sustainability to reforms in government and other institutions steering society.

4th. Ed. 1992, California Institute of Public Affairs, Sacramento, CA, USA, (hbk) 231 pp.; 1st. Ed. 1973.

1442 Union of International Associations (No. 1646)
Yearbook of International Organizations 1991/92

This directory provides comprehensive, up-to-date details of many if not most international organisations. Vol.1, *Organization Descriptions and Index*, contains notes to the user, alphabetical list of organisations, and many sections containing lists of various types of organisations and some miscellaneous information. Types of organisations covered include: federations of international organisations, universal membership organisations, regionally defined membership organisations, and internationally-oriented organisations. Miscellaneous information

includes lists of autonomous conference series and of multilateral treaties and agreements. Each organisation entry has name, contact information, foundation date, aims, structure, consultative status, non-governmental organisation (NGO) relations, activities, events, details of publications, and sometimes membership information. There are 15 appendices. There are indexes of organisations classified by: country of secretariat, countries of location of membership, subject concerns, regional concerns, and type.

28th. Ed. 1991, Union of International Associations, Brussels, Vol. 1 ISBN 3-898-22209-2 (hbk) xviii + 1798 pp., Vol. 2 ISBN 3-898-22210-6 (hbk) 1704 pp.; Vol.3 ISBN 3-898-22211-4 (hbk) xii + 1688 pp.; 1st. Ed. 1905-7.

1443 Woodworth, David & Cynthia Miller
The International Directory of Voluntary Work

This directory gives details of over 500 organisations worldwide, requiring help from all types of people for all types of work, both residential and non-residential.

4th. Ed. 1989, Vacation Work, Oxford, ISBNs 1-85458-001-9 (hbk) & 1-85458-000-0 (pbk) viii + 216 pp.

List of Organisations and Associations

1444 American Association for the Advancement of Science
Largest general scientific organisation in the USA, representing all fields of science and about 300 affiliated scientific societies. It promotes the work of scientists and the improvement of science's effectiveness in promoting human welfare. It has an annual conference, and conducts seminars and colloquia.
Publications: *Science* (No. 1138) and various other periodicals and publications.
Contact: Richard S. Nicholson, Executive Officer
1333 H St. NW Washington, DC 20005, USA
Tel: 202-326-6400

1445 Association for Environment Conscious Building (AECB)
Generates environmental awareness within the building industry
Publications: *Building for a Future* journal.
Contact: Keith D. Hall, Director
Windlake House, The Pump Field, Coaley, Glos. GL11 5DX
Tel: 0453-890757

1446 Association for the Conservation of Energy (ACE)
Research and lobbying group, formed in 1981 by major companies in the energy efficiency industry; promotes awareness of the need for and benefits of energy conservation, and aims to establish a sensible and consistent national policy and programme in the UK.
Contact: Andrew Warren, Director
9 Sherlock Mews, London W1M 3RH
Tel: 071-935 1495, Fax: 071-935 8346

1447 Association of Environmental Consultancies (AEC)
Association, whose members are committed to improving the quality of environmental consultancy in the UK. It has produced codes of practice on environmental assessment and environmental auditing.
Contact: Frank Joyce, Chairman
Priestley House, 28-34 Albert Street, Birmingham B4 7UD
Tel: 021-616 1010, Fax: 021-616 1099

1448 Association of Recycled Paper Suppliers (ARPS)
Aims to help develop more and better products made from recycled paper, and a market for those products.
Contact: ARPS
Bow Triangle Business Centre, Unit 2, Eleanor Street, London E3 4NP
Tel: 081-980 5580, Fax: 081-980 2399

1449 Association of World Federalists
An association standing for world peace, through a reformed and strengthened United Nations, general and complete national disarmament under a world federal authority with limited functions but real powers, and, first of all, a directly recruited impartial, volunteer UN peace-keeping force.
Publications: *World Federalist Bulletin* (No. 1221).
Contact: c/o Federal Trust for Education and Research
158 Buckingham Palace Road, London SW1W 9TR

1450 ATD Third World
International movement supporting the efforts of the poorest people of the world in overcoming poverty and taking an active role in the community.
Publications: Newsletter and books.
ATD Third World
48 Addington Square
London SE5 7LB
Tel: 071-703 3231

1451 British Association for the Advancement of Science (BAAS)
Promotes general interest and understanding in the concepts, language, methods, and applications of science. It has a large annual conference each year, to which members of the public are invited.

Publications: *Science and Public Affairs* journal (No. 1451, copublished with The Royal Society (No. 1618), also*Scope* newsletter for young people.
Contact: Dr Peter Briggs, Executive Secretary
23 Savile Row, London W1X 1AB
Tel: 071-494 3326, Fax: 071-734 1658

1452 British Institute of Energy Economics (BIEE)
Studies energy economics and exchanges information about it.
Publications: *Energy Journal* and newsletter.
Contact: Elizabeth J. Marshall, Secretary
37 Woodville Gardens, London W5 2LL

1453 Business Council for Sustainable Development (BCSD)
Aims to be a truly global task force of about 50 world business leaders, of whom about a third are from developing countries. Played an important part in the preparations for the UNCED (Earth Summit) in Rio de Janeiro in June 1992.
Publications: The book *Changing Course* (No. 815).
Contact: Stephan Schmidheiny
World Trade Center Building, 3rd. Floor, 10 route de l'Aeroport, Geneva, Switzerland
Tel: 22-788-3202, Fax: 22-788-3211

1454 Business in the Community
Partnership of business, local and

central government, chambers of commerce, the trade union movement, and voluntary organisations, working to help industry and commerce contribute to the health of local communities.
Contact: Business in the Community
227a City Road, London EC1V 1LX
Tel: 071-253 3716

1455 Business in the Environment (BiE)

Aims to devise practical steps that will support the UK's progress towards understanding and applying the principles of sustainable development, through action and partnership between business and the community. It was initiated by Business in the Community (No. 1454).
Contact: Elaine Sullivan
41 Threadneedle Street, London EC2A 8AP
Tel: 071-588 6157, Fax: 071-588 9215

1456 Centre for International Environmental Law (CIEL)

Promotes international law as a means of protecting the global environment through teaching, research, and legal advice; provides assistance to non-governmental organisations and the economically disadvantaged through a global network of lawyers.
Contact: CIE:

Kings College London, Manresa Road, London SW3 6LX
Tel: 071-352 8123, Fax: 071-351 6435

1457 Centre for World Development Education

Increases and improves understanding in the UK of world development and interdependence.
Contact: Derek Walker, Director
Regent's College, Inner Circle, Regent's Park, London NW1 4NS
Tel: 071-487 7410, Fax: 071-487 5438

1458 CIEL US

Same description as for CIEL (No. 1456)
Contact: CIEL US
1621 Connecticut Avenue, NW, Suite 300, Washington, DC 20009, USA
Tel: 202-332-4840, Fax: 202-332-4865

1459 Commonwealth Human Ecology Council (CHEC)

Promotes understanding and action in the development of resources for the wholeness of human hope and well-being, through ecological satisfaction policies andprogrammes in Commonwealth and other countries.
Contact: Zena Daysh, Executive Vice-Chairman
57 Stanhope Gardens, London SW7 5RF
Tel: 071-373 6761

1460 Council for Education in World Citizenship (CEWC)

Prepares young people for their responsibilities as citizens in an internationally interdependent world. Encourages the study and understanding of world affairs.

Contact: Patricia Rogers, Head
Seymour Mews House, Seymour Mews, London W1H 9PE
Tel: 071-935 1752, Fax: 071-935 5741

1461 Council on Economic Priorities (CEP)

Runs a Corporate Environmental Data Clearinghouse, that is emerging as a key source of information for environmentalists and investors on the environmental performance of important companies.

Contact: CEP
30 Irving Place, New York, NY 10003, USA
Tel: 212-420-1133, Fax: 212-420-0988

1462 Council for the Protection of Rural England (CPRE)

Works for the English country heritage and for the general conservation and protection of the environment; aspects covered include land use, landscape protection, planning, environmental assessment, water, woodlands, forestry, agriculture, energy, and transport.

Publications: *Agenda 2000*, which summarises CPRE's latest policy recommendations.

Contact: Fiona Reynolds, Director
Warwick House, 25 Buckingham Palace Road, London SW1W 0PP
Tel: 071-976 6433, Fax: o71-976 6373

1463 Department of the Environment (DoE)

UK Government Department with specific responsibilities for the environment and local government. Its sections include: Directorate of Rural Affairs, Health and Safety Executive, Her Majesty's Inspectorate of Pollution, Local Government Directorate, Housing Directorate, Planning Inspectorate, and Inner Cities Directorate. See also No. 1637.

Publications: Many publications, including Nos. 259-262.

Contact: DoE
3 Marsham Street, London SW1P 3PY
Tel: 071-276 3000, Fax: 071-276 0818

1464 Ecological Design Association

Promotes the design of materials and products, projects and systems, environments and communities, which are friendly to living species and planetary ecology.

Contact: Ecological Design Association
20 High Street, Stroud, Glos. GL5 1AS
Tel: 0453-752985, Fax: 0453-752987

1465 Elmwood Institute, The
Ecological think-tank, dedicated to fostering holistic concepts and values for a sustainable future. Aims to: synthesise and refine the emerging new ecological paradigm; promote new perceptions, thinking, and values through public education, discussion, and dialogue; build and nurture an intellectual/activist community.
Publications: *The Elmwood Quarterly*.
Contact: Fritjof Capra
PO Box 57665, Berkeley, CA 94705, USA
Tel: 510-845-4595

1466 Environment Council, The
An umbrella body for UK organisations concerned with the environment and environmental issues. Its Business and Environment Programme helps managers to recognise the opportunities generated by growing environmental pressures on business, and to appreciate the risks of ignoring them.
Publications: *Habitat* newsletter (No. 1193).
Contact: Edwin Datchefski & Val Bowers
80 York Way, London N1 9AG
Tel: 071-278 4736, Fax: 071-837 9688

1467 Environmental Communicators' Organisation (ECO)
Promotes conservationist ideas among professional journalists and broadcasters.

Publications: *Eco Newsletter*.
Contact: Barbara Jefferies, Honorary Secretary
8 Hooks Cross, Walton-at-Stone, Hertford, Herts. SG14 3RY
Tel: 0920-830527

1468 Environmental Data Services (ENDS)
An independent environmental research and information centre, concentrating on the implications of environmental issues for business and industry.
Publications: *The ENDS Report* (No. 1183) and *The Directory of Environmental Consultants*.
Contact: Marek Mayer (Director & Editor)
Unit 24, Finsbury Business Centre, 40 Bowling Green Lane, London EC1R 0NE
Tel: 071-278 7624, Fax: 071-837 7612

1469 Environmental Education Advisors Association (EEAA)
Advises and coordinates environmental education activities in schools and adult education centres.
Contact: Rishard S. Moseley, Honorary Secretary
Education Department, County Hall, Truro, Cornwall TR1 3BA
Tel: 0872-74282

1470 Environmental Law Foundation (ELF)
Provides access to a network of lawyers and other experts, for

individuals, communities, and small businesses to defend themselves and their environment against potentially destructive developments.
Contact: Martin Polden
Rubinstein, Callingham, Polden and Gale, 2 Raymond Buildings, Gray's Inn, London WC1R 5BZ
Tel: 071-242 8404, Fax: 071-831 7413

1471 Environmental Transport Association (ETA)

British mass membership body, covering land-based movement with a clearly defined environmental code of responsibility; provides road rescue services, and promotes environmentally more acceptable forms of transport.
Contact: Chris Bowers, Director
15a George Street, Croydon CR0 1LA
Tel: 081-666 0445, Fax: 081-666 0422

1472 Ethical Investment Research Service (EIRIS)

Provides an information service on the ethical and social aspects of investment, including information on companies' involvement in several relevant aspects.
Contact: EIRIS
504 Bondway Business Centre, 71 Bondway, London SW8 1SQ
Tel: 071-735 1351

1473 Euroenviron

Programme within the European Commission's EUREKA framework, for supporting innovative research and development designed to address some of Europe's major environmental problems.
Contact: Euroenviron
Business and the Environment Unit, Department of Trade and Industry, Room 1010, Ashdown House, 123 Victoria Street, London SW1E 6RB
Tel: 071-215 6527, Fax: 071-215 2909

1474 European Round Table for Industrialists, Working Group on the Environment

Conducts various projects, including analysing the impact on European and world energy use and pollution of application of the best current and prospective technologies, and work on waste disposal. Collects information on work being done by other organisations.
Contact: Lars Buer, Chairman of the coordination group
AB Volvo, Box 1, N-1411, Kolbotn, Norway
Tel: 2-818498, Fax: 2-806877

1475 European Green Table (EGT)

Works on projects at the interface between industry and the environment, including discussion on subjects such as how market signals can be developed to promote environmental excellence and sustainable

development.
Contact: Luis G. Paulsen, Project
Coordinator
PO Box 86, Bryn, 0611 Oslo 6,
Norway
Tel: 2-581811/581800,   Fax:
2-721266

**1476 Franklin Research and
Development Corporation
(FRDC)**
American investment advisers,
specialising in socially responsible investing; a leading provider
of information on US companies.
Contact: Dept. 10, FRDC
711 Atlantic Avenue, Boston, MA
02111, USA
Tel: 617-423-6655,   Fax:
617-482-6179

1477 Friends of the Earth (UK)
A leading British environmental
campaigning organisation, part of
an international group, aiming to
protect the environment and promote sustainable alternatives. It
attempts to develop viable solutions for the problems that it identifies. Its main campaigns focus
on various forms of pollution, but
it also campaigns on various
global, national and local environmental issues. It has annual
'Green Con Awards' to show up
companies claiming to be Green
but not measuring up to these
claims in important respects.
Publications: *Supporters Bulletin*,
newsletter, and various environmental campaign and information leaflets; see also book No.

752.
Contact: Friends of the Earth
26-28 Underwood Street, London
N1 7JU
Tel: 071-490 1555, Fax: 071-490
0881

1478 Friends of the Earth (USA)
A leading American environmental campaigning organisation,
part of an international group,
aiming to protect the environment and promote sustainable
alternatives. It encourages the
involvement of environmental
activists and other citizens concerned about the environment.
Publications: *Friends of the Earth
Magazine* (No. 1185).
Contact: Friends of the Earth
218 D Street SE, Washington, DC
20003, USA
Tel: 202-544-2600,   Fax:
202-543-4710

**1479 Geographical Association
(GA)**
Promotes the study and teaching
of geography at all levels, including environmental education.
Publications: *Geography* journal
(No. 1055) and *GA News* newsletter (No. 1187).
Contact: Miss M. R. Barlow, Senior Administrator
343 Fulwood Road, Sheffield S10
3BP
Tel: 0742-670666

**1480 Global Environmental Man-
agement Initiative (GEMI)**

Aims to foster environmental excellence in business worldwide, especially the application of Total Quality principles to environmental management. Helped to prepare a Charter for Sustainable Development, presented in April 1991.
Contact: George Carpenter, Chairman
1828 L Street NW, Suite 711, Washington DC 20036, USA
Tel: 202-296 7449

1481 Green Alliance
A cross-party ecological initiative, mainly concerned with emphasising the importance of environmental policy in the public and private sectors; runs several programmes, including one on industry and the environment. Focusses on bridge-building and encouraging dialogue, rather than promoting issues. Supports member bodies in the environmental movement, and aims to "ensure that the ecological perspective is injected into the political process".
Publications: Parliamentary newsletter, when Parliament is sitting.
Contact: Tom Burke, Director of Karen Crane, Administrator
49 Wellington Street, London WC2E 7BN
Tel: 071-836 0341, Fax: 071-240 9205

1482 Green College, The
Higher education institution to provide short residential courses on environmental and allied topics, now in process of formation. It is closely associated with GHECO (No. 1483) and GAN (No. 1396).
Contact: Peter de la Cour, Director
154 Buckingham Palace Road, London SW1W 9TR
Tel: 071-730 8868, Fax: 071-730-8858

1483 Greening of Higher Education Council (GHECO)
Promotes curriculum reform and restructuring in almost all higher education disciplines, so that they can address Green issues more effectively; encourages higher education institutions to become models of good Green practice in areas such as energy efficiency, waste management, and recycling. It is establishing a Green Academic Network (GAN) (No. 1396) and runs Green Book Conversations from time to time, where authors of Green books discuss environmental issues with each other and with members of the public.
Contact: GHECO
154 Buckingham Palace Road, London SW1W 9TR
Tel: 071-730 8868, Fax: 071-730-8858

1484 Greenpeace (UK)
A leading British environmental campaigning organisation, part of

an international group, aiming to protect the environment and promote sustainable alternatives. Greenpeace internationally is especially well-known for its dramatic direct action campaigns, for example to save the whales, that have been remarkably effective.
Contact: Greenpeace
30-31 Islington Green, London N1 8XE
Tel: 071-354 5100

1485 Greenpeace (USA)
A leading American environmental campaigning organisation, part of an international group, aiming to protect the environment and promote sustainable alternatives. Greenpeace internationally is especially well-known for its dramatic direct action campaigns, for example to save the whales, that have been remarkably effective.
Publications: *Greenpeace* newsletter (No. 1192)
Contact: Greenpeace
1436 U Street NW, Washington, DC 20009, USA
Tel: 202-462-1177,   Fax: 202-462-4507

1486 Groundwork Foundation
Aims to achieve environmental regeneration around the UK, with backing by government and industry; developing training packages for industry, with a practical focus on actual environmental improvements.

Contact: Groundwork Foundation
Bennetts Court, Bennetts Hill, Birmingham B2 5ST
Tel: 021-236 8565, Fax: 021-236 7356

1487 Henry Doubleday Research Association (HDRA)
Researches and promotes organic growing and environmentally safe, ecologically sound horticulture.
Contact: Alan   Gear,   Chief Executive
National Centre for Organic Gardening,   Ryton-on-Dunsmore, Coventry CV8 3LG
Tel: 0203-303517, Fax: 0203-639229

1488 ICC United Kingdom
British branch of International Chamber of Commerce (ICC) (No. 1499).
Publications: Quarterly newsletter.
Contact: G. N. F. Wyburd, Director
103 New Oxford Street, London WC1AR 1QB
Tel: 071-240 5558, Fax 071-836 5323

1489 Institute for European Environmental Policy (IEEP)
Policy studies institute, based in Bonn, London & Paris, specialising in environmental policy at the European level; performs projects for environmental ministries and the European Community.

Publications: Various reports.
Contact: Nigel Haigh
3 Endslegh Street, London WC1H
0DD
Tel: 071-388 2117, Fax: 071-388
2826

1490 Institute for Planetary Synthesis (IPS), The
Reawakens an awareness of spiritual values in daily life, promotes planetary awareness leading to planetary citizenship, and analyses and helps solve world problems on the baiss of spiritual values and planetary awareness; cooperates with other educational institutions to create a University for Planetary Synthesis.
Publications: Newsletter and other publications
Contact: IPS Geneva
PO Box 128, CH-1211 Geneva 20,
Switzerland
Tel: 022-338876

1491 Institute for Public Policy Research (IPPR)
Promotes research into, and education of the public in, economic, social and political sciences, and in science and technology, relating to public policy. Includes research on the effects of a wide range of factors on public policy, and on the living standards of all sections of the community.
Publications: Many booklets and some books, presenting the results of its research.
Contact: James Cornford, Director

30/32 Southampton Street, London WC2E 7RA
Tel: 071-379 9400, Fax: 071-497 0373

1492 Institute of Economic Affairs Ltd. (IEA)
Promotes extension of the public understanding of economic principles applied to practical problems.
Publications: *Hobart Papers*, also 'occasional papers' and research digests and monographs.
Contact: Lord Harris of High Cross, General Director
2 Lord North Street, London SW1P 3LB
Tel: 071-799 3745

1493 Institute of Energy
Promotes the effective provision, conversion, transmission, and use of all forms of energy, with due regard to the prudent use of resources and the protection of the environment.
Publications: *Energy World* monthly, quarterly journal, *Energy World Yearbook*, and *Fuel & Energy Abstracts*.
Contact: Colin Rigg, Secretary
18 Devonshire Street, London W1N 2AU
Tels: 071-580 7124 (administration) & 071-580 0008 (publications), Fax: 071-580 4420

1494 Institute of Environment Assessment
Independent professional

institute, registering environmental consultancies and companies active in environmental impact assessment.
Contact: Dr Tim Coles, Executive Director
The Old Malthouse, Spring Gardens, London Road, Grantham, Lincs. NG3 6JP
Tel: 0476-68100, Fax: 0476-76476

1495 Institute of Noetic Sciences
Nonprofit research and education organisation, whose purpose is to expand knowledge of the nature and potential of the mind, and apply that thinking to advance human and planetary health and well-being.
Publications: Various publications, including *Noetic Sciences Review* (No. 1113), a *Bulletin*, and No. 436.
Contact: Dr Willis W. Harman, President
475 Gate Five Road, Suite 300, Sausalito, CA 94965-0909, USA
Tel: 415-331-5650, Fax: 415-331-5673

1496 Institution of Environmental Sciences Ltd. (IEnvSc)
Promotes, sponsors, and organises research and interdisciplinary action, consultation, and coordination of all matters concerning environmental sciences.
Publications: *The Environmentalist* journal, 'Proceedings' and newsletter.
Contact: Dr J. F. Potter, Honorary Secretary

14 Princes Gate, Hyde Park, London SW7 1PU
Tel: 0252-515510

1497 Intermediate Technology Development Group (ITDG)
Enhances the productive capacity of poor people in developing countries, through development, adaptation, and discussion of appropriate technologies.
Contact: Frank R. Almond, Chief Executive
Myson House, Railway Terrace, Rugby, Warks. CV21 3H4
Tel: 0788-60631, Fax: 0788-540270

1498 International Centre for the Legal Protection of Human Rights (Interrights)
Provides free advice on all aspects of human rights law, and assistance in taking cases to the European Court of Justice and the Court of Human Rights.
Contact: Interrights
5-15 Cromer Street, London WC1H 8LS
Tel: 071-278 3230, Fax: 071-278 4334

1499 International Chamber of Commerce (ICC)
Represents all aspects of international business on a world scale, and is establishing the international business community's position on key environmental issues. It strives to achieve favourable trade and investment conditions, in which wealth-creating business can grow.

Publications: Environmental guidelines for industry, and briefing documents on such issues as environmental auditing and sustainable development.
Contact: Nigel Blackburn, Director
38 Cours Albert 1er, F-75008 Paris, France
Tel: 1-45623456, Fax: 1-42258663

1500 International Council of Scientific Unions (ICSU)

Encourages international scientific activity for the benefit of humankind, that will serve scientific and technological development and coordinate interdisciplinary scientific research projects and scientific education.
Publications: Various publications.
Contact: Julia Morton-Lefevre, Executive Secretary
51 bd. de Montmorency, F-75016 Paris, France
Tel: 1-45 25 0329, Fax 1-42 88 9431

1501 International Institute for Environment and Development (IIED)

Supports sustainable development through research and study in environmental economics, sustainable agriculture, forestry, human settlements, and climate change. It advises governments and citizens' groups.
Contact: Richard Soundbrook, Executive Director

3 Endsleigh Street, London WC1H 0DD
Tel: 071-388 2117, Fax: 071-388 2826

1502 International Planned Parenthood Federation (IPPF)

Initiates and supports family planning services throughout the world; educates people and governments in the benefits of spacing and planning births for the whole family, especially for mothers and children.
Contact: Halfdan Mahler, Secretary General
Regent's College, Inner Circle, Regent's Park, London NW1 4NS
Tel: 071-486 0741, Fax: 071-487 7950

1503 International Social Science Council (ISSC)

Organisation concerned with social science research worldwide, whose members are international associations in various social sciences.
Publications: *International Social Science Council Bulletin* (No. 1198)
Contact: ISSC
1 rue Miollis, 75015 Paris, France
Tel: 1-45 68 2558, Fax: 1-43 06 8798

1504 International Task Force for the Rural Poor (INTAF)

Identifies and publicises policies, programmes, and projects for integrated education and development, which contribute most to the all-round benefit of

the rural poor.
Contact: Mukat Singh
12 Eastleigh Avenue, South Harrow HA2 0UF
Tel: 081-864 4470, Fax: 071-864 4740

1505 Ipswich Resource Centre
Centre for local meetings and events about current and global issues, including war and peace, the environment, and human and animal rights. It was actively involved in local discussions about the June 1992 Earth Summit.
Publications: *Gadfly* (No. 1186)
Contact: Ipswich Resource Centre
Friends Meeting House, 39 Fonnereau Road, Ipswich IP1 3JH
Tel: 0473-257649

1506 LIFE STYLE Movement, The
A movement of concerned people, who share a conviction that the well-being of all life, especially the human familiy, depends on our living in ways that no longer threaten Earth, our common home. Its members commit themselves to the principle "Live simply that all may simply live".
Publications: *Living Green* (No. 1199)
Contact: Margaret Smith (General Secretary)
1 Manor Farm, Little Gidding, Cambs. PE17 5RJ
Tel: 08323-383

1507 Medical Action for Global Security (MEDACT)
A new organisation of doctors and other health professionals, who are committed to working for a safer, healthier world; formed by the merger of two earlier medically-based organisations against war.
Publications: *Medicine and War* international journal.
Contact: MEDACT
601 Holloway Road, London N19 4DJ
Tel: 071-272 2020, Fax: 071-281 5717

1508 Merlin Jupiter Ecology Fund
First UK Green investment trust, said to be Greener than most of its competitors; an authorised unit trust, investing worldwide in companies making a positive commitment to social well-being and the protection and wise use of the natural environment.
Publications: Regular surveys of environmental performance of key industry sectors.
Contact: Tessa Tennant, Head of Research
197 Knightsbridge, London SW7 1RB
Tel: 071-581 3020, Fax: 071-581 3857

1509 National Association for Environmental Education (NAEE)
National association, with 24 local

associations, for people interested in education and the environment.
Publications: *Environmental Education* journal and newsletter.
Contact: P. D. Neal, Honorary Secretary
University of Wolverhampton, Walsall Campus, Gorway, Walsall WS1 3BD

1510 New Consumer
Organisation encouraging an economy based on service and cooperation, including a clean and healthy environment, an international marketpace that shares and sustains people worlwide, and a more humane and just economic order. It aims to mobilise consumer power for positive economic, social and environmental change, and it runs a research programme to evaluate the products, activities and social responsibility strategies of important British companies.
Publications: *New Consumer* (No. 1201)
Contact: Richard Adams, Director & Managing Editor
52 Elswick Road, Newcastle-upon-Tyne NE4 6JH
Tel: 091-272 1148, Fax: 091-272 1615

1511 Panos Institute, The
Promotes greater awareness and understanding of sustainable development through research and information dissemination. It works in partnership with other organisations dedicated to sustainable development, helping them develop their information activities.
Contact: Jon Tinker, President
9 White Lion Street, London N1 9PO
Tel: 071-278-1111, Fax: 071-278 0345

1512 Pensions and Investment Research Consultants (PIRC)
Advises pension funds and other institutional investors on socially responsible and environmental strategies to pension funds and other institutional investors; has various environmental investment services.
Contact: Stuart Bell
40 Bowling Green Lane, London EC1R 0NE

1513 Policy Studies Institute (PSI)
Indepenedent policy research organisation, contributing to more effective planning and policy making, by studying selected economic, social, and political problems and issues, and bringing together multidisciplinary teams of experts.
Publications: *Policy Studies* (No. 1124), *Cultural Trends*, *PSI Reports and Journal*, also books on its findings, including No. .
Contact: W. W. Daniel, Director
100 Park Village East, London NW1 3SR

Tel: 071-387 2171, Fax: 071-388 0914

1514 Political Studies Association of the United Kingdom
Promotes the study and teaching of the theory and practice of politics.
Publications: *Political Studies* Journal (No.).
Contact: Stephanie Marshall, Honorary Secretary
16 Gower Street, London WC1E 6DP
Tel:071-323 1311

1515 Politics Association
Promotes the study and teaching of the theory and practice of politics.
Publications: *Talking Politics* journal and *Grass Roots* newsletter.
Contact: Stephanie Marshall, Honorary Secretary
16 Gower Street, London WC1E 6DP
Tel: 071-323 1311

1516 Professions for Social Responsibility
Grouping of several participating organisations, that considers the social reponsibilities of the professions in national and global problems and issues, promotes human survival by the peaceful resolution of conflict and the protection of the environment, and fosters the interest and involvement of professional institutions in its aims.

Contact: Dr Jack Fielding, Chairman or Dr Hugh Gordon, Secretary
1 North End, London NW3 7HH
Tel: 081-458 5316

1517 Regional Energy Resources Information Centre (RERIC)
International information centre on alternative energy systems, especially using renewable energy, and their applications worldwide.
Publications: *RERIC International Energy Journal* (No. 1132) and *RERIC News* (No. 1299).
Contact: RERIC
Asian Institute of Technology, Bangkok, Thailand

1518 Research Council for Complementary Medicine, The (RCCM)
Fosters, enables, and sponsors research into the complementary medical therapies, and promotes greater distribution of information on research results to bring about wider options in health care.
Publications: *Research Council for Complementary Medicine Newsletter* (No. 1210) and *The First Ten Years*.
Contact: Jonathan Monckton, Director
60 Great Ormond Street, London WC1N 3JF
Tel: 071-833 8897, Fax: 071-278 7412

1519 Royal Institute of International Affairs (RIIA)

Promotes the study and understanding of all aspects of international affairs.

Publications: *International Affairs* and *The World Today* (No. 1174) journals, *Chatham House* newsletter, and several books, including Nos. 416, 892 & 949.

Contact: Prof. Laurence Martin, Director

10 St. James's Square, London SW1Y 4ZE

Tel: 071-957 5700, Fax: 071-957 5710

1520 Royal Society for the Encouragement of Arts, Manufactures & Commerce (Royal Society of Arts, RSA)

Encourages the arts, manufactures, and commerce through work in design, industry, education, training, and vocational qualifications.

Publications: *Journal of the Royal Society of Arts*.

Contact: Royal Society of Arts

8 John Adam Street, London WC2N 6EZ

Tel: 071-930-5115

1521 Safety and Reliability Society

Provides a forum for exchanging information on safety and reliability engineering; establishes professional and educational standards for safety and reliability engineers.

Publications: Quarterly journal.

Contact: Secretary, N. J. Locke

59 Piccadilly, Manchester M1 2AQ

Tel: 061-228 7824

1522 Save British Science

Communicates to the British public, Parliament, and Government a proper appreciation of the economic and cultural benefits of scientific and technological research and development (R & D); emphasises the national importance of adequate funding of R & D by government and industry.

Publications: Miscellaneous memoranda on R & D.

Contact: Executive Secretary, John Mulvey

PO Box 241, Oxford OX1 3QQ

Tel: 0865-273407, Fax: 0865-511370

1523 Schumacher College

Offers education for the 21st century, including residential courses on ecology, new economics, spirituality, and other contemporary themes.

Contact: The Administrator

The Old Postern, Dartington, Totnes, Devon TQ9 6EA

Tel: 0803-865934

1524 Scientists for Global Responsibility

A recently founded new organisation, formed from the amalgamation of several formerly separate organisations with allied purposes, whose new aims include: responsible use of science and

technology, international peace, justice and security, global survival, disarmament and the elimination of nuclear weapons, and the transfer of military spending to peaceful and sustainable development.
Publications: *Scientists for Global Responsibility* newsletter (No. 1213)
Contact: Kate Maloney, Administrator
Unit 3, Down House, The Business Village, Broomhill Road, London SW18 4JQ
Tel: 081-871 5175

1525 Shared Interest Society Ltd.
Invests its members' money in successful business enterprises working for people in need throughout the Third World. Its typical Third World partners are cooperatives, community businesses, and charities.
Contact: Shared Interest
52 Elswick Road, Newcastle-upon-Tyne NE46 6JH
Tel: 091-272 4979, Fax 091-272 1615

1526 Social Research Association (SRA)
Advances the conduct, application, and development of social science.
Contact: Steven Bernett, Chairman or Norman Clayton, Administrative Officer
9 Windsor Road, London N13 5PP
Tel: 081-886 2052

1527 Society for Computers and Law
Brings together lawyers and computer experts to study problems of current interest.
Publications: *Computers and Law journal and other publications.*
Contact: Ruth Baker, Administrative Secretary
10 Hurle Crescent, Clifton, Bristol BS28 2TN
Tel: 0272-237393

1528 Society for Social Medicine
Society concerned with all aspects of social medicine, including health service provision, prevention of disease, and the health needs of society.
Contact: Mrs K. Dunnett
OPCS, 10 Kingsway, London WC2B 6JB

1529 Society for the Responsible Use of Resources in Agriculture and on the Land (RURAL)
Provides assistance to policy makers in land use, agricultural and rural management; reconciles competing rural interests, formulates views, and gathers and expresses grassroots feeling.

1530 Society of Business Economists (SBE)
Promotes applications of economics in business and industry.
Publications: *The Business Economist* and a monthly newsletter.
Contact: James Hirst
11 Bay Treen Walk, Watford WD1 3RX

Tel: 0923-37287

1531 Society of Metaphysicians
Study of the fundamental principles and applications of neometaphysics to all aspects of life, including the raising of human consciousness.
Publications: *Neometaphysical Digest* (No. 1104) and other publications including No. 980.
Contact: John J. Williamson, President
Archers' Court, Stonestile Lane, The Ridge, Hastings, East Sussex TN35 4PG
Tel: 0424-751577

1532 Strategic Planning Society
Increases the knowledge and understanding of strategic planning.
Publications: *Long Range Planning* (No. 1100) and newsletter.
Contact: G. A. Goodridge, Secretary General
17 Portland Place, London W1N 3AF
Tel: 071-630 7737, Fax: 071-323 1692

1533 SustainAbility Ltd.
Focusses on corporate environmental excellence, sustainable development, Green consumerism, environmental implications and applications of emerging technologies, and sustainable tourism. Carries out research and consultancy projects for a wide range of client. Conducts an annual Green World Survey of environmental campaigning groups, to identify emerging environmental priorities.
Publications: Several books, including Nos. 309-313, are based on its work and outline it.
Contact: John Elkingon, Director
People's Hall, Freston Road, London W11 4BD
Tel: 071-243 1277, Fax: 071-243 0364

1534 Town & Country Planning Association
Promotes a national policy of land-use planning in the UK.
Publications: *Town & Country Planning*.
Contact: David Hall, Director
17 Carlton House Terrace, London SW1Y 5AS
Tel: 071-930 8903, Fax: 071-930 3280

1535 Traidcraft Exchange
Provides an alternative, fair way of trading with the Third World, aiming to sustain both people and the planet, by setting up direct links with producer and marketing cooperatives in the places where products are grown and made. It distributes hundreds of product lines through hundreds of retail outlets.
Contact: Traidcraft plc
Kingsway, Gateshead, Tyne and Wear NE11 0NE
Tel: 091-491 0591

1536 Transport Two Thousand (2000) Ltd.
Coalition of trade associations, trade unions, and environmental groups, forming a political pressure group in fabour of public transport, rail and water freight, cycling and walking.
Publications: *Transport Retort*.
Contact: Stephen Joseph, Executive Director
10 Molton Street, London NW1 2EJ
Tel: 071-388 8386

1537 UK Environmental Law Association (UKELA)
Has about 800 members, including both lawyers and local authorities, environmental organisations, and companies aiming to improve their environmental performance.
Contact: Andrew Bryce
Cameron Markby Hewitt, Sceptre Court, 40 Tower Hill, London EC3N 4BB
Tel: 071-702 2345, Fax: 071 702 2303

1538 United Kingdom Council for Computing Development (UKCCD)
Promotes, coordinates, and directs support by the UK to developing countries to develop their own computing capability.
Publication: *Computing for International Development*.
Contact: A. M. Costain, Chief Executive

Glenthorne House, Hammersmith Grove, London W6 0LG
Tel: 081-741 7305, Fax: 081-741 5993

1539 United Nations Association of Great Britain and Northern Ireland, The (UNA)
Association promoting a deeper public awareness of and support for the role of the United Nations in global affairs, and running specific campaigns to urge a greater commitment to the UN by the British Government of the day. It is interested in international friendship, peacekeeping and nonviolent conflict resolution, arms control and disarmament, human rights, economic and social development, population, and environmental care.
Publications: *New World* newsletter (No. 1205)
Contact: Malcolm Harper, Director
3 Whitehall Court, London SW1A 2EL
Tel: 071-930 5616, Fax: 071-930 5893

1540 United Nations Centre on Transnational Corporations (UNCTC)
Focal point of the United Nations for all matters related to transnational corporations (TNCs); runs several programmes on TNCs and the environment.

Publications: *Criteria for Sustainable Development* (1990).
Contact: UNCTC
United Nations, New York, NY 10017, USA
Tel: 212-963-4689, Fax: 212-963-2146

1541 United Nations Environment Programme (UNEP) Industry and Environment Office (IEO)
Encourages the incorporation of environmental criteria in industrial development plans, facilitates the implementation of procedures and principles for protecting the environment, promotes the use of safe and 'clean' technologies, and stimulates the exchange of information and experience throughout the world.
Publications: Various publications, including *Industry and Environment*
Contact: Jacqueline Aloisi de Larderel, Director
IEO, Tour Mirabeau, 39-43, quai Andre Citroen, F-75739 Paris Cedex 15, France
Tel: 1-40588850, Fax: 1-40588874

1542 Voluntary Service Overseas (VSO)
Sends volunteers to share their skills with people of developing countries; supports development initiatives, which work towards more equal distribution of resources and greater access to land, capital, health care, skills, technology, and education.
Contact: David Green , Director

317 Putney Bridge Road, London SW15 2PN
Tel: 081-780 2266, Fax: 081-780 1326

1543 The WARMER (World Action for Recycling Materials & Energy from Rubbish) Campaign
A very active organisation, committed to ensuring that, if we do throw things away, we encourage the recycling of materials and energy from post-consumer waste, and a safer and healthier environment. Its main concern is that the energy potential in rubbish is fully used, in district heating, electricity generation or refuse-derived fuel.
Publications: *WARMER Bulletin* (No. 1218)
Contact: WARMER
83 Mount Ephraim, Tunbridge Wells, Kent TN4 8BS
Tel: 0892-24626

1544 Wastewatch
Established by NCVO (No. 1412) to become the key agency for coordinating and promoting community-based recycling in the UK; encourages reccycling at a local level, and provides an information service.
Publications: Books and leaflets, designed to help set up rcycling schemes
Contact: Jonathan Wooding
26 Bedford Square, London WC1B 3HU
Tel: 071-636 4066, Fax: 071 436 3188

1545 World Development Movement (WDM)

Democratic, non party political organisation, which successfully campaigns for changes that directly benefit poor people in the Third World. It is especially concerned with cancelling Third World debts, exposing the crippling effects of increasingly unequal trade, and environmental issues.

Publications: *SPUR* newsletter (No. 1215)

Contact: Maria Elena Hurtado, Director

25 Beehive Place, London SW9 7QR

Tel: 071-737 6215, Fax: 071-274 8232

1546 World Vision of Britain

Works to relieve suffering and improve the quality of life amon poor, sick, and underprivileged people worldwide; promotes development education in the UK, and raises awareness of issues affecting Third World poor; responds to situations of injustice, and advocates the care of the powerless.

Contact: Charles Clayton

Dychurch House, 8 Abington Street, Northampton NN1 2AJ

Tel: 0604-22964, Fax: 0604-29317

1547 World Wide Fund for Nature (WWF) UK Branch

The world's largest independent conservation organisation, working worldwide to mobilise support for our planet, its ecosystems and wildlife; promotes sustainable development, conservation of natural resources, and maintenance of biodiversity.

Contact: George Medley, Director

Panda House, Weyside Park, Godalming, Surrey GU7 1XR

Tel: 0483-426444, Fax: 0483-426409

1548 Zoological Society of London

Promotes worldwide conservation of animal species and their habtats by stimulating public awareness and concern; runs London Zoo and Whipsnade Zoo.

Contact: Prof. N. A. Mitchison, FRS, President

Regents Park, London NW1 4RY

Tel: 071-722 3333, Fax: 071-483 4436

7

TRUSTS, FOUNDATIONS AND CHARITIES

Introductory Note

83 trusts, foundations and charities in the UK and USA are listed here. Most of them give substantial grants each year to projects, organisations or other charities, contributing in various ways to a better future for humankind and the planet. In selecting these bodies, I have been guided by the following criteria:

(1) Grants should usually be at least £100,000 per year.

(2) At least one major area of support should for projects addressing a major global or national issue problem area.

(3) At least some newer rather than conventional or traditional needs should be addressed; thus I have not listed most of the bodies supporting orthodox medical research, nor most of the mainstream charities supporting specific national needs like the relief of poverty.

(4) The stated aims should address some specific aspect(s) of life; thus I have excluded bodies indicating only support for 'general charitable purposes'.

(5) The five main UK Research Councils have been included, because they award millions of pounds of grants per year, although they are not technically charities.

I have also included a few foundations, that have *no* large grant-giving programmes, but otherwise do important work relevant to the 21st century. Each of them is indicated by an asterisk before its name; some of them could also be classified as organisations. Some other bodies with charitable status are listed in Chapters 5 and 6, as they are also networks or organisations.

Each entry has as many as possible the following lines: (1) name in bold type (possibly preceded by an asterisk); (2) brief description of aims and areas of support; (3) contact person(s) or organisation; (4) contact address; (5) telephone number and fax number if any.

The main entries are preceded by seven entries describing up-to-date or recent directories and other books about trusts, foundations and charities.

Directories and Other Books on Trusts, Foundations and Charities

1549 Fitzherbert, Luke & Susan Forrester (Eds.)
A Guide to the Major Trusts 1991
This book is a guide for charities seeking grants from major trusts and foundations. Almost all the organisations described here are themselves charities, whose work consists, wholly or partly, in giving grants to other charities or organisations doing charitable work. It contains: (1) introduction; (2) a list of trusts ranked by size, (3) detailed descriptions of trusts in alphabetical name order; (4) a list of medical research charities; (5) local sources of advice to grant seekers; (6) advice on how to apply successfully to charitable trusts. It has a name index.
1991, Directory of Social Change, ISBN 0-907164-61-7 (pbk) 231 pp.

1550 Forrester, Susan (Ed.)
Environmental Grants: A Guide to Grants for the Environment from Government, Companies and Charitable Trusts
This book provides information on sources of funds for environmental work. It gives details of: (1) 300 charitable trusts with an interest in aspects of environmental work; (2) 150 major companies giving charitable donations to environmental groups; (3) grant aid schemes by government departments and statutory agencies; (4) awards and competitions with a financial reward; (5) funding programmes run by environmental organisations themselves; (6) enabling organisations, providing free or low-cost professional services and other support in kind to voluntary and environmental groups. It covers the whole range of environmental concern and activity, including: the natural and built environment, countryside and wildlife conservation, preservation of the architectural heritage, local amenities, regeneration of urban areas, promotion of sustainable development, pollution abatement, energy conservation, and the development of appropriate forms of technology.
1989, Directory of Social Change, London, ISBN 0-907164-47-1 (pbk) vii + 309 pp.

1551 Hodson, H. V. (Consultant Ed.)
The International Foundation Directory
This directory lists foundations in alphabetical country order, and in alphabetical name order for each country. It gives details of each listed foundation. There is a name index of foundations, and a name index of major activities, for each of which relevant foundations are listed in alphabetical order.
3rd. Ed. 1983, Europa Publications, London, ISBN 0-905118-90-1 (hbk) xxviii + 401 pp.; 1st. Ed. 1974.

1552 Humphreys, G. (Ed.)
Charities Digest 1992
This directory provides information about charities in the UK, with pecial emphasis on those working for family and social welfare.
98th. Ed. 1992, Family Welfare Association, London, ISBN 0-900954-48-5 (pbk) 371 pp.

1553 Lee, Stephen, Adrian Randall, Francesca Quint & Fergus Talk (Eds.)
The Henderson Top 1000 Charities 1992: A Guide to UK Charities
This book provides readily comparable information about leading British charities and how they make and spend their money, who is behind them, who is advising them, and who is giving to them. Typical information about a charity, foundation, or trust includes its name(s), aims, address, contact details, status, links, income and expenditure, statement of assets, corporate donors, patrons, advisers, staff, constitution, date of foundation, etc.
1st. Ed., 1992, Hemmington Scott Publishing, London, ISSN 0967-8352 (pbk) 800 pp.

1554 Morgan, M. (Ed.)
Charity Choice 1990: The Encyclopaedia of Charities
This book is a list of British charities, arranged in categories, including: children and youth, family welfare, disabled people, medical, social welfare, ethnic, voluntary organisations, cultural, environment, overseas aid, and finances and services for charities.
3rd. Ed. 1990, Abercorn Hill, London, ISBN 1-870701-02-X (?bk) ? pp.

1555 Villemur, Anne (Ed.)
Directory of Grant-Making Trusts 1991
This book provides a register of grant-making charitable trusts, and gives detailed descriptions of UK-based trusts in alphabetical name order. It introduces a fairly detailed classification of charitable purposes; for each of these purposes, it lists the supporting trusts. It has a geographical index of trusts, together with alphabetical and subject indexes.

12th. Ed. 1991, Charities Aid Foundation, Tonbridge, Kent, ISBN 0-904757-52-8 (hbk) xxvi + 1038 pp.

List of Trusts, Foundations and Charities

1556 Action Aid
Helps identify, fund, and manage long-term integrated rural development programmes, designed to overcome poverty and improve quality of life. It works directly with children, families, and communities in the world's poorest countries. It provides emergency relief when circumstances demand.
Contact: Chris Chataway, Chairman
Tapstone Road, Chard, Somerset TA20 2AB
Tel: 0460-62972, Fax: 0460-67191

1557 Aga Khan Foundation (United Kingdom)
Supports innovative, cost-effective and replicable approaches and programmes of broad use in resolving selected problems in health, education and rural development in different national and cultural contexts.
Contact: Mirza Jahani (Chief Executive Officer)
33 Thurloe Square, London SW7 2SD
Tel: 071-225 2001

1558 Agricultural and Food Research Council (AFRC)
Promotes research in agriculture, fisheries, food, and nutrition.
Contact: Public Relations Office
Polaris House, North Star Abenue, Swindon SN2 1UH
Tel: 0793-413200

1559 Arts Council of Great Britain
To develop and improve the knowledge, understanding, and practice of the arts.
Contact: The Secretary General
14 Grant Seton Street, London SW1P 3NQ
Tel: 0710333 0010

1560 * Association of Charitable Foundations
Promotes philanthropy, good practice in charitable grantmaking, and government understanding thereof.
Contact: Nigel Siederer, Director
34 North End Road, London W14 0SH
Tel: 071-603 1525, Fax: 071-371 1750

1561 BBC Children in Need Appeal, The
Relief of need, hardship, sickness, handicaps, and distress among children and young people.
Contact: British Children in Need
Broadcasting Support Services, PO Box 7, London W3 6XZ

1562 British Academy for the Promotion of Historical, Philosophical and Philological Studies, The

Promotion of education and research in the humanities and social sciences.
Contact: P. W. H. Brown, Secretary, British Academy
20-21 Cromwell Terrace, London NW1 4QP
071-487 5966

1563 * British Nutrition Foundation
Promotes education, research, and information in nutrition and food.
Contact: Dr Daniel M. Conning, Director General or J. P. Wood, Secretary
15 Belgrave Square, London SW1X 8PG
Tel: 071-235-4094, Fax: 071-235 5336

1564 Brookings Institution
Conducts policy-oriented research through its Economic Studies, Governmental Studies, and Foreign Policy Studies programmes; conducts educational conferences on public policy issues for leaders in government, industry, and the professions.
Contact: Brookings Institution
1715 Massachusetts Avenue, NW, Washington, DC 20036, USA
Tel: 202-797-6000, Fax: 202-797-6004

1565 Carnegie Foundation of New York
Promotes the advancement and

diffusion of knowledge and understanding, primarily in education, social welfare, and law, through research and project grants.
Contact: Sar L. Englehardt, Secretary
437 Madison Avenue, New York, NY 10022, USA
Tel: 212-371-3200

1566 Carnegie United Kingdom Trust, The
Aims to improve the well-being of the masses of the people of the British Isles, concentrating on amateur arts, the local environment, and community service, giving priority to agencies benefitting local grassroots action.
Contact: The Secretary
80 New Row, Dunfermline, Fife KY12 7EJ, Scotland

1567 Charities Aid Foundation, The
Promotes and facilitates the distribution of money to general charitable purposes.
Contact: M. J. M. Brophy, Director
48 Pembury Road, Tonbridge, Kent TN9 2JE
Tel: 0732-771333

1568 * Charter 88 Trust, The
Promotes the non-partisan advancement of constitutional and political education generally, and in relation to the rights, liberties, and duties of all citizens. It

publishes the results of relevant research.
Contact: Charter 88 Trust
Exmouth House, 3-11 Pine Street, London EC1R 0JH

1569 Child Poverty Action Group
Promotes action for the relief of poverty amongst children and families with children.
Contact: Fran Bennett, Director
4th. Floor, 1-5 Bath Street, London EC1P 9PY
Tel: 071-253 3406

1570 Christian Action
Supports action to promote human rights and responsibilities, community work, good housing, and the relief of poverty.
Contact: Christian Action
St. Peters House, 308 Kennington Lane, London SE11 5HY
Tel: 071-735 2372

1571 Christian Aid
Primarily supports the furtherance of charitable purposes that tackle and address the root causes of poverty; works directly through local organisations which help the poor to find their own solutions.
Contact: Rev. Michael Taylor, Director
PO Box 100, London SE1 7RT
Tel: 071-620 4444

1572 Churchill Memorial Trust, The Winston
Provides about 100 travelling Fellowships per year, to enable men and women from all walks of life to gain a better understanding of overseas countries, and to acquire knowledge and experience which will make them more effective in their work and in the community when they return; the average tenure of a Fellowship is about eight weeks.
Contact: The Director General
15 Queen's Gate Terrace, London SW7 5PR
Tel: 071-584 3915

1573 * Community Development Foundation (CDF)
Advises both non-governmental organisations and local authorities concerned with community development, training, research, conferences, and publications; has begun to network with Europe, and offers a European service for community development agencies.
Contact: CDF
60 Highbury Grove, London N5 2AG
Tel: 071-226-5375, Fax: 071-704 0313

1574 Economic and Social Research Foundation (ESRC)
Promotes and supports research in economics and the social sciences.
Contact: Public Relations Department
Polaris House, North Star Avenue, Swindon SN2 1UJ
Tel: 0793-413000

1575 * Environmental Law Foundation (ELF)
Provides access to a network of lawyers and other experts, for individuals, communities and small businesses to defend themselves abd their environment against potentially destructive developments.
Contact: Martin Polden
Rubinstein, Callingham, Polden and Gale, 2 Raymond Buildings, Gray's Inn, London WC1R 5BZ
Tel: 071-242 8404

1576 Fairbairn Trust Fund, Esmee
Supports education (mostly at the higher levels), the arts, the natural heritage, social welfare, the environment, and economic and social research.
Contact: The Director
5 Storey's Gate, London SW1P 3AT
Tel: 071-222 7041

1577 Fellowship of Engineering, The (FoEng)
Promotes and encourages excellence in the whole field of engineering, and the advancement of the art, science, and practice of engineering for the benefit of the public.
Contact: Executive Secretary, V. J. Osola
2 Little Smith Street, London SW1P 3DL
Tels: 071-222 2688 & 071-222 3912, Fax: 071-233 0054

1578 * Findhorn Foundation
works for the advancement of spiritual and religious studies and practices worldwide by teaching, example, and demonstration.
Contact: Findhorn Foundation
The Park, Forres, Scotland IV36 0TZ
Tel: 0309-690311

1579 Ford Foundation
Advances public welfare by trying to identify and contribute to the solution of problems of national and international importance.
Contact: Ford Foundation
320 East 43rd Street, New York, NY 10017, USA
Tel: 212-573 5000

1580 * Foundation Center
Gathers, analyses, and disseminates information about philanthropic foundations in the public interest.
Contact: Foundation Center
888 Seventh Avenue, New york, NY 10006, USA
Tel: 212-975-1120

1581 Friends of the Earth Trust
Promotes the beliefs and ideals of Friends of the Earth (No. 1477) through environmental reserach and education.
26-28 Underwood Street, London N1 7JU
Tel: 071-490 1555, Fax: 071-490-0881

1582 Greenpeace Environmental Trust

Supports the aims of Greenpeace (No. 1484) by carrying ut extensive scientific reserach and educationalprojects into world ecology. 30-31 Islington Gree, London N1 8XE
Tel: 971-236 8865

1583 Guggenheim Memorial Fund, John Simon
Aims to improve the quality of education and the practice of the arts and professions, fosters research, and promotes better international understanding; supports the sciences, humanities, and creative arts.
Contact: Stephen L. Schlesinger, Secretary
90 Park Avenue, New York, NY 10016, USA
Tel: 212-687-4470

1584 Gulbenkian Foundation (Lisbon) United Kingdom Branch, Calouste
Promotes charitable purposes, mainly in the arts, education and social welfare.
Contact: The Director
98 Portland Place, London W1N 4ET
Tel: 071-636 5313

1585 Help the Aged
Raises awareness of the needs of the elderly and their contributions to society. It offers both immediate relief and support for long-term development projects overseas.

Contact: Christopher Beer
St. James's Walk, London EC1R 0BE
Tel: 071-253 0253, Fax: 071-253 4814

1586 * Higher Education Foundation
Explores issues of principle and practice in higher education, is concerned about change and its effect on people in higher education, and considers the responsibilities of teachers and taught in higher education.
Contact: Philip Healy, Secretary
Westminster College, Oxford OX2 9AT
Tel: 0865-247644

1587 Homeless International
Assists and supports community-based and non-governmenntal agencies, working to improve the provision of shelter for the poor in the Third World.
Contact: Ruth McLeod, Director
5 The Butts, Coventry CV1 3GH
Tel: 0203-632802, Fax: 0203-632911

1588 IBM United Kingdom Trust
Promotes computer training, the arts, and opportunities for disadvantaged, handicapped and disabled people.
Contact: C. E. Berks, Trust Administrator
IBM South Bank, 76 Upper Ground, London SE1 9PZ
Tel: 071-928 1777

1589 International Foundation
Enables people in developing

countries to improve their standard of life in a sustainable way; gives priority to nutrition, food, development, protection of resources, education, health, and social welfare.
Contact: Chandler McBracken, Chairman of Grants Committee
1375 Route 23, Butler, NJ 07405, USA

1590 * International Science Policy Foundation
Pioneers the presentation and formulation of science and technology poicy for economic and social development; sponsors an annual lecture by a distinguished speaker.
Contact: Dr Maurice Goldsmith, Director
12 Whitehall, London SW1A 2DY
Tel: 071-839 4985

1591 International Voluntary Service
Promotes international understanding through common endeavours, and sets up international work camps of young volunteers as a unique form of voluntary service.
Contact: International Voluntary Service
162 Upper New Walk, Leicester LE1 7QA
Tel: 0533-549430

1592 Kellogg Foundation, W. K.
Supports education, health, and agriculture, but not research.

Contact: J. M. Drevno, Secretary
400 North Avenue, Battle Creek, MI 49016, USA
Tel: 616-965-1221

1593 Kettering Foundation, Charles F.
Supports investigation of questions associated with governing, education, and science, all in relation to one another and in an international perspective, with respect to both formal and informal institutions.
Contact: Robert F. Lehman, Secretary
5335 Far Hills Avenue, Suite 300, Dayton, OH 46429, USA
Tel: 513-434-7300

1594 King George's Jubilee Trust
Advances the physical, mental and spiritual welfare of the younger generation.
Contact: The Director of Administration, The Princes' Trust
8 Bedford Row, London WC1R 4BA
Tel: 071-430 0524

1595 Kleinwort Charitable Settlement, The Sir Cyril
Promotes education, job creation, conservation, the arts, medical research and care, population control, and youth development.
Contact: Kleinwort Benson Trustees Ltd.
PO Box 191, 10 Fenchurch Street, London EC3M 3LB
Tel: 071-956 6600

1596 Kleinwort Charitable Trust, The Ernest
Promotes the conservation and protection of the environment, and helps the elderly and disabled.
Contact: The Secretary
10 Fenchurch Street, London EC3M 3LB
Tel: 071-956 6600

1597 Leverhulme Trust, the
Promotes mainly specific research projects in various fields, and schemes of international academic interchange.
Contact: The Director
15-19 New Fetter Lane, London EC4 1NR
Tel: 071-822 6938

1598 Lloyds Bank Charitable Trust
Provides wide ranging grants, focussed on education, the environment, employment, health, and social welfare.
Contact: A. M. Finch, Community Affairs Manager, Corporate Communications, Lloyds Bank plc
71 Lombard Street, London EC3P 3BS
Tel: 071-626 1500

1599 Medical Research Council (MRC)
Promotes research in medicine and allied biological sciences.
Contact: Public Relations Office
20 Park Crescent, London W1N 4AL
Tel: 071-636 5422

1600 Medical Research Council – Trust Funds
Promotes the development of medical and related research, and the advancement of knowledge leading to improved health care.
Contact: The Accountant, Medical Research Council
20 Park Crescent, London W1N 4AL
Tel: 071-636 5422

1601 Mellon Foundation, Andrew W.
Supports higher education, medical education, and cultural institutions; areas of support include research, population, conservation of the environment, and public affairs.
Contact: J. Kellan Smith, Jr., Secretary
140 East 62nd Street, New York, NY 10021, USA
Tel: 212-838-8400

1602 * National Energy Foundation, The
Promotes better efficiency and effectiveness of energy use, better appreciation of issues relating to energy supply, and better understanding of energy and energy needs.
Contact: The Director
Rockingham Drive, Linford Wood, Milton Keynes MK14 6NG
Tel: 0908-672787

1603 National Environment Research Council (NERC)

Promotes research in the physical and biological sciences, relating to the natural environment and its resources.
Contact: Public Relations Department
Polaris House, North Star Avenue, Swindon SN2 1EU
Tel: 0793-411500

1604 * New Economics Foundation, The (NEF)

Promotes the emergence of a New Economics, based on democracy and sustainability, that provides economic justice, satisfaction of economic needs, and welfare for people and planet.
Contact: Edward Mayo, Director
Universal House, Second Floor, 88-94 Wentworth Street, London E1 7SA
Tel: 071-377 5696

1605 Nuffield Foundation, The

Advances health and social well-being, education, and the care of old people, especially through research in medicine, ageing, and the natural and social sciences.
Contact: Robert Hazel, Director
28 Bedford Square, London WC1B 3EG
Tel: 071-631 0566

1606 Oxfam

Promotes the relief of poverty, distress and suffering in any prt of the world, and education of the public about the nature, causes and effects of poverty; aims to use resources with maximum effectiveness in areas of greatest need.
Contact: The Director
274 Banbury Road, Oxford OX2 7DZ
Tel: 0865-56777

1607 Pilgrim Trust, The

Promotes preservation of the natural heritage and the future well-being of the country; in particular, supports art, learning, museums, libraries, and social welfare.
Contact: The Secretary
Fielden House, Little College Street, London SW1P 3SH

1608 * Population Council

Enhances human welfare, with emphasis on problems of population and reproduction, including contraception; supports social science research on relations between demographic change and social and economic development, and seeks to apply the results to the design of social policies.
Contact: Population Council
One Dag Hammarskjold Plaza, New York, NY 10017, USA

1609 Prince of Wales Charities

Supports organisations whose activities seem particularly enterprising.
Contact: Major General Sir Christopher Airy, KCVO, CBE
The Prince and Princess of Wales Office, St. James' Palace, London SW1A 1BJ

1610 Prince's Trust, The
Provides facilities for people between 14 and 25, for recreation or other leisure occupations, in the interests of social welfare and to improve themselves.
Contact: The Director of Administration
8 Bedford Row, London WC1R 4BA
Tel: 071-430 0524

1611 Project Charitable Trust, The
Advances the education of young people by providing opportunities for service in partnership with people overseas.
Contact: Major N. M. V. Bristol
Breacachadh Castle, Isle of Coll, Argyll PA78 6TB, Scotland

1612 Queen's Silver Jubilee Trust, The
Enables and encourages young people to work together to help others of all ages in the community; in practice, supports voluntary work by groups or individuals.
Contact: The Director of Administration
8 Bedford Row, London WC1R 4BA
Tel: 071-430 0524

1613 * Resources for the Future
Advances research and education in development, conservation and use of natural resources, and improving the quality of the environment; was founded by the Ford Foundation (No. 1529).
Contact: Edward F. Hund, Secretary-Treasurer
1755 Massachusetts Avenue, NW, Washington, DC 20036, USA
Tel: 301-328-5000

1614 Right Livelihood Awards Foundation
Sponsors the Right Livelihood Awards, that have been made annually since 1980, to honour and support work that faces major human and global problems and pioneers solutions to them. For further details of the awards made, see Appendix 2 of *A New World Order* (No. 305).
Contact: Administrative Director, Right Livelihood Awards Administrative Office
PO Box 15072, S-10465 Stockholm, Sweden
Tel: 08-7020340

1615 Rockefeller Foundation
Promotes the well-being of mankind throughout the world, with national and international programmes for the conquest of hunger and associated ills, the solution of population and health problems; especially in the USA, promotes equal opportunities for all, and development of the arts, humanities, and contemporary values.
Contact: Lawrence D. Stigel, Secretary

1133 Avenue of the Americas, New York, NY 10036, USA
Tel: 212-868-8500

1616 Rowntree Charitable Trust, The
Supports work on conflict resolution, justice, equal opportunties, and the unique value of each individual.
Contact: S. Burkeman
Beverley House, Shipton Road, York YO3 6RB
Tel: 0904-617810

1617 Rowntree Foundation, Joseph
Seeks causes of social problems and contributions to their solutions, by funding a programme of useful reserach and pioneering development projects, in housing, social policy, social care, and relations between local and central government.
Contact: Richard Best, OBE, Director
The Homestead, 40 Water End, York YO3 6RB
Tel: 0904-620241

1618 Royal Society, The
Encourages scientific research and its application; promotes international scientific relations and facilitates exchange visits of scientists; has various Research Fellowships and grant schemes.
Contact: The Executive Secretary
6 Carlton House Terrace, London SW1Y 5AG
Tel: 071-839 5561

1619 Save and Prosper Educational Trust
Promotes education, including schools and colleges.
Contact: J. B. Shelley, Director
1 Finsbury Avenue, London EC2M 2QY
Tel: 071-188 1717

1620 Save the Children Fund, The
Promotes the welfare and relief of poverty of children, both in the UK and in the Third World.
Contact: The Save the Children Fund
17 Grove Lane, London SE5 8RD
Tel: 071-703 5400

1621 Science and Engineering Research Council (SERC)
Promotes and supports research in the mathematical, physical and biological sciences, engineering, and technology. It also maintains a fundamental capacity of UK reserach and engineering, and supports relevant post-doctoral education.
Contact: Public Relations Department
Polaris House, North Star Avenue, Swindon SN2 1ET
Tel: 0793-411000

1622 Scott Bader Commonwealth Ltd., The
Assisting distressed and needy people of all nationalities; promoting education, peacebuilding,

and democratic participation; encouraging the careful use and protection of the Earth's resources.
Contact: The Secretary
Wollaston, Wellingborough, Northants. NN9 7RL
Tel: 0933-663100

1623 Shelter – National Campaign for the Homeless
Relieves hardship and distress among the homeless and among those in need living in adverse housing conditions.
Contact: The Director, Shelter Ltd.
88 Old Street, London EC1V 9HU
Tel: 071-253 0202

1624 TSB Foundation for England and Wales
Mainly assists disadvantaged people with education and training; scientific and medical research, especialy on illnesses preventing people from remaining in work; provision of facilities for social and community and social welfare.
Contact: R. W. Jenkins, TSB
25 Milk Street, London EC2V 8LU
Tel: 071-606 7010

1625 Television South Charitable Trust
Advancing education and training in the arts; advancing education, scientific and medical research; providing social welfare
Contact: Liz Delbarre, Appeals Administrator, TVS Trust

60-61 Buckingham Gate, London SW1E 6AJ
Tel: 071-828 9898

1626 * Voluntary Service Overseas
Assists development in the Third World by sending experienced, practical people on two-year projects to share their skills with its inhabitants.
Contact: David Green , Director
317 Putney Bridge Road, London SW15 2PN
Tel: 081-780 2266, Fax: 081-780 1326

1627 War on Want
Relieves poverty and want anywhere in the world, researches into the causes of poverty, and publishes its findings.
Contact: The General Secretary
37-39 Great Guildford Street, London SE1 0ES
Tel: 071-620 1111, Fax: 071-261 9291

1628 Wolfson Foundation
Supports scientific and medical research, and higher education.
Contact: Dr Barbara Rashknow, Director
Universal House, 251-256 Tottenham Court Road, London W1A 1BZ

1629 World in Need
Advances Christianity; relieves sickness, poverty and distress in any part of the world; emphasises

'seed-corn' projects initiating innovative work.

Contact: The Secretary, c/o Andrews Estate Agents

103 High Street, Oxford OX1 4BW

Tel: 0865-794411

1630 * World University Service (UK)

Advances education for development, and the relief of refugee students and staff of universities and other institutions of higher education.

Contact: WUS

20-21 Compton Terrace, London N1 2UN

Tel: 071-226 6747

1631 Y Care International

Supports relief and develops projects run by YMCAs in the Third World. It is part of the worldwide YMCA movement, and is also engaged in development education in the UK.

Contact: Contact: Dr Judith Ennow, Director

640 Forest Road, London E17 3DZ

Tel: 081-520 5599, Fax: 081-503 7461

8

21ST CENTURY PROJECTS

Introductory Note

This short chapter introduces twenty of the considerable number of projects which are oriented to preparing for and improving our long-term future into the 21st century. Cross-references are made to other items where necessary.

Each entry has as many as possible of the following lines: (1) name of project in bold type; (2) brief description of aim, nature and activities; (3) contact person(s) or organisation; (4) contact address; (5) telephone and fax number(s); (6) cross references to other relevant items in *Resources for the Future*.

List of 21st Century Projects

1632 Alternative Futures Project
Founded by several non-governmental organisations to promote a more humane model, based on human solidarity and North-South cooperation. It is now part of the Norwegian National Research Foundation for the Social Sciences and Humanities.
Contact: William Lafferty, Project Director
Sognsvelen 70, 9855 Oslo, Norway
Tel: 2-181170, Fax: 2-182077

1633 Australia's Commission for the Future
Commission established by the Australian Minister of Science in 1985, now acting as knowledge broker between different sectors, focussing on the management of change. It aims to build a world-class futures institution based on information.
Contact: Susan Oliver, Director
98 Drummond Street, Carlton South, 3053 Victoria, Australia

1634 Centre for Our Common Future, The

Centre set up after the presentation and publication of the Brundtland Report *Our Common Future* (No. 988). Its principal purpose is to activate and sustain the global debate on sustainable development, but it also acts as an independent clearing house for the exchange of ideas and activities, providing continual information, advice, and encouragement. It plays a useful role in uniting the various sectors of human activity towards common goals in preparation for the 21st century.
Contact: Warren H. Lidner, Director
Palais Wilson, 52 rue des Paquis, CH-1201 Geneva, Switzerland
Tel: 022-732 7117, Fax: 022-738 5046
Cross-references: No. 864.

1635 Club of Rome, The
International group of leading people in business and public affairs, that has helped to sponsor many important research projects and books addressing human and global problems and issues, well into the 21st century. In particular, it is well-known for its encouragement of *The Limits to Growth* (No. 642) and other global models.
Contact: Bertrand Schneider, President
34 Avenue d'Eylau, F-75116 Paris, France

Cross-references: Nos. 358, 542, 568, 642, 654, 728, 816 & 896.

1636 Countdown 2001
Group of organisations and individuals interested in working towards 'creating the best possible 21st century'. It provides organisational support, including training seminars and workshops on such topics as strategic planning and visioning, creativity, decision making, and education for the future. Its publications include *An Agenda for the 21st Century Project*.
Contact: Dr Sherry L. Schiller, President
110 N. Payne Street, Alexandria, VA 22314
Tel: 703-684-4735, Fax: 703-684-4738.

1637 Darwin Initiative, The
Initiative, founded by the British Prime Minister after the June 1992 Earth Summit (UNCED), to implement the aims of its Biodiversity Convention by using UK scientific and commercial skills to aid biodiversity in developing countries. It works in association with the UK Department of the Environment.
Contact: Kate Mayes
Department of the Environment, 2 Marsham Street, London SW1P 3EB
Tel: 071-276 8386
Cross-references: 1463.

1638 Global Futures Project

School curriculum project, focussing on the entitlement of pupils to prepare for responsible citizenship as future adult members of the community. Helps teachers and pupils to: (1) explore current concerns about the state of the planet; (2) clarify their choice of preferred futures from the personal to the global; (3) envisage just and sustainable alternative futures; (4) develop both critical thinking skills and creative imagination; (5) exercise their rights responsibly as active citizens in the local and global community.

Contact: Dr David Hicks, Director Bath College of Higher Education, Newton Park, Bath, BA2 9BN

Tel: 0225-873701, Fax: 0225-874123 Cross-references: No. 463.

1639 Institute for the Future

Assists organisations, business, industry, and the US Government in conducting long-term futures research. It promotes practical applications of information techniques for improved management and productivity.
Contact: J. Ian Morrison, President
2740 Sand Hill Road, Menlo Park, CA 94025, USA
Tel: 415-854-6322,   Fax: 415-854-7850

1640 Institute for 21st Century Studies

Assists nations to address long-term, worldwide issues, such as population control, national security, agricultural development, and environmental protection, and organises study teams exploring future possibilities in these areas. It works with groups doing national futures studies on alternative futures, and religious organisations interested in global futures. It provides training and computer models for study teams, together with a toolkit of books, handbooks, databases, and software for institutions doing national studies.

Publications: Various publications and software.

Contact: Gerald O. Barney, Executive Director
1611 N. Kent Street, Suite 610, Arlington, VA 22209-2111, USA
Tel: 703-841-0048,   Fax: 703-525-1744

Cross-references: Nos. 50, 363 & 1371.

1641 Institute for Social Inventions

Aims to promote social innovations and inventions, which are new social services or new and imaginative solutions to social problems. It gives an opportunity to tackle such problems before they become crises, by encouraging public participation in continuous problem solving and promoting small-scale innovative experiments. It publishes many examples of social inventions, and runs an annual prize competition.

Contact: Nicholas Albery, Chairman
20 Heber Road, London NW2 6AA
Tel: 081-208 2853, Fax: 081-452 6434
Cross-references: 12 & 1146

1642 Millennium Project, The
Planned joint project of the United Nations University, in collaboration with the Smithsonian Institution, the US Environmental Protection Agency, and the Futures Group. It aims to establish a permanent capability to continuously update and improve humanity's thinking about the future, and make that thinking available through a variety of media. It will create an international system of forecasts, key questions, lessons from history, and potential futres research agandas. It will also evaluate futures research methodology and the potential ofr setting standards, integrate forecasts to describe possible futures, and propose policy choices. The scholars and institutions participating in it will come from many parts of the world and represent a wide range of academic and other disciplines, including futures research.

Contact:Jerome C. Glenn
The American Council for the United Nations University
421 Garrison Street, NW, Washington, DC, USA
Tel: 202-686-5179

1643 New Paradigms Project, The
Project aiming to: (1) explore, develop, and present new paradigms, new ideas, and new possibilities, covering all aspects of life; (2) conduct relevant research and development in various fields; (3) contribute thinking, writing, and preliminary policy proposals, that address human and global problems from now into the 21st century; (4) cooperate with other 21st century studies projects.

Contact: Alan J. Mayne, Coordinator
29 Fairford Crescent, Downhead Park, Milton Keynes, MK15 9AF
Tel: 0908-607022
Cross-references: Nos. 1110 & 1204.

1644 21st Century Studies Programme
Programme for publishing a series of books, based on the premiss that '21st century studies' will become an important discipline through the 1990s into the next century; uniquely devoted to viewing the world in an integrated way, and to improving contemporary situations and policies in many ways. Its books at

present have business, environment, futures, global policy, and reference strands, and there are also plans to republish 'classic' books with important lessons for the future.

Contact: Jeremy Geelan, Editorial Director & Publisher, 21st Century Publishing Programme

Adamantine Press, 3 Henrietta Street, Covent Garden, London WC2E 8LU

Tel: 071-836-4975, Fax: 071-379 0609

Cross-references: No. 1001 for the series, and Nos. 144, 182, 331, 341, 549, 572, 606, 658, 876, 906, 936 & 937 for books in the series.

1645 UNESCO International Commission on Education for the 21st Century

Commission set up by UNESCO to rethink education globally, and ask what kind of education is needed for what kind of society. It strongly emphasises practicality and an action-oriented approach, and aims to stimulate far-reaching public debate on educational reform.

Contact: UNESCO, Place de Fontenoy, Paris 75700, France

1646 Union of International Associations (UIA)

Union with at least the following purposes: (1) facilitating the evaluation of the activities of the worldwide network of nonprosit organisations, especially non-

governmental organisations and voluntary associations; (2) promoting understanding of how international bodies represent valid interests in every field of human activity or belief; (3) enabling these initiatives to develop and counterbalance each other creatively, in response to world problems, by collecting information on these bdies and their mutual relationships; (4) making such information available to them and to others who may benefit from this network; (5) maintaining contact with a wide variety of bodies worldwide, as a foundation for better organisational networking and programme harmonisation in response to increasingly complex world problems. Its publications include: *Transnational Associations* (No. 1160), *Yearbook of International Associations* (No. 1442), and *Encyclopedia of World Problems and Human Potential* (No. 917).

Contact: Anthony J. N. Judge, Secretariat General

rue Washington 40, Brussels 40-1050, Belgium

Tel: 2-640 4109 & 2-640 1808, Fax: 2-646 0525

Cross-references: 917 & 1537.

1647 World Future Society

This society has an extensive future-oriented publications programme, including several important periodicals and books, holds regular conferences, and runs a bookstore. Its membership is mostly American but includes futurists from many other countries.

Contact: Edward Cornish, President
7910 Woodmont Avenue, Suite 450, Bethesda, MD 20814, USA
Tel: 301-656-8274
Cross-references: Nos. 654, 1046, 1047, 1049 & 1050.

1648 World Future Studies Federation

Brings together institutions, scholars, policy makers, and other individuals interested in futures studies, and promotes futures studies and innovative interdisciplinary analyses. It encourages the exchange of information and experience through original research projects, seminars, and conferences.

Current contact: Pentti Malaska, Secretary General
Current address: Turku School of Economics, Rektorinpellonkatu 3, SF-20500 Turku, Finland
Tel: 21-330835, Fax: 21-330755

1649 World Futures

Joint project of the Vienna Academy for Global and Evolutionary Studies, the General Evolutionary Research Group, and Gordon and Breach Science Publishers, New York, for publishing the *World Futures* journal and the *The World Futures General Evolution Studies Book Series*.

Contact: Professor Ervin Laszlo, Rector, European Academy for Evolutionary Management and Advanced Study, Frankfurt, Germany
Cross-references: Nos. 570, 571 & 573.

1650 World Studies 8-13 Project

British curriculum project to make 8 to 13 year olds aware of global problems and issues. Its teacher's handbook has been published and translated into several other languages.

Contact: Miriam Steiner
Faculty of Education, Manchester Metropolitan University, 799 Wilmslow Road, Manchester M20 8RR

1651 Worldwatch Institute

Research organisation with an international approach, aiming to encourage a reflective and deliberate approach to global problem solving. It seeks to anticipate global problems and social trends, and focus attention on emerging global issues. For many years, it has had an extensive programme of publications on a wide range of important human and global problems. Its work is especially oriented to help bring about a sustainable future, with a reasonable quality of life, for humanity and the planet.

Contact: Lester R. Brown, Director

Worldwatch Institute, 1776 Massachusetts Avenue, NW, Washington, DC 20036, USA
Tel: 202-452-1999,

Fax: 202-296-7365
Cross-references: Nos. 119-131,
651, 902, 1176 & 1177.

LIST OF PUBLISHERS OF BOOKS INCLUDED (INCLUDING CURRENT ADDRESSES)

Note that this list includes only publishers, which have published either several of the listed books or a specially important book. A few of the other publishers are organisations listed in earlier chapters.

Academic Press, c/o Harcourt Brace Jovanovich, 1250 Sixth Avenue, San Diego, CA 92101, USA & 24-28 Oval Road, London NW1 7BX

Adamantine Press Ltd., 3 Henrietta Street, Covent Garden, London WC2E 8LU

Addison-Wesley Publishing Co. Inc., Jacob Way, Reading, MA 01867, USA

Albatross Books, PO Box 320, Sutherland, NSW 2232, Australia & PO Box 131, Claremont, CA 91711, USA

Allen & Unwin (*see* HarperCollins)

The Alternative Future Project, Hausmannsgata 27, N-0182 Oslo 1, Norway

Anchor Books (*see* Doubleday)

The Aquarian Press (*see* HarperCollins)

Ashgate Publishing Group, Gower House, Croft Road, Aldershot, Hants. GU11 3HR & Brookfield, VT05036, USA

Avebury (*see* Ashgate Group)

Basic Books, 10 East 53rd Street, New York, NY 10022, USA

BBC Books, Woodlands, 80 Wood Lane, London W12 0TT

Beacon Press, 25 Beacon Street, Boston, MA 02018, USA

Bear & Co., PO Drawer 2860, Santa Fe, NM 87504, USA

Belhaven Press, 25 Floral Street, London WC2E 9DS

Blackwell Publishers Ltd., 108 Cowley Road, Oxford OX4 1JF & 3 Cambridge Center, Cambridge, MA 02142, USA

Bootstrap Press, Council on International and Public Affairs, 777 UN Plaza, Suite 9A, New York, NY 10017, USA

R. R. Bowker, 121 Clandon Road, New Providence, NJ 07974, USA

Bowker-Saur Ltd., 60 Grosvenor Street, London W1X 9DA

Marion Boyars Publisher Ltd., 24 Lacy Road, London SW15 1NL

Brassey's (UK) Ltd., First Floor, 165 Great Dover Street, London SE1 4YA

The British Library, Boston Spa, Wetherby, West Yorkshire LS23 7BQ

Burns & Oates Ltd., Wellwood, North Farm Road, Tunbridge Wells, Kent TN2 3DR

Butterworth-Heinemann & Co. (Publishers) Ltd., Linacre House, Jordan Hill, Oxford OX2 8DP

Calder & Boyars (*see* Marion Boyars)

Cambridge University Press, The Edinburgh Building, Shaftesbury Road, Cambridge CB2 1RP & 32 East 57th Street, New York, NY 10022, USA

Jonathan Cape, 20 Vauxhall Bridge Road, London SW1V 2SA

Cassell plc, Villiers House, 41-47 Strand, London WC2N 5JE

CBD Research, 15 Wickham Road, Beckenham, Kent BR3 2JS

Century-Hutchinson (*see* Random Century)

Chatham House Publications Inc., PO Box One, Chatham, NJ 07928, USA

China Books and Periodicals Inc., 2929 24th Street, San Francisco, CA, USA

Collins (*see* Harper-Collins)

Columbia University Press, 562 West 113th Street, New York, NY 10025

The Conservation Trust, George Palmer Site, Northumberland Avenue, Reading RG2 7PW

Crown Publishers (Division of Random House), 201 East 50th Street, New York, NY 10022

Dartmouth Publishing Co. (*see* Ashgate Group)

Christopher Davies (Publishers) Ltd., PO Box 403, Sketty, Swansea SA2 9BE

Andre Deutsch Ltd., 105-106 Great Russell Street, London WC1B 3LS

Directory of Social Change, Radius Works, Back Lane, London NW3 1HL

Dorling Kindersley, 9 Henrietta Street, London WC2E 8PS

Doubleday & Co., 666 Fifth Avenue, New York, NY 10103, USA

Duke University Press, 6697 College Station, Durham, NC 27708, USA

E. P. Dutton, c/o Penguin USA, 375 Hudson Street, New York, NY 10014, USA

Earthscan Publications Ltd., 120 Pentonville Road, London N1 9JN

The Economist Books Ltd., Axe & Bottle Court, 1 Newcomen Street, London SE1 1YT

Element Books, Longmead, Shaftesbury, Dorset SP7 8PL

Edward Elgar (*see* Ashgate Group)

The Eurospan Group, 3 Henrietta Street, Covent Garden, London WC2E 8LU

Faber & Faber Ltd., 3 Queen Street, London WX1N 3AU

The Findhorn Press, The Park, Findhorn, Forres, Scotland IV36 0TZ

Fontana Books, 77-85 Fulham Palace Road, London W6 8JB

Free Press, c/o Macmillan, 866 Third Avenue, New York, NY 10022, USA

W. H. Freeman & Co., 41 Madison Avenue, 37th Floor, New York, NY 10010, USA & 20 Beaumont Street, Oxford OX1 2NQ

Friedman Publishing Group Inc., 15 W. 26th. Street, New York, NY 10010, USA

Gaia Books, 66 Charlotte Street, London W1P 1LR

Gale Research Co., 835 Penobscot Building, Detroit, MI 48226, USA

Global Action Plan for the Earth, 574 Krumville Road, Olivebridge, NY 12461, USA

Victor Gollancz Ltd. (*see* Cassell plc)

Gordon and Breach Science Publishers, PO Box 786, Cooper Station, New York, NY 10276, USA & PO Box 90, Reading RG1 8LJ

Gower Publishing Co., Gower House, Croft Road, Aldershot, Hants, GU11 3HR & Old Post Road, Brookfield, VT 05036, USA

Green Books, Ford House, Hartland, Bideford, Devon EX39 6EE

Green Print, Merlin Press, 10 Malden Road, London NW5 3HR

Greenwood Press, 88 Post Road West, PO Box 5007, Westport, CT 06881, USA & 3 Henrietta Street, London WC2E 8LU

Grove Weidenfeld, 841 Broadway, New York, NY 10003, USA

Hamish Hamilton, 27 Wright's Lane, London W8 5TZ

Hamlyn, London, Michelin House, 81 Fulham Road, London SW3 6RB

Harcourt Brace Jovanovich, 1250 Sixth Avenue, San Diego, CA 92101, USA & 24-28 Oval Road, London NW1 7DX

Harper & Row (*see* HarperCollins)

HarperCollins, 10 East 53rd Street, New York, NY 10022, USA & 77-85 Fulham Palace Road, London W6 8JB

HarperSanFrancisco, Icehouse 1-401, 151 Union Street, San Francisco, CA 94111, USA

Harvard University Press, 79 Garden Street, Cambridge, MA 02138, USA

Harvester Wheatsheaf, 66 Wood Lane End, Hemel Hempstead, Herts. HP2 4RG

Heinemann (*see* Butterworth-Heinemann)

HMSO Publications Centre, PO Box 276, London SW8 5DT

Hodder & Stoughton Ltd., Mill Road, Dunton Green, Sevenoaks, Kent TN13 2YA

Holt, Rinehart & Winston, 24-28 Oval Road, London NW1 7DX

Houghton Mifflin Co., 2 Park Street, Boston, MA 02108, USA

Hutchinson (*see* Random Century)

Indiana University Press, Tenth & Morton Streets, Bloomington, IN 47401, USA

Institute for Social Inventions, 20 Heber Road, London NW2 6AA

International Centre for Economic Growth, ICS Press, 243 Kearny Street, San Francisco, CA 94108, USA

Intermediate Technology Publications, 103-105 Southampton Row, London WC1B 4HH

IOS Press, van Diemenstraat 94, 1013 CN Amsterdam, Netherlands

Island Press (Division of Center for Resource Economics), 1718 Connecticut Ave. NW, Suite 300, Washington, DC 20009, USA

Johns Hopkins University Press, 700 West 40th, Suite 275, Baltimore, MD 21211, USA

Jossey-Bass Publishers, 350 Sansome Street, San Francisco, CA 94104, USA

Kluwer Academic Publishers, 101 Philip Drive, Assinippi Park, Norwell, MA 02061, USA

Alfred A. Knopf, 201 E. 50th Street, New York, NY 10022, USA

Knowledge Systems Inc., 7777 West Morris Street, Indianapolis, IN 46231, USA

Kogan Page Limited, 120 Pentonville Road, London N1 9JN

Kumarian Press, 630 Oakwood Ave., Suite 119, West Hartford, CT 06110, USA

Peter Lang Publishing, 62 West 45th Street, 4th Floor, New York, NY 10036, USA & Jupiterstrasse 15, Postfach 27, CH-300 Bern 15, Switzerland

Lion Publishing plc, Peter's Way, Sandy Lane Wood, Littlemore, Oxford OX4 5HG

Longman Group UK Ltd., Longman House, Burnt Mill, Harlow, Essex CM20 2JE

Macdonald & Co. (Publishers) Ltd., Orbit House, 1 New Fetter Lane, London EC4A 1AR

The Macmillan Press Ltd., Houndmills, Basingstoke, Hants. RG21 2XS

Macmillan Publishing Co., 866 Third Avenue, New York, NY 10022, USA

M. H. Macy & Co., Box 11036, Boulder, CO 80301, USA

Manchester University Press, Oxford Road, Manchester M13 9PL

McGraw-Hill Book Co., 1221 Avenue of the Americas, New York, NY 10020, USA & Shoppenhangers Lane, Maidenhead, Berks. SL6 2QL

Meckler Publishing, 11 Ferry Lane West, Westport, CT 06880, USA & Artillery House, Artillery Row, London SW1P 1RT

The Mercier Press, PO Box 5, 5 French Church Street Cork, Ireland

Methuen, 7 Kendrick Mews, London SW7 3HG

MIT Press, 55 Hayward Street, Cambridge, MA 02142, USA

Mitchell Beazley, Michelin House, 81 Fulham Road, London SW3 6RB

William Morrow & Co., 105 Madison Avenue, New York, NY 10016, USA

National Academy Press, National Academy of Sciences, 2101 Constitution Ave., Washington DC 20418, USA

NCVO Publications, 8 Regent's Wharf, All Saints Street, London N1 9RL

New Society Publishers, PO Box 582, Santa Cruz, CA 95061, USA

North Atlantic Books, 2800 Woolsey Street, Berkeley, CA 94705, USA

W. W. Norton & Co., 500 Fifth Avenue, New York, NY 10010, USA

OECD Publications, OECD, Paris, France & OECD, 2001 L Street NW, Washington, DC 2036, USA and HMSO, London (*see* HMSO Publication Centre)

Open University Press, Celtic Court, 22 Ballmoor, Buckingham MK18 1XW

Oxfam Publications, 274 Banbury Road, Oxford OX2 7DZ

Oxford University Press, Walton Street, Oxford OX2 6DP & 200 Madison Avenue, New York, NY 10016, USA

Pan Books Ltd., 18-21 Cavaye Place, London SW10 9PG

Pantheon Books (Division of Random House), 201 East 50th Street, New York, NY 10022, USA

Penguin Books Ltd., Bath Road, Harmondsworth, Midd. UB7 0DA, 27 Wrights Lane, London W8 5TZ & 375 Hudson Street, New York, NY 10014-3657, USA

Pergamon Press plc, Headington Hill Hall, Oxford OX3 0BW & Maxwell House, Fairview Park, Elmsford, NY 10523, USA

Pilgrim Books, Lower Tasburgh, Norwich NR15 1LT

Pinter Publishers, 25 Floral Street, London WC2E 9DS

Pion Ltd., 207 Brondesbury Park, London NW2 5JN

Plenum Publishing Corp., 233 Spring Street, New York, NY 10013, USA

Pluto Publishers Ltd., 345 Archway Road, London N6 5AA

Policy Studies Institute (distributed by Pinter Publishers)

Polity Press, 65 Bridge Street, Cambridge CB2 1UR (distributed by Blackwell)

Praeger Publishers, Greenwood Publishing Group, 88 Post Road West, Box 5007, Westport, CT 06881, USA & 3 Henrietta Street, Covent Garden, London WC2E 8LU

Prentice Hall Press (Divison of Simon & Schuster), 15 Columbus Circle, New York, NY 10023, USA

Princeton University Press, 41 William Street, Princeton, NJ 08540, USA

The Random Century Group Ltd., Random Century House, 20 Vauxhall Bridge Road, London SW1V 2SA

The Reader's Digest Association Ltd., Berkeley Square House, Berkeley Square, London W1X 6AB

Rider (*see* Random Century)

Lynne Rienner Publishers, Inc., 1800 30th Street, Suite 314, Boulder, CO 80301, USA & 3 Henrietta Street, Covent Garden, London WC2E 8LU

James Robertson (self-publishing), The Old Bakehouse, Cholsey, near Wallingford, Oxon. OX10 9NU

Routledge, 11 New Fetter Lane, London EC4P 4EE

Routledge & Kegan Paul (*see* Routledge)

Routledge, Chapman & Hall Inc., 29 West 35th Street, New York, NY 10001, USA

Royal Institute of International Affairs, 10 St James's Square, London SW1Y 4ZE

Sage Publications, 2455 Teller Road, Newbury Park, CA 91320, USA & 6 Bonhill Street, London EC2A 4PU

K. G. Saur Verlag KG, Heilmannstrasse 17, D-8000 Muenchen 71, Germany

Scientific American Inc., 411 Madison Avenue, New York, NY 10017-1111, USA

Charles Scribner's Sons, c/o Macmillan Publishing Co., 866 Third Avenue, New York, NY 10022, USA

Secker & Warburg, Michelin House, 81 Fulham Road, London SW3 6RB

Seven Locks Press, PO Box 27, Cabin John, MD 20818, USA

Shambhala Publications, Horticultural Hall, 300 Massachusetts Avenue, Boston, MA 02115, USA

M. E. Sharpe, 80 Business Park Drive, Armonk, NY 10504, USA

Sierra Club Books, 730 Polk Street, San Francisco, CA 94109, USA

Simon & Schuster, 1230 Avenue of the Americas, New York, NY 10020, USA

Sourcebooks Inc., 26 N. Webster Street, Naperville, IL 60540, USA

South End Press, 116 St. Botolph Street, Boston, MA 02115, USA

Sphere Books, London (*see* Macdonald)

Springer-Verlag, Springer House, 8 Alexandra Road, Wimbledon, London SW19 7JZ, also New York, Germany, etc.

Stanford Alumni Assoc., Bowman Alumni House, Stanford, CA 94305, USA

J. P. Tarcher Inc., 9110 Sunset Blvd., Los Angeles, CA 90069, USA

Theosophical Publishing House, 306 West Geneva Road, Wheaton, IL 60187, USA

Thorsons (*see* HarperCollins)

Times Books (*see* HarperCollins)

Trilateral Commission, 345 East 46th Street, New York, NY 10017, USA

UNESCO, 7 Place de Fontenoy, F-75700 Paris, France

Union of International Associations, rue Washington 40, B-1050 Bruxelles, Belgium

United Nations Publications (Unipub), 2 UN Plaza, Room DC2-853, New York, NY 10017, USA

United Nations University Press, Toho Seimau Building, 2-15-1 Shibuya, Shibuya-ku, Tokyo 150, Japan

University Press of America, 4720 Boston Way, Lanham, MD 20706, USA & 3 Henrietta Street, Covent Garden, London WC2E 8LU

Van Nostrand Reinhold International, 2-6 Boundary Row, London SE1 8HN

Viking Penguin (*see* Viking Press)

Viking Press Inc., 40 West 23rd Street, New York, NY 10010, USA

Wadsworth Publishing Co., 10 Davis Drive, Belmont, CA 940022, USA

Westview Press, 5500 Central Avenue Boulder, CO 80301, USA

John Wiley & Sons Inc., 605 Third Avenue, New York, NY 10158, USA & Baffins Lane, Chichester, W. Sussex PO19 1UD

Yale University Press, 92A Yale Station, New Haven CT 06520, USA

Zed Books Ltd., 57 Caledonian Road, London N1 9BU & c/o Humanities Press International Inc., 171 First Avenue, Atlantic Highlands, NJ 07716, USA

PUBLISHER'S NOTE

It has been our overriding priority to include in *Resources for the Future* books that are actually in print. We have also included some important earlier books now out of print. We would be grateful to any reader who would draw to our attention any book, no longer in print, which could usefully be included in the next edition of this volume and/or perhaps be republished in the Adamantine 'Classics for the 21st Century' reprints series. Please write to: The Publisher, 21st Century Studies Programme, Adamantine Press Limited, 3 Henrietta Street, Covent Garden, London WC2E 8LU.

SUBJECT INDEX

Each entry in this index refers to the *reference numbers* of items relevant to its subject, and *not* to page numbers. For entries marked with an asterisk, only some more important items have been selected, as there are too many to list in full.

INDEX BY PUBLISHER

The item numbers, for the publishers whose books were listed most often in Chapter 1, are as follows: